P9-AES-676

The Urban Atmosphere and Its Effects

Air Pollution Reviews – Vol. 1

The Urban Atmosphere and Its Effects

Editors

Peter Brimblecombe
University of East Anglia

Robert L. Maynard
Department of Health, London

ICP

Imperial College Press

Published by

Imperial College Press
57 Shelton Street
Covent Garden
London WC2H 9HE

Distributed by

World Scientific Publishing Co. Pte. Ltd.
P O Box 128, Farrer Road, Singapore 912805
USA office: Suite 1B, 1060 Main Street, River Edge, NJ 07661
UK office: 57 Shelton Street, Covent Garden, London WC2H 9HE

British Library Cataloguing-in-Publication Data
A catalogue record for this book is available from the British Library.

THE URBAN ATMOSPHERE AND ITS EFFECTS
Air Pollution Reviews — Vol. 1

ISBN 1-86094-064-1

Printed in Singapore by World Scientific Printers

CONTENTS

LIST OF CONTRIBUTORS

J.G. Ayres
Department of Respiratory Medicine
Birmingham Heartlands Hospital
Bordesley Green East
Birmingham B9 5SS
United Kingdom

P. Brimblecombe
School of Environmental Sciences
University of East Anglia
Norwich
Norfolk NR4 7TJ
United Kingdom

K. Cameron
Chemicals and Biotechnology Division
Department of the Environment, Transport and the Regions
Floor 3/F8, Ashdown House
123 Victoria Street
London SW1E 6DE
United Kingdom

J. Carrington
School of Environmental Sciences
University of East Anglia

Norwich
Norfolk NR4 7TJ
United Kingdom

J.S. Gaffney
Bld 203 Environmental Research Division
Argonne National Laboratory
Argonne, IL 60439
USA

D.G. Housley
Department of Clinical Biochemistry
Royal Free Hospital, Pond St.
London NW3 2QG
United Kingdom

F. Hurley
Institute of Occupational Medicine
8 Roxburgh Place
Edinburgh EH8 9SU
United Kingdom

M. Jantunen
National Public Health Institute
KTL — Department of Environmental Hygiene
P.O. Box 95
FIN–70701, Kuopio
Finland

M. Johnson
School of Environmental Sciences
University of East Anglia
Norwich
Norfolk NR4 7TJ
United Kingdom

G. LeGouais
School of Environmental Sciences
University of East Anglia
Norwich
Norfolk NR4 7TJ
United Kingdom

N.A. Marley
Bld 203 Environmental Research Division
Argonne National Laboratory
Argonne, IL 60439
USA

R.L. Maynard
Department of Health
Room 658C
Skipton House
80 London Road
London SE1 6LH
United Kingdom

C. Obhrai
School of Environmental Sciences
University of East Anglia
Norwich
Norfolk NR4 7TJ
United Kingdom

R.J. Richards
School of Molecular and Medical Biosciences
University of Wales
Museum Avenue
P.O. Box 911
Cardiff CF1 3US
United Kingdom

L. Rushton
MRC Institute for Environment and Health
University of Leicester
94 Regent Road
Leicester LE1 7DD
United Kingdom

W.S. Tunnicliffe
Department of Respiratory Medicine
Birmingham Heartlands Hospital
Bordesley Green East
Birmingham B9 5SS
United Kingdom

P. Webley
School of Environmental Sciences
University of East Anglia
Norwich
Norfolk NR4 7TJ
United Kingdom

PREFACE

Air pollution is an old problem. During the mid-twentieth century, concern about the effects of air pollutants on health rose to a peak and, in some countries, measures were introduced to control emissions. In the last decade, interest and concern have surged again and a great deal of new research has been undertaken. This has led to a literature explosion which resulted in problems such as rapid updating of standard accounts and difficulties for students trying to identify the leading edge of the field. The results of recent research have also posed problems for those involved in developing policies for controlling the levels of air pollutants. The application of time-series methodology to the analysis of the day-to-day effects of air pollutants and the maturing of the Harvard Six Cities cohort study have led to the assumptions about thresholds of effects and safe levels of exposure being challenged. Demands for tighter standards and continued reductions in levels of air pollution have intensified.

This volume, and those that follow it, has been designed as a bridge between student-level accounts of air pollution science and the original literature. Thus, it is hoped that the volumes will be useful to postgraduate students beginning work in this field, to students reading for their Master's degrees and to undergraduates interested in pursuing aspects of their studies in depth. Air pollution scientists, doctors and those involved in policy development should also find these volumes of use and interest.

That this volume is devoid of all errors seems unlikely, given the speed of development of the field. Therefore, constructive criticism and correction of errors are invited.

Producing a book, let alone a series, is a team effort and the editors wish to thank the authors who have contributed chapters, colleagues who have advised on the contents, and especially the staff at Imperial College Press — past and present — who have made this an enjoyable task.

R.L. Maynard
P. Brimblecombe
J.G. Ayres
R.J. Richards
M. Ashmore

CHAPTER 1

URBAN AIR POLLUTION

Peter Brimblecombe

Our air is unfit to breathe. At least that is a perception one gains from popular literature, the media and the World Wide Web. Political interest parallels this and acid rain, global warming and the ozone hole remain sensitive issues involving the atmosphere on a larger scale. The last decade of the 20th century has seen a return to concern about the urban atmosphere with amendments to the US Clean Air Act, a National Strategy in the UK and a wide ranging European Directive.

1. Cities at the Millennium

This century has been one of enormous urbanisation with predictions suggesting that almost half the world's population will live in cities by the new millennium. At the end of this century, almost a hundred cities will have populations greater than three million. This implies a significant increase in human exposure to urban air pollutants. The potential for increased exposure that results from urbanisation can be balanced by improvements in air quality, but population growth often seems greatest in cities with only poorly developed regulatory procedures.

The problems and difficulties that cities pose have not prevented a significant reappraisal of their role within environmentally aware development. Victorians frequently regarded cities as a necessary evil and promoted return-to-nature as an ideal, deriving such views from the Romantic philosophies of the 18th and 19th centuries. Perhaps it was such negative visions of cities as unhealthy anomalies on the landscape that aided the decline in their quality through the 20th century.

Cities are necessarily the focus of many environmental problems, given their density of energy use and occupation. The early 20th century did not see cities develop with any special regard to the environment. Declining environmental quality has also been in evidence in recent years with the unplanned growth in many developing countries. Suburbanisation has frequently led to very large dispersed cities that encroach on the surrounding countryside. Suburbs have often imposed long travel times, usually by private automobiles, and frequently left urban cores impoverished. Open dispersed cities can give a low density of pollution sources and access to parks and amenities, but increasingly such development is recognised as environmentally costly. Suburbanisation thus adds to the loss of land and increases energy use (most obviously in the transport sector). Traditional views saw urban parkland as a way of diluting pollutants, but in the modern urban atmosphere, it has become evident that biogenic hydrocarbons from vegetation can contribute to photochemical smog (Geron *et al.*, 1995; Benjamin, 1998).

Our contemporary reappraisal of cities increasingly sees them as assets that go far beyond their role as the productive base for economic growth and improved living standards. They can be readily viewed in terms of their cultural and intellectual diversity, opportunity, liberalism, etc. In the environmental sense, cities can provide a route to efficient resource use.

Current initiatives have started to place cities in a more ecological context. The UK's Natural Environmental Research Council has developed a programme called URGENT which attempts to shift its research focus towards the urban environment. This parallels other programmes; such as the World Health Organisation's interest in *Healthy Cities* and the OECD Environment Group on Urban Affairs' work on *The Ecological*

City. These developments show an institutional recognition of the city as a type of ecosystem.

Sustainable cities are increasingly seen as the ecologically appropriate direction for humanity. The idea of *sustainability* has been enthusiastically embraced. It has become a key guide to socioeconomic development (e.g. DG XI, 1993) despite a lack of clarity about application to some important issues. Sustainability in forestry and fishing has accustomed us to think in terms of replacement levels, overpopulation, etc. Broader definitions of *sustainability* vary, but all embody the idea that actions undertaken now should not limit options in the future. It has often included issues of social justice, which have been seen as an important source of degradation. In sustainable cities, economic development would continue in harmony with cultural aspirations and environmental quality.

2. Sources of Pollution

Cities are a source of air pollution, but in reality air pollution is the result of a combination of factors. It is the sum of these factors that gives air pollution particular characteristics in a given city. Fuel use is a prime source of air pollution in contemporary cities although there are cities where wind-blown dusts or the long range transport of pollutants, produced at great distance, are dominant sources.

In the initial stages of historical growth the fuel adopted in a city is typically that which is conveniently available. This was usually wood, but as cities grow, they deplete nearby forests and fuel must be imported. The fuel adopted then becomes a matter of choice influenced by economics, efficiency, regulation or convention.

Our use of fuel introduces a broad social control on the nature of urban air pollution. Most obvious are the weekly cycles imposed by the pattern of work. Holiday weekends can also be a major source of traffic generated pollutants where vehicle numbers are large while in some remote areas, a weekly feast day can produce pollution episodes. Although the diurnal production of air pollutants may be socially driven, these are usually modified by diurnal changes in meteorology and photo-chemistry.

Air pollution has a very long history extending back thousands of years (Brimblecombe, 1987a; 1987b). However, beyond a few nations of Europe, it was the developments of the 19th century that were significant in creating urban wide air pollution problems (Brimblecombe, 1998).

2.1. Odour

Although combustion is usually seen as the dominant issue, odour was a major problem in cities of the past and in many ways it remains a contemporary issue as local environmental officers frequently find. Historically, rotting organic matter and sewage was an ever-present urban problem. It was so serious in 17th century Paris that silverware rapidly tarnished. Police regulations in Papal Rome discouraged cooking as there was an objection to the odours. Tripe and cabbage were not prepared in the home for fear of prosecution, so tripe was cooked in enormous cauldrons set up in front of the church of San Marcello and cabbages boiled near the Piazza Colonna. Coffee was roasted at the foot of the column of Marcus Aurelius (Brimblecombe, 1998). In modern times, the production of chocolate, beer or coffee brings frequent complaint. At the margins of cities, agricultural odours are important sources of complaint.

3. Solid Fuels and Primary Pollutants

It is more typical especially when viewing the pollution of the urban atmosphere as a historical change to consider the changes that fuel use has imposed upon the air.

3.1. Historical Development

The earliest change in fuels to dramatically affect urban air pollution was the shift from wood to coal. This occurred in medieval times in

Britain after the depletion of conveniently usable wood supplies. The rapid growth in coal use was more noticeable than the gradual increase in smokiness resulting from a rise in population. The unfamiliar smell of coal smoke led to early fears about health risks through the belief that disease was carried in malodorous air (miasmas).

Thus, citizens of 13th century London became aware of problems of air pollution. This early use of coal was mostly in industrial processes such as cement making and metallurgy. Legislation prescribed a return to wood, but its high price compared with coal meant that control measures ultimately failed. It was not until the 16th century, with the widespread construction of chimneys, that the coal began to be used domestically (Brimblecombe, 1987a).

The transition to coal was virtually complete in London by the early 17th century, but it was delayed elsewhere. The intensity of the energy requirements meant that industrialising cities required coal, so its use grew rapidly in the 19th century. The development of the steam engine gave a particular focus to smoke pollution and protest against it forced engineers to examine the way in which smoke was generated. Benjamin Franklin advised on smoke from the earliest Boulton and Watt steam engines (Marsh, 1947) and aided the concept of "burning your own smoke". This control philosophy remained central to smoke abatement well into the present century.

Civic administrators realised that they needed to respond to the nuisance and health risk created by industrial smoke pollution. The Manchester and Salford Police Act of 1792 meant authorities emphasised chimney construction "to burn the Smoke arising". Throughout Europe, much legislation developed but often lacked technological and administrative instruments to make it workable (Brimblecombe, 1998).

At the beginning of the 20th century, smoke abatement legislation had failed to deliver smoke free air. At a local level attention shifted between the many smaller sources and the problems were treated in a case-by-case manner. Occasionally the air quality improved, but this was not usually the result of clear solutions. It was typical for new technology or plant to remove troublesome and emotive local issues. Pollution problems were not so much solved as superseded, but these indirect improvements came with imperceptible slowness (Brimblecombe & Bowler, 1992).

The lack of regular air pollution monitoring meant that the effectiveness of legislation remained almost impossible to assess. The most widespread early observations were simple descriptions of smoke from chimneys, although the deposit gauge (essentially a large rain gauge) was regularly used in the first half of this century and charted the decline of coarse particulate material deposited from the air of large cities. In some cities, such as London, the concentrations of primary pollutants were occasionally monitored with simple pumps and these hinted at long term declines in sulphur dioxide and smoke. In other cities, improvements in these pollutants have been more recent.

Despite modest controls on the emissions of combustion-derived air pollutants in the first half of the 20th century, these were insufficient to prevent air pollution disasters: Meuse Valley, Belgium 1930 (Firket, 1931), Donora Pennsylvania 1948 (Halliday, 1961) and the devastating London smog of December 1952. This winter event caused as many as four thousand excess deaths in London and led to a public outcry and strong political reaction. The subsequent inquiry helped the development of the Clean Air Act of 1956. Thus, the 1952 Smog and its resultant legislation are often viewed as a turning point in dealing with the problems of urban air pollution. However, the effectiveness of the Act has often been questioned (Elsom, 1992). Some have felt (e.g. Brimblecombe, 1987a) that the Act simply reinforced shifts to less smoky fuels that were well underway, while others were critical of its inability to cover anything but smoke emissions. Despite these arguments, the years that followed have seen a continued reduction in domestic coal use (as it was replaced by gas and electricity). These declines have been balanced by the growth in emissions from liquid fuels, mostly those used in road transport. Thus reductions in traditional pollutants (SO_2 and smoke), unfortunately, drew attention away from new pollutants in the urban atmosphere.

3.2. Air Pollutants from Solid Fuels

The combustion of fuel has always been a major source of air pollution. Hydrocarbon fuels would be expected to oxidize to give water and carbon dioxide. However, combustion is frequently limited by availability of oxygen so carbon monoxide and smoke (carbon and pyrolised fuel) are also typical products. When coal is burned on a low temperature grate, combustion is ineffective and leads to the release of a range of polycyclic aromatic hydrocarbons (PAHs). These include carcinogens such as benzo(α)pyrene. As the domestic use of coal has declined and it has been burnt in more efficient furnaces, especially using pulverised fuel, the emission of unburnt hydrocarbons from solid fuels has been much reduced. Nevertheless, carbon-containing air pollutants remain the most abundant produced by combustion.

High temperatures from combustion process encourage the formation of oxygen atoms that can enter chain reactions known as the Zeldovich cycle:

$$O + N_2 \rightarrow NO + N$$

$$N + O_2 \rightarrow NO + O$$

besides rapid pathways relying on the CH radical (from hydrocarbons) which yields HCN:

$$CH + N_2 \rightarrow HCN + N$$

which can also be oxidised to NO. In addition, nitrogen compounds present as impurities in the fuel can also be oxidised to yield nitrogen oxides.

Impurities in fuel are thus important in generating air pollutants during combustion. Sulphur is a major impurity in coal, partly present as the mineral pyrite FeS_2. This is converted to sulphur dioxide on combustion:

$$4FeS_{2(s)} + 11O_{2(g)} \rightarrow 8SO_{2(g)} + 2Fe_2O_3$$

Sulphur is also found as organosulphur compounds, but there are many other impurities in coal. Halogens are sometimes abundant and chlorine is released as hydrochloric acid. This is especially common in refuse derived fuels (RDF) which contain plastics such as PVC (polyvinylchloride). More innocuous volatile compounds released in coal burning are the boron oxides, probably volatilised as H_3BO_3, although this becomes partially associated with particles (Anderson, 1994).

Problems with disposal of solid waste have created the need for more incineration, although there has been a growth in public opposition to incinerators. The chlorine compounds within refuse can be readily converted to dioxins and other halogenated aromatic compounds, which can be released to the atmosphere without adequate controls. Even well-designed and operated incinerators remain unpopular, although it is hard to see how the problem of waste disposal will not be severe enough to create continued pressures to use them.

3.3. Particles

Besides these gases, there are large amounts of ash and grit produced by coal burning. These materials presented a considerable problem in the past. Fly ashes contain oxides of calcium particularly which contribute to atmospheric alkalinity (Hedin, 1994). Additionally, a whole range of metallic oxides is present. Modern coal-fired furnaces adopt grit arrestors, cyclones and electrostatic precipitators to remove this particulate material from the exhaust gases.

The particles are also sites for accumulation of compounds of moderate volatility. As mentioned before, inefficient coal combustion gave high yields of PAHs in the past and these were involatile enough to be associated with particles. There are observations that airborne concentrations of PAH have been on the decline (Lawther & Waller, 1978). In rural UK, the decline has been noticeable among the lighter species such as phenanthrene (Jones *et al.*, 1992).

3.4. Reactions

The gas phase oxidation of sulphur dioxide in the atmospheres of coal burning cities is generally slow, so heterogeneous reactions become important. Sulphur dioxide dissolves in aqueous droplets, but is only slowly oxidised by dissolved oxygen to sulphuric acid. In the presence of iron or manganese, a faster catalysed oxidation becomes important. At acidities typical of atmospheric aerosols, sulphur dioxide will be present mainly at the bisulphite ion (HSO_3^-), thus the oxidation is represented:

$$HSO_3^- + 1/2O_2 \xrightarrow{Fe/Mn} SO_4^{2-} + H^+$$

Hydrogen peroxide and ozone can be more effective oxidants, but they are not likely to be present at significant concentrations under winter smog conditions.

3.5. Gradual Improvements

Atmospheres dominated by coal smoke are far less common than in the past. They may still be found on occasions particularly outside Europe and North America and are still characterised by high concentrations of sulphur dioxide and particulate material. In recent years, Belfast has been noted for this traditional type of air pollution, but it is mostly China and the former Warsaw Pact countries which have experienced some of the worst smoke problems.

Generally, emissions of SO_2 have become dominated by small number of large sources, often major power stations and industries remote from urban centres. Total sulphur emissions have declined, so in line with this urban concentrations have become lower. However, short-term concentrations of SO_2 can still be high as plumes from distant sources, although only occasionally fumigating a location, can contribute to high concentrations.

4. Liquid Fuels and Photochemistry

4.1. History

The growth in emissions from coal combustion in the 19th century is paralleled by the rise in emissions from petroleum combustion in the

20th century. This has also been accompanied by a shift to mobile sources of air pollutants.

The novel characteristics of air pollution from this new source first emerged in the Los Angeles basin in the 1940s. Here, an urban structure encouraged a low population density. Dominance of the automobile over public transport, weak winds, an inversion that traps pollution over the city against a mountain barrier and long hours of sunshine allow for photochemical reactions.

The war years (1941–45) saw the development of a persistent pall of haze over the city. The public wanted a culprit and much of the blame for the smog was directed initially towards gas companies, oil refineries and chemical plants. This situation was so novel that the pollutants were not recognised. Early attempts at control failed, so the problem was generalised and not associated with these identifiable sources. Professor Raymond Tucker, a St. Louis based air pollution expert, investigated the smog and recommended more controls on industries in 1947, but absolved automobiles of key responsibilities. However, by the 1950s, it was clear that traditional primary pollutants, such as SO_2, were not responsible for plant damage. Arie Jan Haagen-Smit showed that oxidative reactions of gasoline vapours yielded smog which led to the controversial assertion that NO_2 and petroleum vapours produce ozone on irradiation. Ultimately, Stephens' group used long path length infra-red spectroscopy to confirm the suggestion of Haagen-Smit (Stephens, 1987). The oil industry had resisted the shift in blame, but the complicity of the motor vehicle was apparent by the mid 1950s (Krier & Ursin, 1977).

Los Angeles offered such favourable conditions for the formation of photochemical smog, that it seemed a unique type of pollution. Nevertheless, it was obvious that the precursors would be commonly present in modern cities (Leighton, 1961), so by the late 1960s, measurements in Europe soon identified photochemical products. The following decade saw high summer ozone concentrations of almost continental dimensions (e.g. Cox *et al.*, 1975; Guicherit & Van Dop, 1977). The heat wave of 1976 produced an impressive episode across southern England that particularly captured the public imagination. An increasing frequency of ozone episodes allowed these to be linked with forest damage along with acid rain. In the 1980s, the ozone problems of a more urban scale were recognised in Mediterranean cities, such as Athens.

The early legislation in the UK such as the Clean Air Act 1956 tackled the primary pollutant smoke, while in the US, the Clean Air Act 1970 had to consider photochemical pollutants. These raised more difficult control issues and non-attainment of the standards was so widespread that it created a perception of failure. However, the desire to achieve higher air quality was reasserted under the amendments that resulted in the Clean Air Act 1990. When these were passed, 140 million people were living in non-attainment areas, with respect to ozone, but the situation has actually much improved. Initially, 91 areas were designated "non-attainment" but this was reduced to less than forty embracing a population under 100 million people by mid-1998. Areas of non-attainment required programmes such as car-pooling, clean fuel and auto inspection and maintenance.

The problem of photochemical smog requires particular attention to VOC emissions and concern over fuel evaporation, volatility and composition. The photochemical ozone creation potential (POCP) of volatile organic compounds (VOCs) has become increasingly important as a consideration of their impact on the atmosphere. The reactivity of the individual organic compounds in the atmosphere is important in the production smog. A high POCP means that some VOCs, at relatively low concentration, can play a dominant role in the oxidation process.

4.2. Sources from Petroleum Burning

As with coal burning, incomplete combustion of petrol produces a large amount of carbon monoxide. Soot is not a major product of petrol engines, but is characteristic of diesel-engined vehicles that have grown common in Europe. Nitric oxide in cities is now mostly produced by automobiles and they are also the dominant source of carbon monoxide.

Liquid fuels typically yield volatile hydrocarbons (with increasing interest in the reactivity of aromatic and oxygenated compounds). Evaporative losses of fuel from engines during storage and transfer, were often underestimated in the earliest attempts to control photochemical smog, but they are now matters of great concern.

Automotive fuels are refined and so tend to have relatively low levels of non-combustible impurities; fuel oils for heating, by contrast, can contain much higher concentrations. Vanadium is often particularly enriched as vanadyl porphyrin complexes in oil, so the combustion of heavy oils can yield fly ashes that contain oxides such as V_2O_5, V_2O_4 and V_2O_3 besides the mixed oxide 2NiO. V_2O_5 that contains another nickel element enriched in oil.

4.3. Reactions

The most characteristic compound within photochemical smogs is ozone. It can be generated by the reactions:

$$NO_2 + h\nu(\text{less than } 310\,\text{nm}) \rightarrow O(^3P) + NO$$

$$O(^3P) + O_2 + M \rightarrow O_3 + M$$

$$O_3 + NO \rightarrow O_2 + NO_2$$

These equations can be imagined as represented by a pseudo-equilibrium constant relating the partial pressures of the two nitrogen oxides and ozone.

$$K = [NO].[O_3]/[NO_2]$$

However, as an equilibrium, it is not a net producer of ozone in the atmosphere because the reactions both produce and destroy ozone. However, if hydrocarbons are present, we can get more effective production of oxidants:

$$OH + RH \rightarrow H_2O + R$$

$$R + O_2 \rightarrow RO_2$$

$$RO_2 + NO \rightarrow RO + NO_2$$

Thus we have used the oxidant RO_2 to oxidise NO to NO_2 rather than ozone. Thus, the photolysis of NO_2 now produces extra ozone. If R is the ethyl group (CH_3CH_2), we could write subsequent reactions:

$$CH_3CH_2O + O_2 \rightarrow CH_3CHO + HO_2$$
$$HO_2 + NO \rightarrow NO_2 + OH$$

thus the oxidation of a hydrocarbon has given rise to an aldehyde (here acetaldehyde, or ethanal). Note that the OH radical is regenerated, so it can be thought of as a kind of catalyst. Aldehydes undergo further attack by OH radicals:

$$CH_3CHO + OH \rightarrow CH_3O + H_2O$$
$$CH_3CO + O_2 \rightarrow CH_3\,COO_2$$
$$CH_3COO_2 + NO \rightarrow NO_2 + CH_3CO_2$$
$$CH_3CO_2 \rightarrow CH_3 + CO_2$$

However, CH_3COO_2 may react as below:

$$CH_3COO_2 + NO_2 \rightarrow CH_3COO_2NO_2$$

leading to the formation of the eye irritant, peroxyacetylnitrate PAN.

Some compounds such as acetylene (ethyne) react so slowly in smog situations that they become a useful marker in the long-range transport of photochemically polluted air.

In recent years it has become clear that winter conditions with high concentrations of NO must have alternative oxidation routes for the production of NO_2. Some, but probably not all, of this oxidation can be explained by the reaction:

$$O_2 + 2NO \rightarrow 2NO_2$$

4.4. Particles

The shift from solid to liquid fuel as a source of energy has tended to reduce the smoke concentrations in the urban atmosphere. Modern coal-fired furnaces have also achieved a remarkable reduction in the emission of particles through careful combustion control, cyclones and electrostatic precipitators.

In recent years, there has been a sharp increase in the use of the diesel automobile engines in some countries. They have been widely adopted in Europe and have made significant penetration into the private automobile fleet. There has been much debate about air pollution from diesel engines and although there are still questions about their relative merits as they produce large quantities of particulate material especially when under load. Their presence in cities now means that diesel-engined vehicles contribute to an important fraction of the fine particulate. It is also likely that diesel soot may be reponsible for much of the soiling properties of the air in European cities.

Particulate material is also generated from reactions in the atmosphere. The inorganic component tends to consist of sulphates and nitrates, from the oxidation of the sulphur and nitrogen oxides. Organic secondary particles are less well known, but the dicarboxylic acids, particularly oxalic, are easily generated in photochemically active urban atmosphere and can be found associated with the fine fraction of the aerosol.

5. Improving Air Quality

5.1. Urban Air Pollution in the 20th Century

The 20th century has forced us to recognise differences between the air pollution arising from solid and liquid fuels. This has allowed us to recognise two archetypes for polluted urban atmospheres: the London and Los Angeles smogs. These typify coal-burning as *primary* and auto-derived *secondary* pollution. Such a simple defining typology is obviously inadequate to describe the enormous differences between cities. Yet modern tabulations such as those of the Dobris (Sluyter, 1995), or GEMS reports (e.g. WHO/UNEP, 1992), offer us little coherent synthesis as a route to interpretation.

In the future we may be confronted by quite different situations or ones of greater complexity. The air pollution of Fairbanks, has been offered as an example of a quite novel kind of pollution, where the absence of solar radiation and an inversion at the snow surface create special conditions (Benson, 1986); or Christchurch, New Zealand, an example of the dominance of solid fuel heating in winter. Industrial emissions can become more complex, but can be balanced by friendlier technologies (such as the use of aqueous solvents) to limit the impact on air quality.

5.2. Control Technology and Strategy

Attempts at controlling urban air pollution before the 20th century promoted legislation that gave little thought to how emissions reduction might be achieved or how their effectiveness might be monitored. The early part of the present century saw the development of a range of technological fixes, although clearly, proper training of technicians and stokers was important adjunct to smoke abatement.

The early stages of control related to combustion improvements or emission control such as designing efficient boilers and furnaces that "burned their own smoke". A little later, there was interest in end-of-pipe technologies, filter bag houses, cyclones, electrostatic precipitators and scrubbers (Hesketh, 1991). In the mid 20th century, the control of automobile emissions has centred on the catalytic convertor, which has been widely adopted in the developed world. Automobile engines have improved greatly and lean-burn engines have shown great promise, but the effectiveness of the catalytic converter has recommended its use for automobiles. The poor performance of the catalytic converter when the engine is cold, typically when driving cycles are short has raised concern. A reduction in diesel emissions is particularly desirable in Europe where the fleet has increased dramatically.

Fuel change has also been seen as an important approach from the earliest times when civic administrators insisted on a return from coal to wood. Today much attention is given to the composition of automotive fuels — the volatility is clearly relevant to the presence of VOCs in the atmosphere, while the types of volatile hydrocarbons are important fuels that yield organic compounds with a low potential for the creation of ozone (i.e. low in aromatics and olefins which have high POCPs), along with a low S content are desirable (as in reformulated gasolines). More exotic fuels, natural gas, alcohols and oxygenates continue to attract attention, but are often restricted to commercial fleets of vehicles (where the provision of special refuelling sites is possible). Real success is likely to come from the introduction of cleaner fuels usable by conventional engines (Lloyd, 1997).

However, strategies for improving air pollution have to be broader than this simple control on emissions. They have to consider a range of instruments to implement policy such as: pollution taxes, cost benefit analysis and air quality management (e.g. Elsom, 1992). The economic instruments have an interesting place, but traditionally emission control has been the primary method of control. Such policies seek to confront the source directly. In the case of primary pollutants, it meant that a large chimney stack or an obvious source could become the target for emission control in a neighbourhood. In simple situations, such as the control of the Alkali Industry in the UK last century (Brimblecombe, 1987a) this could be successful, at least in terms of the legal requirements of the controlling legislation.

However, the novel complexity that results from photochemical pollution serves to blur the link between the sources of pollution and the resultant polluted atmosphere. This transition has served as a catalyst for change in control strategy. Simple emission controls are not necessarily effective because there are situations where reduction of a pollutant such as NO may well enhance the concentration of ozone, for example. Thus, effective approaches have to consider managing air quality. Air quality management (AQM) has become more common throughout the world. In Europe it is embodied in the emerging European legislation 96/62/EC. In AQM the development of air quality standards, emission inventories, air pollution monitoring and modelling, are all integrated into the development of effective policy (Elsom, 1992).

5.3. The Public

The public has become more concerned about the state of the environment in the 20th century and governments are required increasingly to provide information about pollution. Strategies (e.g. DoE, 1996) and legislation (e.g. OJ, 1996) insist on the importance of communicating with the public. Thus in the late 20th century, air pollution monitoring networks have become as much an instrument of planning and regulation as they are providers of public information.

The complexity of contemporary monitoring networks means that they provide an enormous mass of numerical data. Sometimes, the sheer wealth of information has increased distrust, with the mass of misunderstood numbers seen as an indication of poor air quality, or, that governments are attempting to confuse their citizens with indigestible science. Systems of reporting, such as the air pollution indices in the US or France or the classification bands of the UK attempt to simplify data for the public. Such simplifications can distort information, as the presentation of this complex real-time data is no small problem.

The regulatory desire for more detailed spatial structure to air pollution data has increased the requirement on the number of monitoring sites. People also want to know about the air pollution at where they live or work. The high cost of dense monitoring networks has led to development of computation techniques or indicative measurements that are use inexpensive integrating methods (e.g. diffusion tubes), grab samples or mobile monitors (e.g. Brown, 1993; Alst *et al.*, 1998).

5.4. Broadening the Vision

The maintenance of air quality requires more than the simple application of regulatory strategy. The application of the UK Clean Air Act 1956 is often seen as an example where merely one problem was solved; that of smoke only unmasked others. A continuing vigilance is required to recognise the growth of novel problems because a city and its environment is ever changing. Legislative frameworks are time consuming to develop, so are often overtaken by more rapid changes in our atmosphere. Scientific understanding of fine particles in urban air was growing while legislators in Europe and the US were developing regulations. This means that drafts of the European Union Directive 96/62/EC were awkward in its definition of particulate matter (OJ, 1996). Although good legislation tries to be forward-looking, it is based on experience of the past and cannot necessarily reflect recent novel postulates. As an example, hydrogen sulphide is not included in the emerging EU regulations. Yet this gas is emitted by the growing fleet of catalytic converter-equipped cars while we have little understanding of the significance of claims that low levels of hydrogen sulphide have long-term neurotoxicological effects (Watts, 1998).

Our consideration of air quality and its control has to be seen as part of the way we think about the structure of cities. The way in which transport evolves and even the requirements for transport is important. Good public transport may be relevant, but decreasing the need for transport is also a useful strategy. Carless cities have been proposed, and although these sound attractive, they may be enthusiastic aspirations rather than workable entities occupied by people.

Urban design can also consider structures that minimise air pollution by correct spatial structure. Architectural layouts can align urban canyons to lessen pollutant accumulation, or choose urban vegetation with a low release of photochemically active compounds.

Lifestyle has also changed and the patterns of our lives are varying. Our exposure to air pollutants indoors or in the workplace can become an increasingly alarming phenomenon. Exposure to tobacco smoke, radon and consumer products are all important contributors to our burden of indoor air pollutants.

References

Anderson D.L. *et al.* (1994) *Atmospheric Environment,* **28**, 1401–1410.

Alst R. *et al.* (1998) *Guidance Report on Preliminary Assessment under EC Air Quality Directives,* DGXII.

Brown R.H. (1993) *Pure and Applied Chemistry,* **65**, 1859–1874.

Benjamin M.T. & Winer M.W. (1998) *Atmospheric Environment*, **32**, 53–68.
Benson C. (1986) In *Arctic Air Pollution*, ed. Stonehouse B. pp. 69–84, Cambridge University Press.
Brimblecombe P. (1987a) *Atmospheric Environment*, **21**, 2485.
Brimblecombe P. (1987b) *The Big Smoke*. Methuen, London.
Brimblecombe P. & Bowler C. (1998) *J. Air Waste Management Association*, **42**, 1562–1566.
Brimblecombe P. (1998) In *Urban Air Pollution — European Aspects*, eds. Fenger J., Hertel O. & Palmgren F. Kluwer, Dordrecht, in press.
Cox R.A. *et al.* (1975) *Nature*, **255**, 118–121.
DG XI, (1993) *Towards Sustainability*. DG XI, Commission of Eurpean Communities, Brussels.
DoE, (1996) *The United Kingdom National Air Quality Strategy*. Department of Environment, London.
Elsom, D.M. (1992) *Atmospheric Pollution*, 2ed. Blackwell, Oxford.
Firket M. (1931) *Bulletin de L'academie Royale de Medicine de Belgique*, **11**, 126.
Geron C.D. *et al.* (1995) *Atmospheric Environment*, **29**, 1569–1578.
Guicherit R. & Van Dop H. (1977) *Atmospheric Environment*, **11**, 145–156.
Halliday E.C. *et al.* (1961) In *Air Pollution*. pp. 9–37, WHO, Geneva.
Hedin L.O. *et al.* (1994) *Nature*, **367**, 351–354.
Hesketh H.E. (1991) *Air Pollution Control*. Technomic, Lancaster Pennsylvania.
Jones K.C. *et al.* (1992) *Nature*, **356**, 137.
Krier J.E. & Ursin E. (1977) *Pollution and Policy*. University of California Press, Berkeley CA.
Lawther P.J. & Waller E. (1978) *J. Envir. Health Persp.* **Feb.**, 71–73.
Leighton P.A. (1961) *Photochemistry of Air Pollution*. Academic Press, NY.
Lloyd A.C. (1997) In *Air Quality Management*, eds. Hester R.E. & Harrison R.M. pp. 141–156, Royal Society of Chemistry, London.
Marsh A. (1947) *Smoke*. Faber and Faber, London.
OJ, (1996) *Official Journal of the European Communities*. No. L 296/61.
Sluyter R.J.C.F. (1995) *Air Quality in Major European Cities*. Rijksinstituut voor Volksgezondheid en Milieu, No. 722401004.
Stephens E.R. (1987) *EOS*, **68**, 89–93.
Watts S.F. (1998) *Atmospheric Environment*, **32**, in press.
WHO/UNEP (1992) *Urban Air Pollution in Megacities of the World*. Blackwell, Oxford.

CHAPTER 2

TRENDS IN AIR POLLUTION RELATED DISEASE

W.S. Tunnicliffe & J.G. Ayres

1. Background

The burgeoning interest in the health effects of air pollution has resulted in an explosion of publications, many considering changes in health outcomes over a period of time. When attempting to measure the health effects of air pollution, which are in general small, confounding factors need to be accounted for with care. For example, in panel studies, where a sample of (usually susceptible) subjects are followed over a relatively short period of time, allowance has to be made for changes over that time period of the health outcome itself for other reasons, such as day of the week or seasonal effects. These problems are well discussed by Roemer *et al.* (1998) and are not considered further here. However, for studies over longer periods, perhaps years, underlying trends over longer time periods in a particular health outcome need to be considered when attempting to identify, either qualitatively or quantitatively, the proportion of that trend which might be attributable to exposure to an environmental agent or agents such as are found in ambient air. Examples of studies of this type are the US Six Cities Study (Dockery *et al.*, 1993) and the American Cancer Society Study (Pope *et al.*, 1995), both of which followed populations over years producing similar, although not identical findings of an effect of air pollutant exposure on mortality from cardiopulmonary disease and bronchial carcinoma. Such studies need also to consider long-term confounders which may both in themselves have a bearing on general disease susceptibility that may also enhance the effects of pollutant exposure, the most obvious example, being dietary antioxidant consumption.

This is an important principle as incorrect conclusions may be drawn if these factors are not taken into consideration when interpreting results from specific studies, although such mistakes are unlikely to be made by the researchers themselves, more by the media and the general public. An example of where an inappropriate attribution of a trend in a condition to changes in air pollution was inferred, was the marked increase in hospital admissions for asthma, particularly in children, seen in the UK in the 1980s (Committee on the Medical Effects of Air Pollutants, 1995). This was matched by similar trends in other countries (Committee on the Medical Effects of Air Pollutants, 1995). However, measured air pollutant levels showed little change over this period and as there has subsequently been a decline in UK asthma admissions since 1990, with air pollutant levels remaining much the same, it is difficult to attribute these changes in asthma health outcomes directly to varying conditions in air pollution.

This chapter comprises a justification of those diseases considered to be pollution-related, followed by a disease by disease breakdown of the relevant health outcomes (where adequate data exist) comparing possible trends between countries. The aim has been to produce a broad picture rather than a detailed analysis of all data which may be in existence. We have not considered sick building syndrome nor have we considered any occupational exposures, concentrating entirely on those diseases which have been associated with exposure to outdoor air pollutants.

1.1. Definitions

A pollution related disease has been defined here as:

- A condition whose incidence (or prevalence) or frequency of exacerbation has been identified as being related to variations in pollution levels in ambient air, or
- A condition where there is sufficient concern that the incidence (or prevalence) of the condition might be affected by air pollutants but where direct evidence is lacking.

1.2. Disease Inclusions

A number of conditions have been recognised to be associated with exposure to polluted air, variations in exacerbation rates usually relating to day to day changes in air pollutant levels. Most are predictable as they are diseases of the respiratory tract, while other conditions have emerged where an identified association could at first sight be regarded as more surprising, such as ischaemic heart disease (IHD) and sudden infant death syndrome (SIDS). The conditions considered in this chapter are summarised in Table 1 which broadly cover diseases of the upper and lower respiratory tract, cardio- and cerebro-vascular disease, leukaemia and sudden infant death syndrome. While it is likely that there may be other conditions which are also affected by exposure to air pollution we have considered here only those conditions which fulfil the above definitions.

1.3. Health Outcomes

Health effects related to air pollution have been identified across the full range of health outcomes, from changes in symptoms on a day to

Table 1. Diagnostic categories of pollution related diseases.

Upper respiratory tract	croup rhinitis
Lower respiratory tract	asthma chronic obstructive pulmonary disease bronchial carcinoma
Cardiovascular	ischaemic heart disease
Cerebrovascular	stroke
Blood	chronic myeloid leukaemia acute lymphatic leukaemia
Other	sudden infant death syndrome (SIDS)

day basis, to mortality. Consequently, within the diagnostic categories outlined in Table 1, we have where appropriate, considered mortality, hospital admissions, incidence of attacks in primary care, and prevalence. We have considered trends over time and where possible, break down these trends in terms of age and sex where data are available. As there is no information on trends over time in day to day changes in symptoms or lung function at a population level, we have not included these outcomes.

2. Upper Respiratory Tract Disease

2.1. Rhinitis

Definition

There are no widely agreed criteria for the diagnosis or classification of rhinitis (Mygind, 1985), the symptoms of sneezing, rhinorrhea and/or nasal blockage being common also to viral upper respiratory tract infections. Although a classification into allergic and non-allergic types is commonly employed, the alternative, division into seasonal or perennial depending on the pattern of the occurrence of symptoms, is also frequently employed. The degree to which either is used depends on how actively the physician pursues an allergic cause. The term hay fever usually refers to seasonal (summer) allergic rhinitis. Whatever the label applied, this pattern of symptoms contributes a significant burden of disease in primary care but this will also be added on in epidemiological terms, by the likely high frequency of unreported symptomatology. The threshold for consultation due to these symptoms will vary for individual, temporal and geographical reasons. Factors that may act at this level need to be carefully considered when examining reported differences and/or changes in incidence and prevalence rates of rhinitis in populations.

Risk factors

There are a number of recognised and further possible risk factors for rhinitis which will impact on comparisons of trends between countries. Hay fever often runs in families, a tendency which can be attributed to genetic factors rather than a shared familial environment. There is very little information about racial differences in the prevalence of hay fever, but some of the reported differences are significant. For instance in Britain, general practice consultation rates for allergic rhinitis have been found to be 60–80% higher for Afro-Caribbean people and around 50% lower for the southern Irish than for native Britons (Gillam *et al.*, 1989). In New Zealand, the period prevalence of hay fever is 20% to 30% higher among the children of Pacific islanders than among children of Maori or European origin (Pattermore *et al.*, 1989). Two other strong and important inter-related factors are those of birth order and family size. First born children have the highest risk of developing hay fever, with the risk falling progressively for each subsequent sibling (Strachan, 1989; Butland *et al.*, 1997) possibly due to the increased incidence of viral infections in larger families. To an extent, this finding demonstrates a possible effect of socio-economic status on the diagnosis of hay fever. There is little evidence of a true social class gradient for allergic rhinitis, although a study from south-west London suggested that more privileged individuals were more likely to attract a diagnostic label of hay fever from their doctors than those of poorer socio-economic background (Sibbald & Rink, 1991a). Dietary factors and smoking may also impact on the epidemiology of hay fever but evidence is limited.

Incidence

Data on the incidence of rhinitis are scarce. A four-year follow-up of hay fever in 6563 people of all ages in Tecumseh, USA, (Broder *et al.*, 1974) revealed an overall incidence of 2.16% in males and 1.67% in females. The peak incident rates were amongst adolescents of both sexes although the onset of hay fever in girls tended to be later than in boys. The overall remission rate over the same period was 10% for males and 5% for females, with rates higher in older individuals, in those with a recent onset of symptoms, and in those with associated perennial symptoms. The presence of asthma had no demonstrable effect on remission. There are no adequate data reflecting the lifetime experience of any one generation or birth cohort with regards to hay fever or the other forms of rhinitis. If, as is contested, the incidence of atopy is increasing over time, the Tecumseh data, due to its cross-sectional nature, may be confounded by such a cohort effect.

Prevalence

There is also a paucity of data on prevalence. One community-based study in London (UK) has reported on the epidemiology of both seasonal and perennial rhinitis (Sibbald & Rink, 1991b). Of the 5349 subjects, age range from 16–65, the overall prevalence of rhinitis was 24%. An estimated 3% had seasonal symptoms alone, of whom 78% were atopic on skin prick testing. 13% had perennial symptoms alone, (50% atopic) while the remaining 8% had perennial symptoms with seasonal exacerbations, of whom 68% were atopic. It is uncertain if this pattern of disease applies to other populations.

In the UK, there is a clear age-related pattern to hay fever, being unusual amongst children aged less than five years old, highest rates occurring amongst those in their twenties and thirties, and with declining rates as age progresses (Ayres, 1986). The peak prevalence in Denmark (Pedersen & Weeke, 1981) is similar to the UK, but slightly older (24 years) in the USA (Broder *et al.*, 1974) and in Australia (Australian Bureau of Statistics, 1991). There are no appreciable sex differences in the overall prevalence of hay fever, although the ratio of affected males to females has been reported to vary with age. Very little is known about the age or sex distribution of perennial rhinitis.

Estimates of the prevalence of hay fever amongst schoolchildren and adults around the world have shown ranges of between 0.5% and 28% for children and 0.5% and 15% for adults (Montgomery, 1983). However, direct international comparison of these rates is not justified as the rates are derived from the variable application of a variety of case definitions, and the wide age make-up of the sample populations studied. Some geographic differences in the period prevalence of hay fever have been reported from Australia: cumulative lifetime prevalence of hay fever in children aged eight to ten years was significantly higher in dry inland areas (31%) than in damp coastal regions (21%) (Britton *et al.*, 1986). In the UK, general practice hay fever consultation rates tend to be higher in the south than the north (Strachan *et al.*, 1990). These differences have been attributed to geographic variations in the aeroallergen load, this influence of aeroallergen probably being of the most importance in early childhood. Data concerning the effects of urbanisation are contradictory.

Time trends

Anecdotally, hay fever was rare in Britain before the industrial revolution. More recently it has been labelled as a "post-industrial revolution epidemic" (Emanuel, 1988). The consensus is that the prevalence of allergic rhinitis continues to rise. Possibly, the best evidence for this comes from the examinations of Swedish army recruits performed in

1971 and 1981. Allergic rhinitis was diagnosed in 4.4% of conscripts in 1971 and 8.4% of conscripts in 1981 (Aberg, 1989). Consistent with this trend are the rising consultation rates for hay fever in Britain; 5/1000 in 1955–56, 11/1000 in 1970–71 and 20/1000 in 1981–82 (Fleming & Crombie, 1987).

2.2. Croup

Definition

Croup is a disease of infants characterised by inspiratory stridor, a barking cough and hoarseness due to airway obstruction in the glottic and subglottic area (Skolnik, 1989). It is usually a self-limiting illness, but is also a frequent cause of hospital admission, occasionally requiring intubation and ventilatory support (Geelhoed, 1996). The epidemiology of croup is bedeviled by the lack of a consistent case definition. Historically, the label was attached to infants with diphtheria who developed upper airway obstruction due to membrane formation. The term pseudo-croup was applied to another condition, thought to be inflammatory, affecting infants without membrane formation. As diphtheria vaccination coverage increased, the prefix pseudo tended to be dropped (De Boeck, 1995). Another historical area of confusion lay in the classification of other disorders affecting the glottic and supra-glottic area including epiglottitis and bacterial tracheitis, as croup, and this confusion persists (Couriel, 1988; Grad & Taussig, 1990). The syndrome remains poorly defined for the purposes of comparative epidemiology and considerable caution needs to be exercised when examining reported rates even from the same institution at different times.

Aetiology

While a number of viral causes have been identified, human parainfluenza virus 1 (HPIV-1) is the aetiologic agent most commonly associated with croup in developed countries (Denny *et al.*, 1983), being detected in one fifth of all cases of croup and accounting for around 40% of cases where a specific agent is identified. In the developing world, measles virus and, in communities where immunisation is not universal, diphtheria remain important causes of the croup syndrome.

Incidence

Comprehensive incidence data by country are not available. However, it is established that the majority of cases of croup occur within the first three years of life, with the highest incidence in the second year. In developed countries, hospitalisation for croup tends to be twice as common among boys than girls up to the age of five years; a similar though less pronounced pattern has been reported for children in developing countries. The origin of the gender differences is uncertain. In the USA, between 1% and 5% of all children up to the age of three years require outpatient evaluation for croup, and it accounts for between 10% and 35% of lower respiratory tract illness seen in paediatric practice in America. There, between 1% and 6% of children seen in the primary care setting with croup are hospitalised, as opposed to around 30% of cases of croup seen in paediatric emergency rooms. The mean annual number of croup hospitalisations for the years 1979–93 was 41 000 (range 27 000 to 62 000 per year). Ninety-one percent of these hospitalisations were for children aged less than five years old (Marx *et al.*, 1997). Comparable data for the UK are not available.

HPIV-1 is associated with epidemics of acute respiratory illness every other year. It has produced national epidemics in the autumn of odd-numbered years since 1973 in the United States, and the virus is infrequently detected during the intervening periods.

Time trends

There are very few data that adequately reflect time trends in croup incidence. Recently, the first national review of trends in hospitalisation for croup in the United States of America was reported (Marx *et al.*,

1997). Over the 15-year survey of hospital discharge data for the years 1979 to 1993, croup hospitalisations and hospitalisations for all causes for children aged under 15 years declined. The magnitude of this decline is not reported.

3. Asthma

3.1. Definition

Asthma is characterised by variable airway obstruction that is on the whole reversible, airway inflammation, and increased airway responsiveness to a variety of stimuli (Sheffer, 1991). It usually manifests as an episodic condition, with exacerbations being interspersed with symptom free periods. Only a minority displays symptoms continuously. It is a matter of considerable concern in most developed countries; in Britain around 10% of children have a diagnosis of asthma and a further 5% have symptoms suggestive of asthma. It is estimated that between 4% and 6% of adults have diagnosed asthma and an unknown, but probably substantial proportion remain undiagnosed. Acutely, it can cause considerable distress and may be life threatening. As a chronic disease it interferes with normal activities of living, impacting on the education, employment and lifestyle of sufferers and their families. The total cost of asthma to the UK is estimated to approach £1 billion per annum.

There are problems in determining the prevalence of asthma in epidemiological studies (National Institutes of Health, 1995a). The two approaches that are commonly used, questionnaires and tests of lung function have specific problems associated with each. In the former, while responses to questions on asthma and wheeze have been found to be broadly reliable, they may lack precision in differentiating asthma from other respiratory conditions particularly in the very young and the elderly. Bias in applying a diagnostic label may also be a problem. For example, for adults over 40 years of age with similar symptom patterns, women tend to be diagnosed to have asthma more often than men (Dodge & Burrows, 1980). Equally, a single measurement of lung function can be misleading. Serial measurements are more useful, although it is difficult to acquire them reliably at a population level. The measurement of airway responsiveness may be more valuable particularly in children subject to exercise challenge. Among current symptomatic asthmatics, a histamine $PC_{20} \leq 8$ mg/ml has been found to be 100% sensitive, but only has a positive predictive value of 29% (Cockcroft *et al.*, 1992). These issues need to be borne in mind when comparing asthma statistics from differing populations or even within the same population at different time points.

3.2. Asthma Causation

Our current understanding of the causes of asthma is incomplete but a combination of host and environmental factors appear to be involved in both the development of asthma (primary sensitisation) and in the subsequent intermittent production of symptoms (provoking/trigger factors).

Host factors

The clustering of asthma in families and data from twin studies suggest the role of specific genes related to asthma. Atopy, gender (asthma being more common in boys but roughly equal in adulthood), birth factors and perinatal factors (neonatal lung disease, first born and season of birth) have all been shown to impact on asthma prevalence (Gergen & Weiss, 1995). Some, therefore, may potentially affect assessment of prevalence rates for asthma between countries where these factors differ materially (e.g. average family size).

Environmental factors

The most important environmental factor in asthma is allergen exposure (Gergen & Weiss, 1995), both inhaled (e.g. house dust mite and grass pollen) and to a lesser extent ingested. Exposure to allergens will vary both geographically and over time. Exposure to tobacco smoke is also of considerable importance; children whose parents smoke (in particular the mothers) appear to have a greater risk of subsequently developing asthma than children of non-smoking parents (Gortmaker *et al.*, 1982). This effect also appears to occur *in utero* (Taylor & Wadsworth, 1987). Passive smoking is also associated with increased asthma severity (Murray & Morrison, 1988) and may be associated with the earlier onset of asthma (Weitzman *et al.*, 1990). Active smoking increases the risk of developing some forms of occupational asthma (Chan-Yeung & Malo, 1995), occupational sensitisers being major causes of asthma in adults (National Institutes of Health, 1995b).

Migrant studies suggest that immigrants tend to acquire the prevalence of asthma in the area to which they have moved (Gergen & Weiss, 1995). There is no clear pattern of asthma with respect to urban or rural areas. Aspects of the home environment also appear important including the presence of dampness and overcrowding (Brunekreef *et al.*, 1989). It is difficult to differentiate between specific effects attributable to these factors and the effects of socio-economic status (Watson *et al.*, 1996). All these factors need to be considered when comparing trends in asthma over time, between studies.

3.3. Burden of Disease

Incidence

The incidence of asthma varies with age. For all age groups combined, the incidence of new cases of asthma has been estimated to vary from 2.65 to 4/1000 per year; among children under five years of age, the incidence ranges from 8.1 to 14/1000 per year for boys and from 4.3 to 9/1000 per year for girls (Broder *et al.*, 1974; Dodge & Burrows, 1980). The incidence of asthma after the age of 25 years has been estimated to be around 2.1/1000 per year (McWhorter *et al.*, 1989).

There are few valid studies examining changing trends in the incidence of asthma, particularly in age groups other than schoolchildren. In a study of the population of Rochester, Minnesota, between January

1964 and December 1983, the annual age- and sex- adjusted incidence of definite plus probable asthma increased from 183 to 284/100 000. The increase was predominantly in the 1 to 14 years age group (Yunginger *et al.*, 1992).

Point prevalence and trends

There are no comprehensive data on the prevalence of asthma over time (Gergen & Weiss, 1995). Caution must be exercised in comparing reported prevalence data, as confounding variables (e.g. public awareness, cultural attitudes, variations in access to medical care, medical labelling) can impact on observed differences. Asthma prevalence is also highly dependent on the definition used. Table 2 (adapted from Evans III *et al.*, 1987) presents asthma prevalence data from two national surveys that attempt to monitor the health of the US population, the National Health and Nutrition Examination Survey (NHANES) and the

Table 2. Estimates of asthma prevalence: results from two national population-based surveys.

Survey and questions asked	Rate per 100 population
National Health and Nutrition Examination Survey	
(Persons aged 6 months to 74 years, 1976 to 1980)	
Did a doctor ever tell you that you had asthma?	6.2
Do you still have asthma?	3.0
During the past 12 months, not counting colds or the flu, have you frequently had trouble with wheezing?	6.5
Has a doctor ever told you that you had asthma and/or asthma?	10.5
Do you still have asthma and/or wheezing?	7.7
National Health Interview Survey	
(All ages, 1979 to 1981)	
During the past 12 months, did you have asthma?	3.1

National Health Interview Survey (NHIS) conducted from 1976 to 1981. The apparent prevalence of asthma ranges from 3 to 10.5% depending on the question asked. Until the relatively recent initiation of the European Respiratory Health Survey (ERHS) and the International Study of Asthma and Allergies in Childhood (ISAAC) there were few international studies of asthma prevalence employing the same study methodologies across different populations adding further to the complexities of data interpretation.

Table 3 [adapted from Global Initiatives for Asthma (National Institutes of Health, 1995a] displays a summary of some studies of the

Table 3. Prevalence of asthma in children in studies using airway hyperresponsiveness as a test.

Country	Year	Number	Age (years)	Current asthma %	Ever asthma %	Wheeze ever %	Hyperresponsiveness %	Atopy %
Australia	1982	1487	8 to 10	5.4	11.1	21.7	10.1 (H)	29.3
	1986	1217	8 to 11	6.7	17.3	26.5	10.0 (H)	31.9
	1991	1575	8 to 11	9.9	30.8	40.7	16.0 (H)	37.9
New Zealand	1981	813	9	11.1 #	27		22.0 (M)	45.8
	1988	1084	6 to 11	9.1	14.2	27.2	20.0 (H)	
	1989	873	12	8.1 #	16.8	26.6	12.0 (E)	
England	1980	1613		8.0 #		14.8	? (H)	
Wales	1989	965		5.3 #	12	22.3	8.0 (E)	
Germany	1990	5768	9 to 11	4.2 #	7.9		? (C)	?
Denmark	1987	527	7 to 16	5.3			16.0 (H)	31
Spain	1990	2216	9 to 14	?		?	6.9 (E)	
Indonesia	1981	406	7 to 15	1.2	2.3	14.5	2.2 (H)	24.1
China	1988	3067	11 to 17	1.9	2.4	6.3	4.1 (H)	?30
Papua New Guinea	1985	257	6 to 20	0	0	1.7	1.0 (H)	17
Kenya	1991	402	9 to 12	3.3	11.4		10.7 (E)	
Australia (indig. aborigines)	1991	215	7 to 12	0.1	0	1.4	2.8 (H)	20.5

Current asthma: airway hyperresponsiveness plus wheeze in the last 12 months, # indicates a figure calculated from published data.
Ever asthma: asthma ever diagnosed.
H: Histamine, M: Methacholine, E: Exercise, C: Cold,
Atopy: positive skin prick test response
All figures are a percentage of the population tested

Table 4. Prevalence of asthma in children in studies using alternative approaches to the detection of asthma.

Area	Date	Number	Age (years)	Current %	Cumulative prevalence %	Definition of asthma	Method
UK							
Birmingham	1961	49 273	5–15		1.76	"Recurrent wheezing and dyspnoea"	
	1971	20 958	5–18	(2.3)	(4.2)	"Ever had asthma?"	School medical examination
				3.2	5.7	"Wheezing and shortness of breath"	
				5.4	9.9	Both	
	1976	12 733	5–16	(2.6)		"Ever had asthma?"	School medical examination
				6.3		Asthma and wheezing, and shortness of breath	
Oxford	1958	14 005		(1.1)	(1.6)	"Asthma" mentioned in records	School records
Aberdeen	1969	2511	10–15	(3.6)	11.5	"Recurrent dyspnoea of an obstructive type"	Random selection questionnaires
Nationwide	1978	13 509	11	(2.0)	(3.5)	"Ever had asthma?" or	One week birth cohort
				2.9	8.8	"Ever had wheezy bronchitis?"	
				4.9	12.3	Both	
US							
Colorado	1964	2235		(2.8)		"Difficulty in breathing with wheezing"	Questionnaire
Tucson	1980	437	5–14	(8.5)		"Ever had asthma?" and "Have you seen a doctor about your asthma?"	Random selection of population questionnaires
Tecumseh (Michigan)	1974	3761	0–15	7.9	12.5	Wheeze and shortness of breath	Population survey questionnaire
				(4.3)	(6.8)		
Nationwide	1973	66 711	< 17	(3.1)		"In the past year did anyone in the family have asthma?"	Questionnaire
Sweden	1954	60 063	7–14	(1.4)		Not given	School questionnaire
Switzerland	1979	2451	15	(0.89)	(1.9)	Not given	Interview examination
Australasia							
Melbourne (Aus)	1969	30 000	7–10	(3.7)	(11.4)	Wheezy breathlessness	Random sample questionnaire, examination
					19.1		

Table 4 (*Continued*)

Area	Date	Number	Age (years)	Current %	Cumulative prevalence %	Definition of asthma	Method
Lower Hutt (NZ)	1969	952			(7.1)	"Has this child suffered from asthma?"	Questionnaire of school population
Sydney (Aus)	1980	6566	mean 13.9		(8.8)	"Asthma that has required treatment by a doctor or hospital"	Schoolchildren random sampling questionnaire
Israel	1973	14 941	< 14	(0.4)		More than 5 asthmatic attacks- most recent within 2 years	General practitioner lists, hospital records, interview
Japan Tokyo	1966	113 112	6–12	(0.7)		Not given	Randomly selected schoolchildren
Singapore	1972	21 406		(1.5)	(3.9)	Asthma diagnosed by family doctor	School survey questionnaire
India Patna	1966	4413	0–9	(0.2)		Not given	Population survey questionnaire
Zimbabwe (semiurban)	1980	4293	5–19		1.2	Variable wheezy breathlessness	Population survey questionnaire
Tanzania (rural)	1977	242	mean 12.5	7.9		Recurrent wheezy breathlessness	Schoolchildren questionnaire
Turkey Ankara	1969	1163	6–13		2.1	Wheeze and difficulty in breathing	Medical history and examination
Maldives	1979	2903	< 15	20.7		Reversible paroxysmal airways obstruction	Whole population survey, questionnaire
Tokelau Islands	1978	706		(1.3)		Probable history and signs of asthma	Interview and examination
Western Caroline Islands	1978	123	15	49.0		Recurrent wheeze and dyspnoea	Questionnaire and examination
Papua New Guinea South Fore	1983	1081	< 20	0.6		History of intermittent breathlessness and bronchodiltor response/histamine provocation	Population study questionnaire
Tahiti	1982	3870	14–18		11.5	"Have you had attacks of asthma?"	Population study questionnaire

Figures in brackets reflect crude morbidity rates.

prevalence of asthma in children which used airway hyperresponsiveness as a test. There are many data available for Australasia and the United Kingdom, but fewer for other regions.

Table 4 (adapted from Gergen & Weiss, 1995) summarises other estimates of the crude prevalence of asthma in children adopting alternative definitions of the diagnosis of asthma.

Some broad conclusions can be drawn from these data; the point prevalence of current asthma for children in 'industrialised' countries ranges from between 0.7% in Tokyo, Japan, to 8.5% in Tucson, Arizona, USA and averages around 5% worldwide. Rates for "non-industrialised" countries vary from high levels in the Western Caroline Islands, 49% to 0.6% in South Fore, Papua New Guinea. Broadly, "industrialised" countries tend to have higher prevalence rates in children than "non-industrialised" countries.

Table 5 [adapted from Global Initiatives for Asthma (National Institutes of Health, 1995a)] and Fig. 1 show estimates of the prevalence

Table 5. Changes in the prevalence of asthma or symptoms in populations studied with the same method on two occasions.

Country	Year	Number	Age (years)	Current asthma %	Diagnosed asthma %
Australia	1982	769	8 to 12	6.5	12.9
	1992	795	8 to 11	9.9	19.3
New Zealand	1975		12 to 18		26.2
	1988		12 to 18		34
Wales	1973				6
	1988	965			12
United States	1971–74	Large	6 to 11		4.8
	1976–80	27 275	6 to 11		7.6
France	1968	814	21		3.3
	1982	10 559	21		5.4
Tahiti	1979	3870	16		11.5
	1984	6731	13		14.3

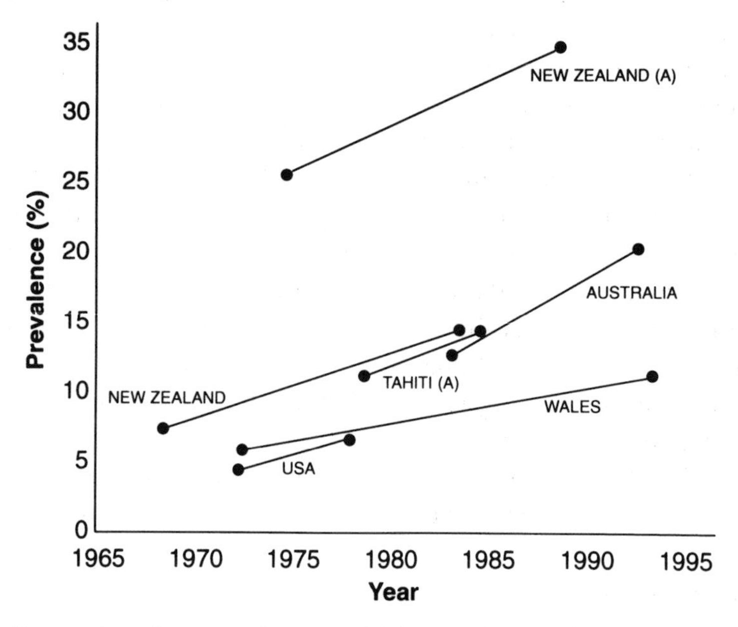

Fig. 1. Changes in asthma prevalence in children over time where comparable methodology is employed using paired studies.

of asthma or symptoms in populations studied with the same method on more than one occasion. In addition, in the USA the NHIS found that reported asthma amongst 6–16-year-olds increased by 28% from 3.2% in 1970 to 4.1% in 1978–80 (Halfon & Newacheck, 1986). Further NHIS analyses comparing the 1981 and 1988 Child Health Supplement reported that asthma increased from 3.1% to 4.3% among 0 to 17 year olds (Weitzman *et al.*, 1992). When broken down by race, the increase was noted in white but not black children. The NHANES study found that the reported prevalence of "ever asthma" amongst 6 to 11 year olds increased by 58% from 4.8% in NHANES I (1971–74) to 7.6% in NHANES II (1976–80) (Gergen *et al.*, 1988). Comparing across studies, therefore, can be difficult even where there is internal consistency in methodology over time.

Increases in prevalence have been reported from various countries including England. In Sweden, the prevalence of asthma increased 47% (from 1.9% to 2.8%) among 18-year-old military recruits between 1971 and 1981 (Aberg, 1989), whereas in Finland, prevalence increased by around 500% (from 0.29% to 1.79%) among 19-year-old military recruits between 1966 and 1989 (Haahtela *et al.*, 1990). In New Zealand, an identical questionnaire administered to 11- to 13-year-olds in 1969 and 1982 revealed an increase in prevalence of asthma by around 90% (from 7.1% to 13.5%) (Mitchell, 1983). The populations attending the schools became poorer and more Polynesian over this time. Two surveys of 7-year-old Melbourne (Australia) schoolchildren 26 years apart reported an increase in ever asthma or ever wheezing with bronchitis or cold by 141% (from 19.1% to 46%) (Robertson *et al.*, 1991).

In England, the data are conflicting. Burney *et al.* (1990) examined the reported incidence of asthma or bronchitis in the preceding 12 months and ever wheezing over a 13-year period (1973–86) among children aged between 5 and 12 years old. Over this period the prevalence of asthma increased approximately 138% among boys and 378% among girls with similar changes in wheezing. Anderson felt there was little evidence for a trend in the prevalence of reported wheezing in the previous year having reviewed the results of prevalence studies for the UK from 1964 and 1986 (Anderson, 1989). Other community-based sources of data (General Household Surveys (GHS) of 1988 and 1989 and the most recent Morbidity Study in General Practice 4 (MSGP4)) suggests a slight increase over time (Central Health Monitoring Unit, 1995).

The data from the Health and Lifestyle Survey conducted in 1984/5 (HALS1) and 1991/2 (HALS2) gives us a further perspective on changes in asthma prevalence in the UK (Central Health Monitoring Unit, 1995). Unlike the GHS in which a new sample is selected each year, the HALS2 sample consisted of those of the HALS1 sample who could be traced and agreed to co-operate. In 1984/5, the survey covered persons aged 18 and over; in 1991/2, because the same individuals were involved, the youngest participants were 25. The HALS1 sample consisted of around 9000 persons and HALS2, of over 5000 of them. In HALS1, 5% of women and 6.5% of men said they had ever had asthma;

the corresponding figures for HALS2 were 8.6% and 9.1%, respectively. These increases occurred in virtually all age/sex groups. This represents good evidence for a rise in asthma in both children and adults, over the 1980s, in the UK.

Asthma severity

Few of the serial prevalence studies of asthma have gathered comparable data on asthma severity. Data are available from two UK studies: the comparison between the 16 year olds from the national birth cohorts of 1958 and 1970, and from the replicate surveys of 8 year old children in Croydon, South London, in 1978 and 1991. In the former, the prevalence of frequent attacks of asthma or wheezy bronchitis (defined as attacks occurring more than once a week) more than trebled from 0.2% in 1974 to 0.7% in 1986 (Anderson *et al.*, 1994). Although this was a significantly greater proportionate increase than for annual period prevalence, the proportion of children who had missed school due to wheezing illness over the past year was little changed (1.5% in 1974 compared to 1.7% in 1986). More extensive information is available from the Croydon studies. The prevalence of frequent attacks of wheezing was almost the same in each survey while the prevalence of sleep disturbance due to wheezing, restriction of activities at home and the prevalence of speech limiting attacks of wheeze had almost doubled. In contrast, prolonged school absence due to wheezing fell by half (Anderson *et al.*, 1994).

Mortality

Fortunately, asthma deaths are rare and consequently are a poor indicator of asthma incidence or prevalence. Adequate mortality statistics are also available for relatively few countries and are rarely available for different populations within countries. Despite these limitations, there is considerable interest in recent trends in asthma mortality as they are believed to include a number of preventable deaths and consequently have been suggested as surrogates for the quality of primary and secondary health care provision within communities.

Death certification provides the source for nearly all the major studies of asthma mortality. The limitations of these data are now being understood and include the decreasing accuracy of asthma as a cause of death for those aged over 35 at the time of death, the effect of the change in ICD coding rules in 1979, the low percentage of deaths that are validated by autopsy, and the regional and temporal habits of certifying doctors within and between countries (Pearce *et al.*, 1995). Despite the limitations of these data, some general conclusions can be drawn. There is a large variation in asthma mortality by country (Fig. 2 adapted from Sears 1991); during the period 1985–87, rates varied from a high 9/100 000 in West Germany to a low 1.5/100 000 in Hong Kong for persons of all ages. Increased mortality appears to be associated with age, ethnicity and poverty at the population level (Pearce *et al.*, 1995). Age-specific seasonal variations in asthma deaths have also

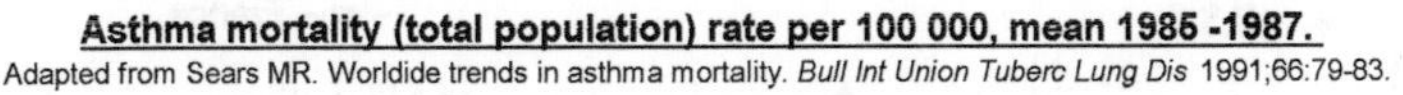

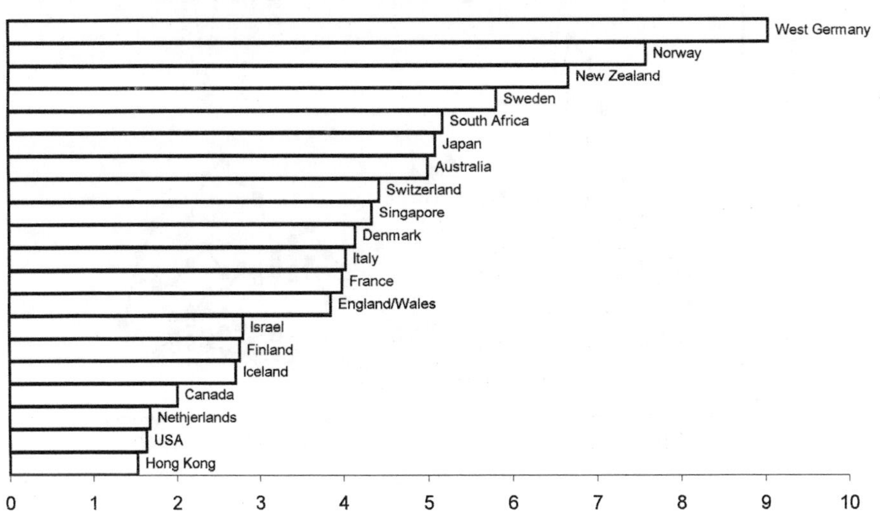

Fig. 2. International comparisons of asthma mortality rates.

been noted. For instance, during the epidemic of asthma deaths in the UK in the 1960s, mortality was highest in the third quarter of the year (Inman & Adelstein, 1969).

At the level of the individual, a history of severe disease, lack of access or reluctance to seek care, sub-optimal pharmacotherapy, depression and family disturbance, and rapidity of onset of attack are all associated with increased risk for asthma death (Pearce *et al.*, 1995).

Figures 3 and 4 [adapted from the Global Initiatives for Asthma (National Institutes of Health, 1995a)] demonstrate international trends in asthma mortality from the 1960s onwards, for 5- to 34-year-olds and for all ages, respectively. Rates for Canada and the USA have been generally lower than most other countries for this period, although within the United States there are wide variations in asthma mortality rates. In the 1960s, there was a rise in death rates in New Zealand,

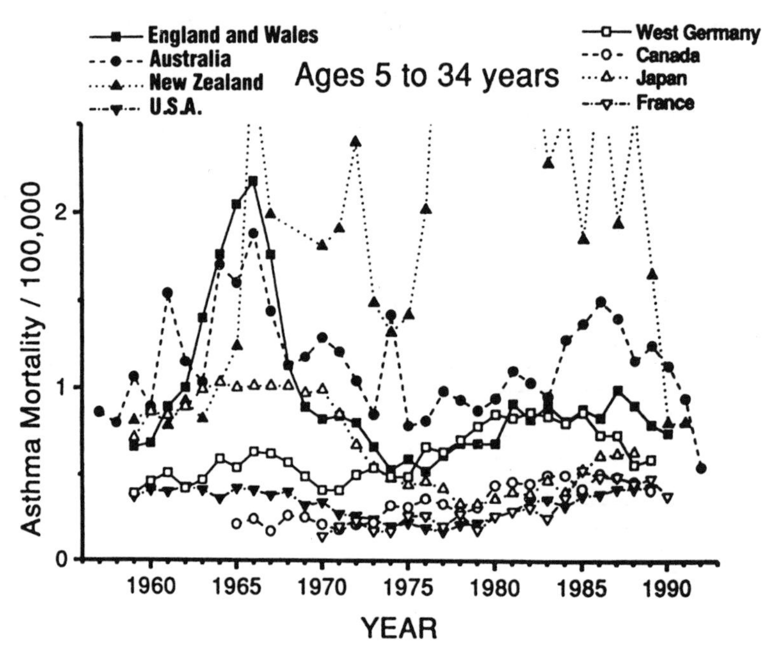

Fig. 3. Trends in asthma deaths over time by country (age 5–34).
(Printed with permission from Prof. Stephen T. Holgate)

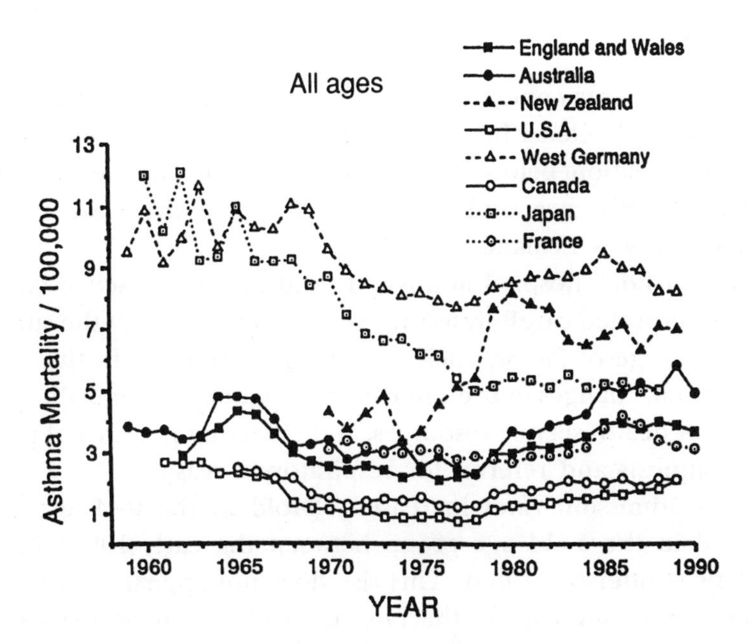

Fig. 4. Trends in asthma deaths over time by country (all ages). (Printed with permission from Prof. Stephen T. Holgate)

Australia and the United Kingdom; a decade later, a second epidemic of deaths was seen in New Zealand. In the late 1970s and early 1980s, there was a general increase in asthma mortality rates in most countries. Since then, rates in England and Wales appear to have fallen for all age groups except those aged over 85. A similar downward trend has been seen for the 5–54-year-old age group in New Zealand. Other countries have had different experiences; in Scotland, rates for the 5–44-year-olds have been relatively stable whereas the French have seen rates peak for both the under and over 35-year-old age groups between 1985 and 1987 which have been attributed to influenza epidemics.

Why some countries appear to have managed to cut their asthma mortality rates, and others not, remains controversial. In the United Kingdom, the increasing awareness of asthma by patients and doctors, and the encouraging increasing uptake of prophylactic medication have been offered as possible explanations (Campbell *et al.*, 1997).

Health service use

Disease-specific data concerning health service utilisation for countries other than the United Kingdom is scarce, and we have concentrated on this information below. Nonetheless, there is good evidence for increasing health care utilisation for asthma in the USA, Canada, Australia and New Zealand.

Trends in both hospital admissions and general practice contacts must be interpreted carefully as they are as likely to reflect the summed effect of a range of factors such as shifting patterns of health service use in general, changes in preferences for inpatient or domiciliary care, rationing of health service resources, and changing patterns of presentation, diagnosis and referral for respiratory disease.

Asthma admission rates increased 13-fold in the 0–4 age group and 6-fold in the 5–14 age group between the early 1960s and the mid 1980s (Anderson, 1989). This rise does not appear to have been explained by an increase in the rates of readmission or a substantial shift in disease severity prompting admission. Similarly, only a small proportion of the rise has been accounted for by changes in diagnostic labelling. Over the same period, the proportion of all paediatric admissions due to asthma also rose. Several forces may have tended to increase the likelihood that acute episodes of wheezing would have resulted in hospital admission over this period, including a tendency for members of the public to use accident and emergency rather than their general practitioner as the point of first access to care for their sick children.

Trends in hospital admissions for adults over the same period have been less marked, and their direction has varied over time, declining slightly in the early 1970s before resuming an upward trend in the 1980s (Committee on the Medical Effects of Air Pollutants, 1995). The reasons behind these trends remain speculative. In Scotland and Wales, the upward trend for all age groups continued through the 1980s (Mackay, 1992), although recently, Scottish rates have levelled off and Welsh rates may even be in decline. The apparent plateau in the English rates is consistent with these patterns.

The National Morbidity Surveys conducted in 1955–56, 1970–72, 1981–82 and 1991–92 suggest a marked increase in the proportion in all age groups consulting a general practitioner for asthma over this time, probably not due to changes in the catchment population or to diagnostic transfer (Central Health Monitoring Unit, 1995). Weekly returns are available for the limited number of general practices participating in the Royal College of General Practitioners' sentinel practice scheme. These demonstrated a striking increase in the mean weekly attack rate for acute asthma in all age groups from 1976 to 1992 particularly in the under-fives (Central Health Monitoring Unit, 1995). Trends in acute bronchitis also rose over the same time period, suggesting that diagnostic transfer is unlikely to be responsible for the trend (Central Health Monitoring Unit, 1995).

3.4. Conclusion

In the past three decades, data from a wide variety of sources have supported the contention that asthma prevalence is rising. It is highly unlikely that this world-wide pattern has a single cause. Environmental factors are clearly important, with the increased prevalence of asthma being largely accounted for by increased expression of IgE dependent hypersensitivity (Holgate, 1998). This expression can be linked to a variety of stimuli acting on differing populations with allergen exposure and factors operating in pregnancy and in the first three years of life being very relevant. Any inference concerning the potential role of air pollution in the rising prevalence of asthma needs to be considered against this complex background.

4. Chronic Obstructive Pulmonary Disease (COPD)

4.1. Definition

There is no generally agreed, precise definition of chronic obstructive pulmonary disease (COPD) but it may be understood as a disease state characterised by airflow obstruction, particularly of the small airways, in the presence of chronic bronchitis and/or emphysema, sometimes accompanied by airway hyper-reactivity and may be partially reversible (American Thoracic Society, 1995). Generally, individuals with asthma who demonstrate incomplete reversibility of airflow limitation should not be considered to have COPD, but a degree of misclassification is inevitable. Some clinicians would demand a history of tobacco smoke exposure to apply a diagnostic label of COPD.

Chronic bronchitis

Chronic bronchitis (CB) became the focus of epidemiological interest following the demonstration of the acute effects of the London smog episode of 1952 on respiratory mortality (Ministry of Health, 1954) resulting in it becoming one of the first diseases to be defined for the purpose of epidemiological research. It was defined by the MRC as the presence of a chronic productive cough for at least three months of the year in two successive years (Medical Research Council Committee, 1965); it was considered "obstructive" or "simple" depending on the presence or absence of breathlessness on exertion. Later, longitudinal studies of working men demonstrated that mucus hypersecretion and progressive airflow obstruction while linked through the risks associated with tobacco smoking, were independent disorders with distinct natural histories (Fletcher *et al.*, 1976).

Tobacco smoking has been identified as the major risk factor for the future development of CB while a number of other less important associations have been demonstrated. The combination of multiple risk factors, and incomplete understanding of aetiology has made it impractical to use aetiology as a defining characteristic for CB.

Emphysema

Emphysema is defined as a condition of the lung characterised by abnormal, permanent enlargement of airspaces distal to the terminal bronchiole accompanied by destruction of their walls and without obvious fibrosis (Snider *et al.*, 1985). This leads to airflow limitation consequent to the loss of alveolar attachments to bronchioles and small bronchi. The small airways are the major site of increased resistance in patients with COPD, due possibly to small airway inflammation (Hogg *et al.*, 1968), present in both patients with predominant emphysema and predominant chronic bronchitis. Tobacco smoking is again recognised as the major risk factor associated with this disease. In the clinical setting (and therefore with respect to epidemiology) the diagnosis is reached on the basis of clinical and physiological features. The development and application of quantitative CT scanning has facilitated the process of diagnosis at the level of the individual, but has yet to be applied in the field of population studies.

4.2. *Development of Impaired Lung Function*

The occurrence of severe disability and death in chronic lung disease is related most closely to symptoms of breathlessness and to levels of ventilatory function measured up to 20 years previously. Traditional epidemiological studies of risk factors for COPD have focused on measures of ventilatory decline in adult life and have provided attributable risk estimates for active cigarette smoking in adult life of around 80%. More recently, it has become appreciated that there are likely to be early life predictors of adult COPD risk. To date, little attention has been paid to these predictors or the possible influence of changing environmental factors on their distribution within the population over time.

Reduced forced expiratory volume in one second (FEV_1) is the cardinal feature of airflow obstruction and is the most important predictor of the subsequent development of COPD. Normally FEV_1 increases with age until adulthood, then plateaus and subsequently declines linearly with age. Theoretically there are 3 separate processes which could result in a reduced FEV_1 in later adult life:

- failure to achieve normal FEV_1 growth in childhood followed by a normal rate of decline in adulthood

- truncation of the plateau phase followed by a normal rate in decline of FEV_1 in adulthood
- an accelerated loss of FEV_1 in adult life

A combination of these three could also occur, which may in part explain some of the complex clinical heterogeneity of the natural history of COPD. When considering the possible role of air pollutant exposure when assessing trends in COPD (either prevalence or exacerbations), these factors need to be considered and, where possible, allowed for. Unfortunately, sufficient information is rarely available for such considerations to be adequately dealt with.

Factors predicting submaximal growth in FEV_1

Cross-sectional and longitudinal studies have shown a 5–7% reduction in FEV_1 attainment by the age of 14 in children exposed to environmental tobacco smoke (ETS) (Tager *et al.*, 1983). More recently, in infants as young as 1 month of age, investigators have demonstrated reductions in maximal flow at functional residual capacity in infants of mothers who smoke (Hanrahan *et al.*, 1992; Young *et al.*, 1991) more so in females than males (Tager *et al.*, 1993), suggesting an *in utero* effect of smoke exposure. Children who begin to smoke actively at age 15 years and who continue to smoke, achieve only about 92% of their expected FEV_1 by the age of 20 years (Tager *et al.*, 1985), the total individual consumption needed to achieve this reduction being in the order of one pack-year. Increased airway responsiveness may also be associated with lower maximal attainment of FEV_1 in children (Weiss, 1995).

It has been suggested that lower respiratory tract illness (before the age of two) is associated with reductions in maximally attained FEV_1, although the potential role of recall bias in explaining these findings has always been problematic. Recently, cohort studies of children, followed from an early age, have confirmed this association (Gold *et al.*, 1989), more so for boys than girls and being independent of exposure to ETS and of a physicians diagnosis of asthma. Symptomatic childhood asthma is associated with a reduction in FEV_1 of up to 15% by the age of 15 years (Weiss *et al.*, 1992). The reduction appears to be proportional to the severity of asthma as defined by symptoms. Again, gender differences are evident, with greater effects in girls with mild asthma than in boys.

Factors associated with premature onset of lung function decline

Active smoking is an important factor in the early onset of lung function decline. For instance, the FEV_1 plateau phase evident in normal adult males was markedly attenuated in cigarette smokers and in those with chronic respiratory symptoms (Tager *et al.*, 1988). Lung growth improves in individuals who stop smoking between the ages of 15 and 20 years (Xu *et al.*, 1994). This appears to be the only period in life where smoking cessation actually improves lung function as opposed to simply attenuating its rate of decline.

The presence of wheezing at 14 years of age and its persistence to 28 years of age is associated with decrements in lung function of approximately 20% predicted by the end of the plateau phase in asthmatic subjects (Roorda *et al.*, 1994).

Factors associated with accelerated decline in lung function in adulthood

Cigarette smoking is the most important risk factor for this disease in adults in developed countries causing both mucus hypersecretion and progressive airflow limitation. There is a strong dose-response relationship between the amount smoked and the rate of decline in FEV_1 in population terms; active smokers losing around 50 ml per year of FEV_1 as compared with around 30 ml per year in non-smokers (Pride & Burrows, 1995), albeit with appreciable between individual variation (Fletcher & Peto, 1977). While high tar cigarettes are more likely to cause mucus hypersecretion, the tar content of cigarettes has not been shown to be a determinant of rate of decline in lung function.

There is a weak but statistically significant association between ETS exposure and lung cancer among non-smokers. Given that active smoking is as strongly associated with death from chronic respiratory disease as it is to death from lung cancer, a similar weak relationship might be expected between passive smoke exposure and the development of COPD. The findings of the limited numbers of studies to date addressing this issue are consistent with such a relationship, but none, either alone or pooled are statistically significant.

The occupation of individuals used in studies of the effects of air pollution is important and is only partly addressed in some studies. Many dusty occupations are associated with the development of mucus hypersecretion. Recently, evidence has been accumulating supporting an association between dust exposure and reduced FEV_1, but it has been more difficult to determine whether persistent airway obstruction develops to a greater degree than would be predicted by smoking history and socio-economic factors alone. Longitudinal studies of decline in FEV_1 together with quantitative measures of dust exposure in gold and coal miners have established a small additive effect of dust exposure and smoking in accelerating decline in FEV_1 (Marine *et al.*, 1988). Cotton workers, grain handlers, cement workers and farmers also appear to be at risk of progressive airway obstruction. It is methodologically difficult to determine the relatively subtle influences that occupational factors may have in COPD, but if anything, it is likely that they are being underestimated.

Severe deficiency of the antiprotease α_1-antitrypsin, associated with the PiZZ phenotype, is an important risk factor for the development of emphysema and associated airflow limitation. The current best evidence suggests that intermediate deficiency of α_1-antitrypsin associated with the heterozygous state is not associated with an excess risk of COPD.

The frequency of infective episodes tends to increase with the severity of airway obstruction and infection is the commonest precipitant of acute on chronic respiratory failure and/or death in advanced COPD. Although trials of continuous winter prophylactic antibiotics over several years in men with chronic bronchitis have shown no significant slowing rate of decline of lung function, such information is important when assessing time trends in COPD. However, once obstruction has developed, mucus hypersecretion and/or infections may accelerate further decline in FEV1. In addition, in the presence of defects in lung defence mechanisms such as complete or subclass hypogammaglobulinaemia, frequency of infections does appear to be related to sustained decline in FEV_1.

The role of airway hyperresponsiveness continues to be controversial. It has however been difficult to determine whether it is a true risk factor for accelerated decline in FEV_1 or whether it simply follows as a result of the development of smoking-related airway disease and decreased lung function. The American Lung Health Study demonstrated a strong linear relationship between bronchial hyperresponsiveness and accelerated decline in FEV_1 reinforcing its potentially important role (Anthonisen *et al.*, 1994).

4.3. Prognosis

Age and baseline FEV_1 are the strongest predictors of mortality in COPD. Less than 50% of patients with an FEV_1 less than 30% predicted, survive for five years or more (Anthonisen, 1989). Mortality is also related to reduced body weight, but because this is in turn related to reduced FEV_1, it is not clear if this has an independent effect on survival in COPD.

4.4. Burden of Disease

Despite the importance of COPD, very little comprehensive data are available on the burden of this disease borne by societies around the world. Adequate data are really only available for the USA and the UK.

Prevalence

Table 6 summarises the prevalence of symptoms of chronic cough and phlegm production reported in surveys of general population samples in Great Britain over the past 40 years. In general, the prevalence of

Table 6. Prevalence of chronic cough and phlegm in population-based studies in Great Britain 1950–90.

Area	Date	Age	Men					Women				
			No. of men	% Chronic cough	% Chronic phlegm	% Persistent cough + phlegm	% Current smoker	No. of women	% Chronic cough	% Chronic phlegm	% Persistent cough + phlegm	% Current smoker
Leigh [urban England]	1954	55–64	84	30	33	18	87					
Glamorgan [rural Wales]	1955	55–64	86	31	28	26	74	92	13	13	8	24
Annandale [rural Scotland]	1956	55–64	87	29	23	20	80	92	15	14	11	22
Rhondda Fach [urban Wales]*	1958	55–64	88	42	35	29	73	173	16	16	10	17
GB	1958	45–54	422	43	34		77	346	20	15		37
		55–64	420	53	49		73	344	22	21		26
GB	1965	35–54	2936			17	57	3115			9	44
		55–64	1355			25	56	1591			9	25
GB	1972	37–54	3610	28		17		3716	18		12	
		40–59	3832	29		19		4031	20		13	
		55–64	1836	32		24		2045	21		14	

Table 6 (*Continued*)

Area	Date	Age	Men					Women				
			No. of men	% Chronic cough	% Chronic phlegm	% Persistent cough + phlegm	% Current smoker	No. of women	% Chronic cough	% Chronic phlegm	% Persistent cough + phlegm	% Current smoker
GB [selected towns]	1978	40–59	7735			16	41					
London [suburban practice]	1985	40–59	414	20	18			482	12	8		
		55–64	197	27	27	21		245	15	12	9	
Scotland	1987	40–59	4114	16	17		39	5220	12	11		38
S. England [rural popn.]	1990	35–54	1123	16		8		1241	11		7	
		56–64	539	19		15	29	716	13		8	23

*Non-miners only.

chronic cough and phlegm in men appears to have declined in line with a decreasing proportion of active smokers. There has been little change in women. In the late 1980s, 15–20% of middle-aged men and about 8% of middle-aged women in Britain reported chronic cough and phlegm.

Data concerning ventilatory function are considerably more limited; The Health and Lifestyle Survey assessed a representative sample of 2484 men and 3063 women aged 18 to 65 years by turbine spirometry in their homes (Cox, 1987). Overall, 10% of men and 11% of women had significant airflow obstruction, the proportions with poor function

Table 7. Proportion of persons consulting a general practitioner annually for chronic respiratory disease by age and sex, Great Britain 1955–56, 1970–71, 1981–82.

Diagnosis	Age	Men (per 1000)			Women (per 1000)		
		1955–56	1970–71	1981–82	1955–56	1970–71	1981–82
Chronic bronchitis*	45–64	32.7	29.6	12.3	12.9	12.0	6.7
	65–74		73.8	37.9		23.5	13.6
	75+		70.7	47.5		23.2	12.3
	65+	72.0			32.3		
Emphysema & COPD**	45–64	3.1	3.4	6.5	0.2	0.5	3.0
	65–74		11.1	26.2		1.5	7.8
	75+		5.0	31.2		0.4	7.2
	65+	4.4			0.8		
Asthma	45–64	9.0	8.1	13.9	10.6	9.7	18.1
	65–74		10.7	21.5		9.1	18.7
	75+		6.6	16.2		5.7	12.9
	65+	7.9			8.9		
Chronic bronchitis, emphysema, asthma & COPD	45–64	44.8	41.1	32.7	23.7	22.2	27.8
	65–74		95.6	85.6		34.1	40.1
	75+		82.3	94.9		29.3	32.4
	65+	84.3			42.0		

*Excludes bronchitis unspecified but includes bronchitis with mention of emphysema.
**Emphysema without mention of bronchitis in 1955–56; includes COPD in 1981–82 only.

increasing with age especially among smokers. In British general practice, there has been a modest decline in the proportion of middle-aged men and a slight increase among middle-aged women consulting for chronic respiratory disease between the 1950s and the 1980s (Strachan, 1995) (Table 7). Among both men and women over the age of 65 years of age, the overall proportion consulting with chronic respiratory disease has changed little, but there is evidence of some shifts in diagnostic preference away from chronic bronchitis towards other terms. The extent to which these shifts reflect real changes in the distribution of disease as opposed to diagnostic fashion is unclear.

Mortality

Interpreting mortality statistics relating to chronic respiratory disease is difficult. A substantial proportion of people have evidence of 'chronic respiratory disease' (CRD) and consequently a large proportion of people die "with" CRD but it is often debatable if an individual dies "as a consequence" of CRD. While recent changes in the rules for coding underlying cause of death (WHO Rule 3) have tried to address this problem, ambiguity and inconsistent application of the changes is inevitable.

Also, it is now accepted that reduced levels of FEV_1 predict subsequent mortality from a range of non-respiratory diseases, including coronary heart disease and stroke. These associations are even apparent in lifelong non-smokers and are independent of smoking habit. Significant numbers of deaths in part attributable to COPD are consequently likely to be concealed among these deaths certified to other causes.

Another problem that needs to be borne in mind when comparing mortality statistics over time and between countries is the differing use of diagnostic labels. Historically in Britain, chronic bronchitis (ICD9 491) was the cause certified in the majority of people dying from chronic respiratory disease. Emphysema, asthma and bronchitis unspecified (ICD9 492, 493 and 490, respectively) were each coded in only a small proportion of cases. A separate ICD code for "chronic airway

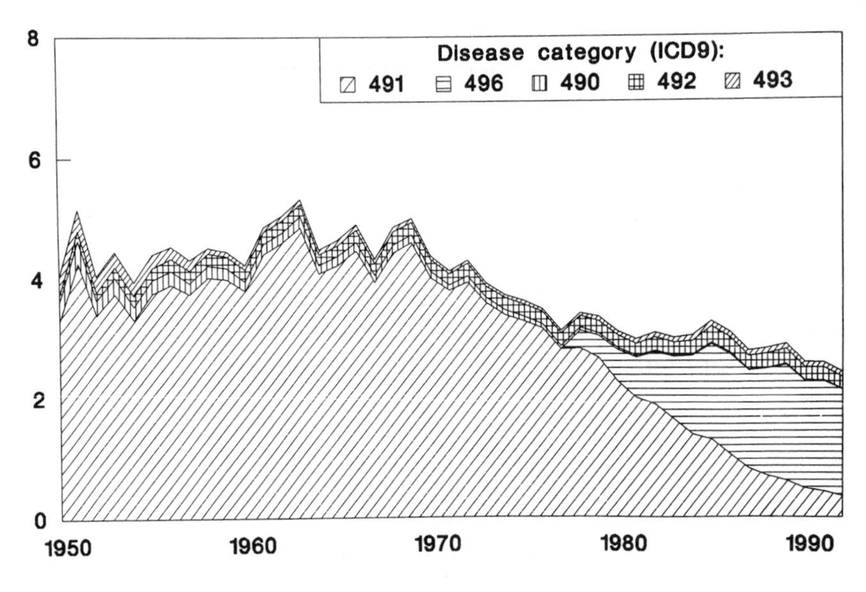

Fig. 5. Deaths in males for chronic respiratory disease over time in the UK.

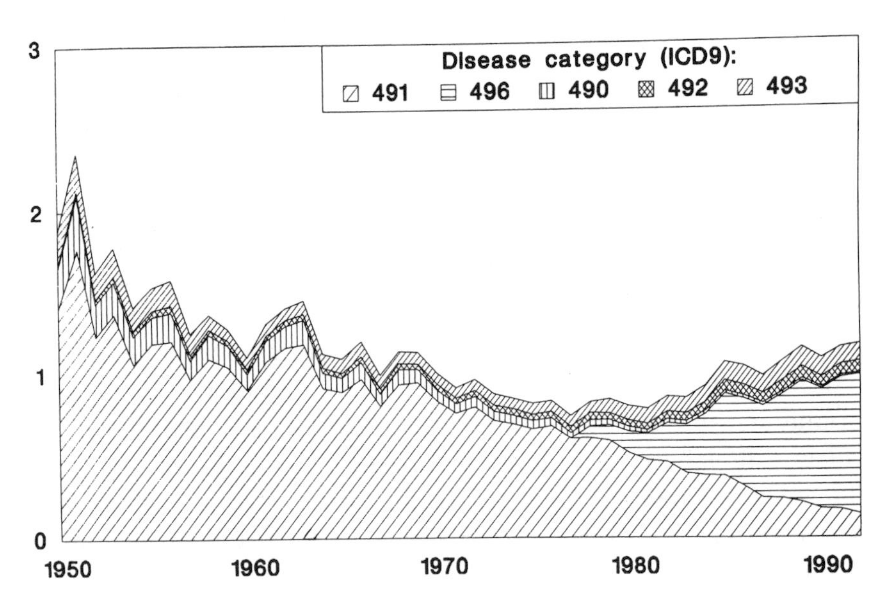

Fig. 6. Deaths in females for chronic respiratory disease over time in the UK.

obstruction not elsewhere classified" was introduced in 1978 (ICD8 519.8, ICD9 496). Since then it has been used increasingly and over the same period there has been a substantial decline in mortality attributed to chronic bronchitis, suggesting diagnostic transfer is at work (Figs. 5 and 6).

In 1992, in England and Wales, there were nearly 27 000 deaths due to CRD, of which 19 963 were due to COPD, accounting for 6.4% of all deaths in males and 3.9% of all deaths in females. In 1995, the total number of deaths certified due to chronic obstructive lung disease was 21 199 (40.9 deaths per 100 000 of the total population).

Health service impact and employment

Chronic respiratory disease has a considerable impact in terms of lost working time. Due to changes in the rules for self certification and sick pay benefits, adequate data are only available for the UK up to 1983. Before the changes, around 25 million days per annum were lost among men (9% of all certified sickness absence) and over 3 million days among women (3.5% of all certified sickness absence). Similar data for other countries are not readily available so this endpoint, if being considered for studies in air pollution would need to be gathered prospectively specifically for the purpose of the study.

4.5. Trends

Time trends

Mortality rates in England and Wales for chronic respiratory disease broadly correspond to the changes in disease prevalence, with declining rates among older women in the 1950s and 1960s and among middle aged men in the 1970s and 1980s (Figs. 7 and 8). These rates are likely to have been subject to non-linear cohort effects, mortality rates showing a marked peak for men born around 1900 and for women born around 1925 (Barker & Osmond, 1986). These effects closely follow those for lung cancer mortality and are thought to correspond

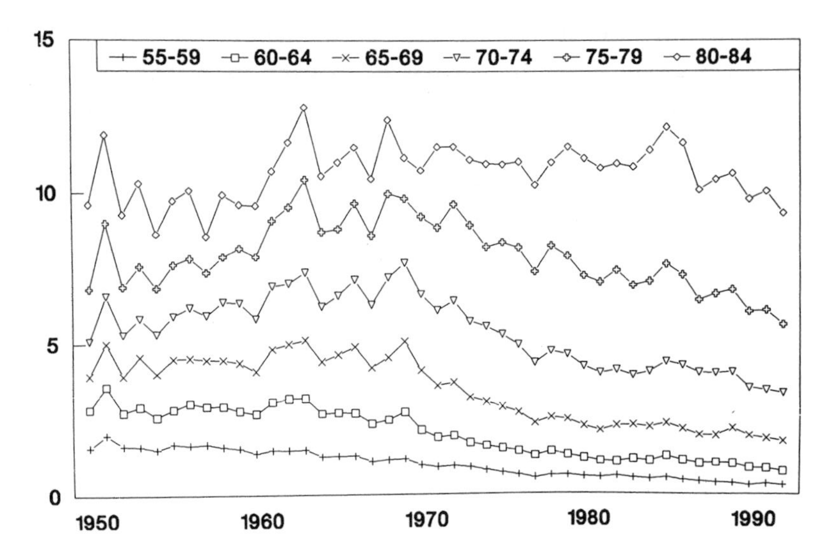

Fig. 7. Age-specific deaths from chronic obstructive pulmonary disease in males over time in the UK.

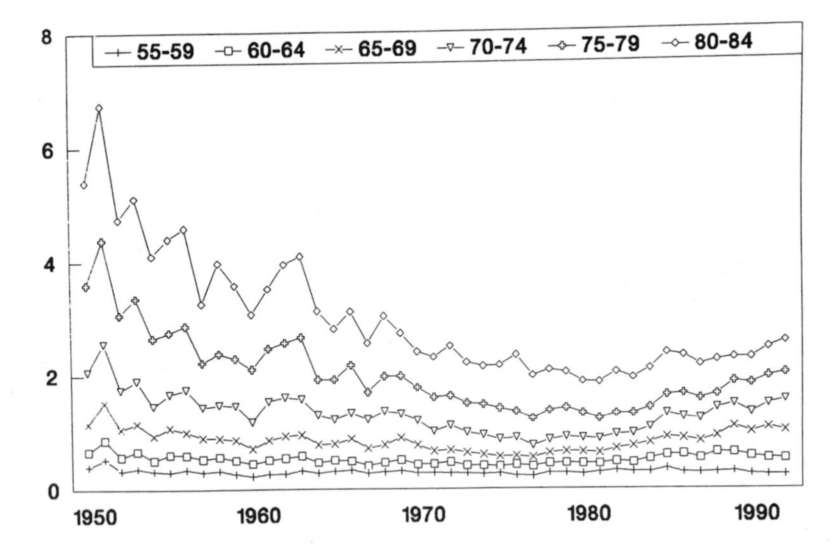

Fig. 8. Age-specific deaths from chronic obstructive pulmonary disease in females over time in the UK.

to inter-generational differences in lifetime cigarette consumption at various ages. These effects are a little at odds with those shown for period of death, which have shown a general decline since the 1940s, but the background of a general decline in death rates for all age groups needs to be borne in mind when interpreting these long-term trends.

International variations

International comparisons of deaths attributable to COPD is difficult chiefly due to the differing certification habits that exist between countries at one time and within any country from one time to another. For example, in 1984, the proportion of all chronic respiratory deaths attributed to COPD in USA was around two-thirds as opposed to almost none in Poland. Some general patterns for mortality from all chronic respiratory disease do however emerge; in an international comparison (Thom, 1989) conducted with figures from 1984, high rates were seen in Great Britain, Eastern Europe and Australasia, intermediate rates in Western Europe and North America, and low rates in Southern Europe, Scandinavia, Israel and Japan. British rates were generally only exceeded by Romania (both sexes), Germany (men), and New Zealand (women).

The high rates of mortality in Britain are thought not to be solely attributable to diagnostic variations; the few international comparisons of symptom prevalence and ventilatory function (chiefly between GB, Norway and USA) suggest that international differences in smoking habit largely explain variations in the prevalence of chronic phlegm, but that British men have lower ventilatory function after controlling for smoking (Reid & Fletcher, 1971).

Geographic variations within England and Wales

Age adjusted death rates from chronic respiratory disease vary considerably between small areas in Great Britain. There are two major underlying trends; generally higher in towns, especially in major conurbations, and a regional trend independent of urbanisation, with lower rates in the south-east, rising to higher rates in the north-west, with high rates in South Wales and Scotland. Prevalence studies of morbidity broadly follow the same pattern, in contrast with studies in individuals where cough and chronic phlegm production are poor predictors of mortality from chronic respiratory disease. There is some evidence to suggest that these regional variations in chronic respiratory disease in adults, in part reflects differences in their environmental experience of childhood, with place of birth (rather than death) being an important correlate.

Social class

In Great Britain, there is a powerful inverse association between socio-economic status and chronic respiratory disease. These trends apply in women as well as men, suggesting that specific occupational exposures play only a small explanatory role. The gradient is similar to that seen for mortality from lung cancer, suggesting that lifetime differences in smoking habits between the socio-economic groupings largely explain the differences (Lee *et al.*, 1990). On the other hand, there is some evidence for this association, independent of current smoking habit and again, differences in childhood socio-economic status appear to be important determinants in the association.

5. Bronchial Carcinoma

5.1. Causation

Lung cancer is the most common cancer in the Western world, with over 130 000 deaths per year occurring in the USA and over 150 000 per year in Europe (Mannino *et al.*, 1998). Worldwide, over a million deaths per year occur from this disease, two thirds in developed countries. The main risk factor for lung cancer is cigarette smoking although occupational agents, for example, asbestos, ionising radiation, polyaromatic hydrocarbons and chrome exposure, are recognised to exert important effects. Occupational lung cancer is probably under-recognised, although it has been estimated that as many as 15% of all lung cancers in the USA are occupationally related (Vineis *et al.*, 1988). It is only recently that concern about the role of air pollution as a factor for lung cancer has been expressed although the effect is small compared to actively inhaled cigarette smoke. Lung cancer incidence is closely matched by deaths as the majority of patients are incurable at the time of diagnosis, the overall five-year survival rate being around 5% (all histological types included). Consequently, mortality data are the best source for following trends. There have been no significant changes in the treatment of lung cancer over the last five decades which would have impacted on mortality.

5.2. Trends

At the beginning of the 20th century, lung cancer was extremely rare, but by the late 1940s, the incidence had risen dramatically. Rates continued to rise in both sexes in Western countries although over the last decade or so the rise in men has plateaued and in some countries has begun to fall. In women, however, the rise continues inexorably upwards in most areas (Thom & Epstein, 1994) and it is estimated that, in the UK, lung cancer will exceed breast cancer as the commonest cancer in women by the year 2000 (Gillis *et al.*, 1992) and has already done so in the United States. The reversal of the trend in men has been seen most clearly in the UK, Finland and New Zealand (Table 8). In a few countries, flattening of or reversal in trend has been seen in women, notably in Australia, Ireland, Israel, Japan and New Zealand. Table 8 depicts summaries of the rates for men and women for lung cancer for the years 1950–54 compared to 1984–87 for a range of countries for which there are reliable statistics. Lung cancer mortality rates are continuing to rise in the older age groups but are falling in the younger groups, a cohort effect due to changes in smoking habits. Rates in American blacks are higher in

Table 8. Mortality rates (per 100 000) from lung cancer for different countries over a 37-year period with % changes over specified times (M = male, F = female). Data are taken from Thom and Epstein (1994).

Country		1950–54	Highest rate 1950–87 (year range: 84–87 unless stated)	1984–87	% change 50s to 80s	% change max. year to 80s
Australia	M	45	111 (79–83)	110	+144	–4
	F	7	28	28	+300	0
Austria	M	110	120 (60–64)	102	–7	–9
	F	13	13	20	+54	0
Belgium	M	65	190	190	+125	0
	F	8.5	13	13	+53	0
Canada	M	42	130	130	+210	0
	F	8	38	38	+375	0
Czechoslovakia	M	125	200	200	+60	0
	F	11.5	20	20	+74	0
Denmark	M	43	120	120	+179	0
	F	10	52	52	+420	0
England/Wales	M	110	190 (65–69)	140	+27	–26
	F	15	50	50	+233	0
Finland	M	110	170 (70–74)	120	+9	–29
	F	11	16	16	+45	0
France	M	33	110	110	+233	0
	F	8	11	11	+37	0
Germany (FR)	M	63	115 (79–83)	112	+78	–3
	F	9.5	16	16	+68	0
Hungary	M	100 (65–69)	170	170	+70	0
	F	15	30	30	+100	0
Ireland	M	38	110	110	+190	0
	F	11	42	42	+282	0
Israel	M	50 (65–69)	60 (75–78)	58	+16	–3
	F	14.5	19 (70–74)	18	+24	–5

Table 8 (*Continued*)

Country		1950–54	Highest rate 1950–87 (year range: 84–87 unless stated)	1984–87	% change 50s to 80s	% change max. year to 80s
Italy	M	30	130	130	+333	0
	F	6	28	28	+367	0
Japan	M	9	60	60	+567	0
	F	3	17	17	+467	0
Northern Ireland	M	62	125	125	+102	0
	F	10	40	40	+300	0
Norway	M	10	75	75	+650	0
	F	5	21	21	+320	0
Netherlands	M	70	190 (75–78)	180	+157	−5
	F	7	20	20	+186	0
New Zealand	M	55	110 (75–78)	105	+91	−4.5
	F	7	35	35	+400	0
Poland	M	38	160	160	+321	0
	F	8	20	20	+150	0
Portugal	M	20	60	60	+200	0
	F	5	9	9	+80	0
Scotland	M	110	200 (70–74)	180	+64	−20
	F	19	70	70	+268	0
Spain	M	45 (60–64)	95	95	+111	0
	F	8.5	9.5 (70–74)	8	−6	−16
Sweden	M	25	60	60	+140	0
	F	9	22	22	+144	0
Swetzerland	M	62	120 (75–78)	110	+77	−9
	F	5.5	16	16	+191	0
United States	M	54	130 (79–83)	120	+122	−8
	F	9	50	50	+455	0
Yugoslavia	M	52	110	110	+112	0
	F	9	17	17	+89	0

males compared to whites but are similar in black and white women (Ries *et al.*, 1991).

5.3. Geographical Distribution

The wide geographical spread of mortality rates, present both in the early 1950s and at the present time cannot be entirely explained on the basis of differences in cigarette smoking although the changes in rates over latter years do match reductions in the prevalence of cigarette smoking during that period. Under-recognition of occupational causes of lung cancer may explain part of these differences but exposure to poor urban air quality over years may have played a role. However, those in their 70s who are dying today from lung cancer will have been exposed to a different air pollutant mix in their younger years compared to that experienced today, making source attribution of air pollution to lung cancer rates difficult. However, in Third World countries where the cigarette manufacturers have been concentrating their efforts, lung cancer rates may follow changes in cigarette consumption. Recent work from South Korea shows an increase in age adjusted annual mortality rate per 100 000 for lung cancer from 3.7 in 1980 to 17.8 in 1994, a 380% rise (Jee *et al.*, 1998) with projected further increases over the next decade to 65.4 for men and 15.1 for women.

5.4. Histological Type

There has been an apparent increase in lung cancer in non-smokers hidden within these trends, particularly in the adenocarcinoma cell type (Stockwell *et al.*, 1990). However, there has also been an increase in adenocarcinoma in smokers and the possibility has been raised that this is associated with increased nitrate content of the newer generation of cigarettes (Wynder & Muscat, 1995). Trends in the type and content of cigarettes smoked needs to be taken into consideration when assessing air pollution as a possible contributor but reinforces the importance of qualitative differences in air pollution with respect to causal associations.

5.5. Summary

The dominant factor driving all changes in incidence of lung cancer over time remains cigarette smoking, and the effect of changes in other potential aetiological factors may prove difficult to identify. The importance of occupational exposures (past and present), passive exposure in non-smokers and type of cigarette smoked will all need to be accounted for when determining the contribution of urban pollution to the incidence of this disease.

6. Ischaemic Heart Disease

6.1. Causation

In Europe, North America, Australia and New Zealand, ischaemic heart disease (IHD) is the leading cause of mortality in men over the age of 45 and in women over the age of 65. The risk factors are well recognised (hypertension, obesity, hypercholesterolaemia, cigarette smoking) but other factors may also prove to be of importance such as antioxidant consumption. For instance, the high rates of IHD in Eastern European countries seem to be associated with diets severely deficient in antioxidant foods (Ginter, 1995). There has been considerable effort to reduce mortality from IHD in many countries over the last few decades and these have been to a greater or lesser degree successful. This has been partly due to preventive approaches encouraging changes in life style but also to changes in management such as the advent of thrombolytic therapy in the treatment of acute myocardial infarction (Schroder *et al.*, 1997) and coronary angioplasty for localised disease. Consequently, there are existing trends which are due to recognised interventions, whose rate of change may occur at different rates over time and between countries, which need to be determined when considering the impact of air pollution exposure on overall trends. Air pollutants have been shown to be associated with both mortality and hospital admissions for IHD and for congestive cardiac failure in a number of studies (Schwartz & Dockery, 1992; Schwartz, 1997) although there are no data on the effect of day to day changes in particulate pollution on patients with angina. The most useful indicators for the effects of air pollution in heart disease are therefore, mortality and hospital admissions.

6.2. Trends and Geographical Distribution

Mortality from IHD shows a wide variation world wide with England and Wales amongst those with the highest rates. Although trends have been downwards in many countries, higher rates of decline are not necessarily seen in those countries with initially higher rates. Using data summarised by Thom and Epstein (1994), various trends can be identified. In men, IHD mortality has declined in most countries (Table 9). Declines began earlier in some European countries (e.g. Belgium, Denmark) but a little later (1970s) in others such as England and Wales, Israel and Norway. Increases in rates were seen in Eastern European countries at least up to the late 1980s. For women, declines in IHD deaths occurred in all countries except Poland, Czechoslovakia and Yugoslavia and the decline in mortality started at an earlier time (1950s) than in men. Recent (1984–87) data from the USA show a decline in in-hospital mortality from IHD by 5.1% per annum and in community mortality by 3.6% per annum (Rosamond *et al.*, 1998). The effect was slightly greater in white compared to black individuals. This occurred against a background of a slight increase in rates of hospitalisation for IHD over the same period, a time when thrombolytic therapy for myocardial infarction had begun to be used to a much greater extent.

6.3. Summary

These considerable geographical variations in IHD rates will be due to the constellation of factors which are recognised pre-disposing

Table 9. Mortality rates (per 100 000) from IHD for different countries over a 37-year period with % changes over specified times (M = male, F = female). Data are taken from Thom and Epstein (1994).

Country		1950–54	Highest rate 1950–87 (year range: 50–54 unless stated)	1984–87	% change 50s to 80s	% change max. year to 80s
Australia	M	700	730 (65–69)	400	−43	−45
	F	300	300	160	−47	−47
Austria	M	410	480 (75–83)	400	−2.5	−20
	F	210	210	170	−25	−25
Belgium	M	400	500 (65–69)	330	−17.5	−34
	F	240	240	205	−14.6	−14.6
Canada	M	600	630 (60–64)	400	−33	−36.5
	F	210	210	130	−54	−54
Czechoslovakia	M	400	500 (84–87)	500	+25	–
	F	300	300	220	−27	−27
Denmark	M	390	500 (65–69)	410	+5	−18
	F	230	230	135	−41	−41
England/Wales	M	505	600 (70–74)	500	−1	−17
	F	260	260	180	−30.8	−30.8
Finland	M	710	810 (65–69)	600	−18.3	−25.9
	F	370	370	200	−45.9	−45.9
France	M	305	305	210	−31	−31
	F	180	73	73	−59.4	−59.4
Germany (FR)	M	320	470 (70–74)	410	+28	−12.8
	F	205	205	160	−22	−22
Hungary	M	360	600 (84–87)	600	+75	–
	F	270	270	225	−16.7	−16.7
Ireland	M	550	600 (75–83)	590	+7.3	−1.7
	F	410	410	220	−46.3	−46.3
Israel	M	490	505 (70–74)	400	−18.4	−20.8
	F	300	320 (60–64)	200	−33.3	−37.5

Table 9 (*Continued*)

Country		1950–54	Highest rate 1950–87 (year range: 50–54 unless stated)	1984–87	% change 50s to 80s	% change max. year to 80s
Italy	M	310	400 (65–69)	300	–3.3	–25
	F	305	305	170	–44.3	–44.3
Japan	M	200	200	130	–3.5	–35
	F	140	140	75	–46.4	–46.4
Northern Ireland	M	600	720 (75–78)	620	+3.3	–13.9
	F	400	400	250	–37.5	–37.5
Norway	M	300	490 (70–74)	450	+50	–8.2
	F	180	190 (60–64)	115	–36.1	–39.5
Netherlands	M	290	420 (70–74)	390	+34.5	–7.1
	F	200	200	120	–40	–40
New Zealand	M	600	700 (65–69)	500	–16.6	–28.6
	F	310	310	180	–41.9	–41.9
Poland	M	320	500 (84–87)	500	+56.3	–
	F	180	205 (75–83)	190	+5.6	–7.3
Portugal	M	310	310	220	–29	–29
	F	205	205	100	–51.2	–51.2
Scotland	M	630	700 (60–64)	610	–3.2	–12.9
	F	350	350	260	–25.7	–25.7
Spain	M	220	290 (75–78)	240	+9.1	–17.2
	F	120	125 (65–69)	100	–16.7	–20
Sweden	M	390	460 (75–83)	410	+5.1	–10.9
	F	240	240	120	–50	–50
Switzerland	M	330	405 (60–64)	300	–9.1	–25.9
	F	200	225 (55–59)	100	–50	–55.6
United States	M	780	780	460	–41	–41
	F	380	380	200	–47.4	–47.4
Yugoslavia	M	320	500 (84–87)	500	+56.3	–
	F	290	290	290	0	0

influences for IHD; the best recognised being smoking, blood pressure and cholesterol although these three together will only account for one quarter of the variance of mortality differences (Epstein, 1996). Adding BMI and other dietary factors will improve this but further factors such as air pollution may be identifiable as contributory factors, provided an appropriate index of long-term exposure can be produced.

7. Stroke

7.1. Measures of Stroke

Stroke varies in severity from transient loss of power or sensation due to cerebral embolism to a fatal cerebral haemorrhage. At the milder end of the spectrum, many patients are not admitted to hospital and epidemiological data for these patients are not routinely available. Air pollution changes have been shown to be associated with hospital admission for stroke, so these more severely affected patients can be identified and trends followed. However, death from stroke may follow many days or weeks after the original event and consequently mortality is not the best index of the effects of air pollution at least on a day-to-day basis. The most relevant marker for determining the effects of day-to-day changes in air pollution on stroke is the incidence of all strokes in the community regardless of severity or immediate outcome (e.g. death or hospital admission). Such data are, however, not routinely available and so the most useful outcome for assessing day to day effects are hospital admissions. Confounders need special attention in this condition, notably weather conditions (e.g. temperature) and factors such as population density and socio-economic factors, both risk factors for stroke (Starr *et al.*, 1996). One factor which has probably affected estimates of stroke incidence is the marked improvement in imaging techniques (CT, MRI) which has enabled the diagnosis to be made with more certainty (Brown *et al.*, 1996) although this is more likely to apply to those individuals who are admitted to hospital.

7.2. Seasonal Factors

It has been widely reported that there is a seasonal pattern to the incidence of stroke with rates being higher in the winter (Jakovljevic *et al.*, 1996) although this may be simply a feature of colder ambient temperatures (Rothwell *et al.*, 1996). This reinforces the importance of dealing adequately with meteorological parameters when estimating air pollution effects. In the United States, stroke incidence age-adjusted rates of around 145 per 100 000 have been reported for the 1980s (Brown *et al.*, 1996) and in the MONICA study, they ranged from 61 (women in Friuli, Italy) to 388 (men in Novosibirsk, Russia) in the countries involved (Thorvaldsen *et al.*, 1996). Rates have in general fallen slightly over time, although in many countries the falls are not significant, in contrast to the changes in mortality.

7.3. Trends and Geographical Distribution

A review of mortality rates for stroke in the 35 to 74 age group (Thom & Epstein, 1994) reveal slightly higher rates in men, for most countries falling between 52 per 100 000 and 120 per 100 000, the exceptions being Czechoslovakia, Hungary, Portugal and Yugoslavia with rates between 170 and 250. For women, rates are higher in the same four countries (ranging from 102–160) while for the remaining countries for whom there are good data rates range from 32 to 95 per 100 000. In most countries, mortality from stroke has fallen remarkably over the last four decades, the steepest declines being seen in women compared to men. In eastern European countries, however, the trends are smaller or flat and in countries such as Hungary, Poland and Yugoslavia, rates have, in fact, increased to some extent over this period.

7.4. Summary

While the incidence of stroke has fallen only slightly, if at all, over the last 20 years or so, age-adjusted mortality is falling worldwide, particularly in women. United States data reveals that, on a cohort basis, each successive cohort since 1890 has improved their life expectancy with respect to stroke (Feinlieb *et al.*, 1993). Again, as with IHD, trends in stroke due to other factors must be allowed for when considering the impact of air pollution.

8. Leukaemias

8.1. Risk Factors

Chronic myeloid leukaemia (CML) and acute lymphocytic leukaemia (ALL) have both been associated with exposure to the genotoxic carcinogens benzene (CML) and 1,3-butadiene (ALL) in occupational environments. The UK Expert Panel on Air Quality Standards have taken the view that, being genotoxic carcinogens present in outdoor air, no safe air quality standard can be set and, further, that exposure of populations to these carcinogens will be conferring an increased risk of leukaemia on those populations even though the risk is regarded as immeasurably small (Expert Panel on Air Quality Standards, 1994a; 1994b). We have therefore included these conditions in this chapter.

The leukaemias are classified into those deriving from the lymphoid cell series and those from the myeloid cell series, both acute and chronic forms of which ALL and CML concern us here. Chronic leukaemias are almost exclusively found in adults with the lowest rates found in the 20–30-year age group, increasing exponentially with age thereafter. Acute leukaemias are seen mostly in children but also in adults above the age of 60 years, with much lower frequencies in the intervening ages. However, there is a paucity of registries across nations which would allow assessments of trends over time and geographical comparisons to be made with certainty.

8.2. Trends

Mortality from all leukaemias in the USA has fallen considerably in children over the last 20 years whereas mortality from chronic leukaemias has remained unchanged over the same period. Similar trends have

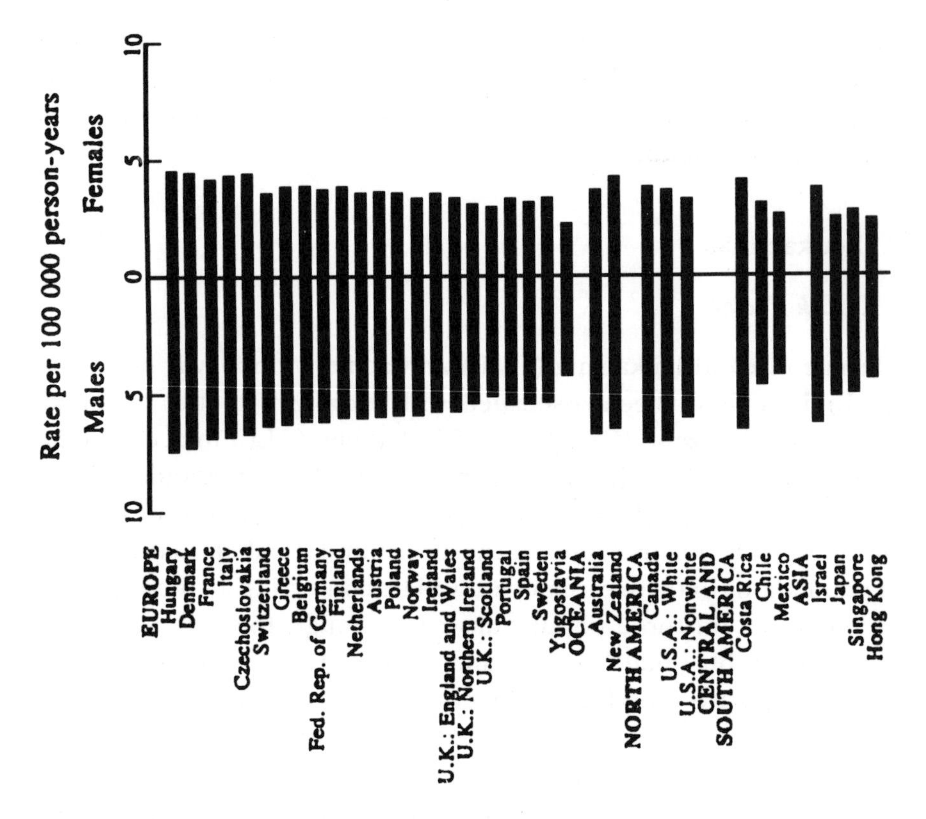

Fig. 9. International variation in leukaemia mortality (age-adjusted, world standard) by sex, 1983–87.

been seen in other westernised countries but we can find no information on trends in Third World countries. There is, however, information on mortality rates between countries in cross sectional assessments at least for the period 1983–87 (Groves *et al.*, 1993). These data reveal modest differences in age-standardised mortality with values for males ranging from 3.5 per 100 000 person years in Mexico to 7.5 per 100 000 person years in Hungary (Fig. 9). The rates are lower in women (2 to 4.5 for Mexico and Hungary) but with a similar pattern across countries as seen in males. Rates are highest in the elderly, US data showing an approximate rate of 30 per 100 000 in the over 65s in 1990 compared to 10 per 100 000 in the 45–64-year age group. The US data also suggest a difference in mortality according to racial origin, with rates in the elderly increasing in non-Caucasians over the last 40 years from a lower rate than their Caucasian counterparts to similar levels in the 1990s. This would suggest that one or more environmental factors have exerted an effect in the non-Caucasian population which were not having such an effect in 1950.

8.3. ALL

The range of incidence rates for ALL internationally is smaller than for all leukaemias, ranging from around one to around two cases per 100 000 person years in males and slightly less in women. However, there appears to be some variation by race as rates in the Latino population in Los Angeles are twice that of the black population in the same city. Incidence rates have increased over time (Fig. 10) but are higher in males and in Caucasian populations. This is the only leukaemia whose incidence has increased over time but survival has improved very dramatically over the same period due to much improved treatment regimes. It is therefore important to consider incidence rates rather than mortality when considering air pollutants in the aetiology of this condition. The reason for the rise in ALL is unknown.

8.4. CML

Age adjusted incidence rates for CML have shown a slight decline over time in Caucasian populations (Fig. 11) with females showing higher rates, although there is a suggestion that black subjects, while exhibiting lower overall rates, show no downward trend over time. The reason for this difference is unknown but is not felt to be due to exposure to recognised sources of ionising radiation. This leukaemia shows the smallest international variation with rates always less than 2 per 100 000 person years, suggesting that environmental factors are less important for this leukaemia than for others, notably chronic

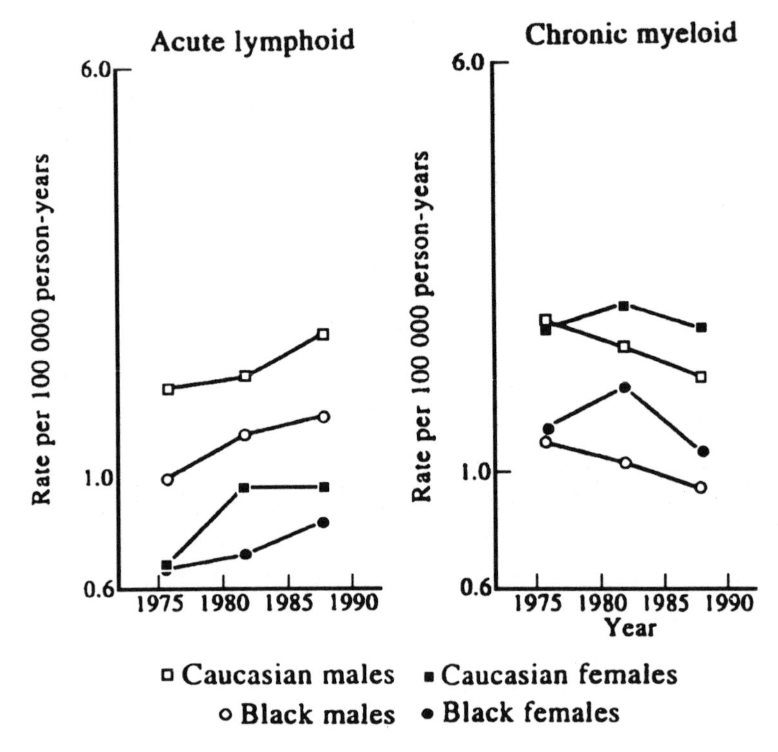

Figs. 10 & 11. Trends in age-adjusted (world standard) incidence rates for ALL and CML by race and sex 1973–90. □ Caucasian males; ■ Caucasian females; ○ black males; • black females.

lymphocytic leukaemia (CLL) where rates range from around 1 per million in China to 60 per million in male Canadians. However, there is no suggestion from occupational cohorts exposed to genotoxic carcinogens at higher levels that CLL is in any way related to such exposures.

8.5. Summary

Although there are limited data in this group of diseases, a number of observations are of interest and merit further investigation with respect to causality (Groves *et al.*, 1993). Amongst these are the increase in incidence of ALL over time, the higher rates in Spanish and Latino populations and the bimodal age distribution. For CLL, the declining incidence over time and very low rates in oriental populations needs to be explained as does the marked male predominance and its virtual absence below the age of 30 years. CML has shown a decline over the last few decades, more so amongst Caucasian compared to African American groups, whereas across nations, there is a lack of variation in rates. Its virtual absence in childhood remains unexplained. While it is likely that genetic factors are responsible, at least in part, for some of these observations, the role of environmental factors would seem to be crucial, which would include exposure to air pollutants whether indoor or outdoor.

9. Sudden Infant Death Syndrome

9.1. Causation

Sudden infant death syndrome (SIDS) has been defined as "The sudden death of an infant under one year of age which is unexpected by history, and in which a full post-mortem examination fails to demonstrate an adequate cause of death" (Beckwith, 1970). However, this definition has come under criticism because of the variation in extent of a post mortem examination and differences in opinion attached to microscopic inflammatory changes seen in the lungs of some of these infants (Mitchell & Becroft, 1997). In reality, the diagnosis is by no means always based on post-mortem findings, and data on the epidemiology of SIDS have largely accepted that a basic diagnosis based on the unexplained death of an infant is acceptable and have simply used ICD-9 code 798.0 in analysis.

9.2. Trends and Seasonality

When considering the trends in SIDS over the last two decades, one major factor has affected the trends, namely the dramatic effect that

Table 10. Changes in incidence of SIDS before (pre-) and after (post-) the adoption of the "face-up" sleeping position. Numbers are rates per 1000 live births.

Country	UK	US	Australia	Scandinavia	Austria
Pre-	2.0	1.4	2.0	1.5	2.0
Post-	0.9		1.2	0.5	1.0

adoption of the "face-up" sleeping position has had in effectively halving the incidence of this cause of death (Table 10). Consequently, assessment of trends in SIDS over time with respect to air pollutant exposure will be difficult as the contributory factor is likely to be very small, although geographical patterns according to air pollutant exposure may prove more amenable to investigation.

There is a consistent seasonality in the incidence of SIDS with the highest rates in the winter months (Douglas *et al.*, 1996). Prior to the fall in incidence due to the "face upwards" sleeping position, rates were approximately twice as high in the winter as in summer in both the UK and Australia (Fig. 12). The sharp fall seen in the early 1990s followed a more gradual decline over years probably attributable to alteration in other risk factors such as overwrapping and parental cigarette smoking. For instance, in the USA, the incidence of SIDS fell by around one fifth between 1980 and 1988 (*Statistics Bulletin,* 1993). Rates both before and after general uptake of the "face-up" sleeping position are broadly similar across a range of countries (Table 10) although within this range, some differences in trends can be seen, notably, the relative sparing of Finland from the earlier high rates and the increase followed by a fall in Sweden (Vege & Rognum, 1997).

9.3. Ethnicity

Such data as are available for differences in trends across different ethnic groups come from the United States which reveal a disproportionately higher rate among black infants compared to white (*Statistics*

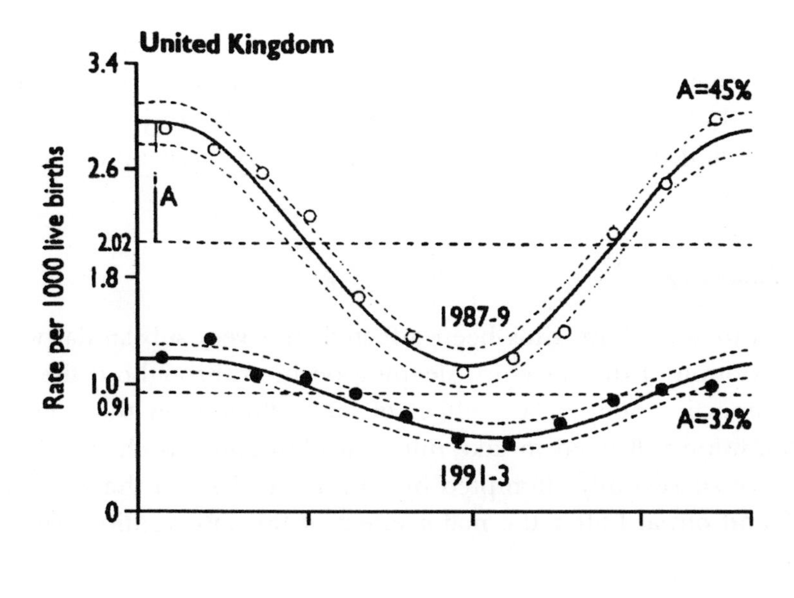

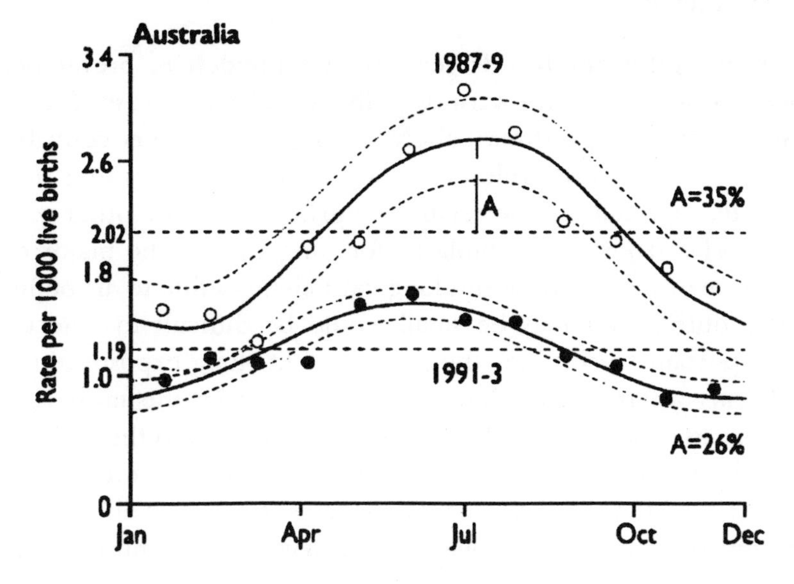

Fig. 12. SIDS in Britain and Australia, 1987–89 and 1991–93. Incidence in 1987–89 (∘) compared with that in 1991–93 (•) in both countries. Discontinuous lines indicate 95% confidence intervals. Discontinuous horizontal lines give mean values. A = amplitude.

Bulletin, 1993). However, the difference between the black and white groups seems to be narrowing with rates falling by 19% between 1980 and 1988 in black infants compared to a fall of just 4% for the same period in white infants. The reasons for these differences are unknown.

9.4. Summary

Trends in SIDS have thus been affected in a very substantial way by the adoption of the "face-up" sleeping position throughout the world and apportionment of any effect of air pollution on the incidence of condition will need to take into consideration this change. It may be more successfully attempted by limiting analysis to the years from 1990 and onward after the major effect of this intervention has been seen.

10. Summary

There are different trends seen in the incidence, prevalence or exacerbation rates of the various pollutant-related diseases over time. In some cases (e.g. asthma), there are multiple factors contributing to such trends and the role of air pollutant exposure is thus difficult to define, especially as the relative contribution of the other relevant factors is far from clear. Similarly, for lung cancer, the major causal factor remains as cigarette smoking, and changes due to air pollution, whose contribution may be small, could be difficult to detect. It is also important to consider the pattern of life-long exposure in those conditions where a cumulative exposure may be the most relevant exposure to consider, as is the case for lung cancer and the leukaemias. It may be possible to begin to unravel these complex interactions by considering differential rates and exposures seen across different countries, but retrospectively collected data may not be robust enough to use in a situation where all co-exposures and confounders need to be carefully assessed.

If a pollution control measure has been introduced in a country (e.g. the banning of the sales of coal in Dublin, Ireland) or where a major change in pollutant exposure has occurred, (e.g. reunification of West and East Germany), determination of subsequent changes in health outcomes may contribute to identifying the role of air pollution on health outcomes over both short and longer time scales. Such natural experiments need to be identified and, where identifiable in advance, studied in a planned way.

It has been hypothesised that certain chronic conditions have common causes which would thus respond to specific control or treatment strategies. This has been elegantly addressed by Thom and Epstein (1994) who estimated the degree of concordance between separate trends in mortality from common diseases across a wide spectrum of countries, including in their analysis, separate time lags to identify the time taken for a change in behaviour or treatment to impact. They found, amongst other findings, a very high concordance between deaths from lung cancer and for IHD in men but not in women, possibly a reflection of the different patterns of smoking behaviour over time. Compared to IHD there is low concordance for stroke under the age of 64 years but high concordance over that age in both sexes, perhaps reflecting the combined effects of nutrition, smoking and alcohol. Conversely, there are completely divergent associations in trends between lung cancer and stroke probably because cigarette smoking is so much more important a factor for lung cancer than for stroke. With respect to the contribution of air pollution to the prevalence of chronic disease states (as opposed to the effect of day to day changes on daily events), this approach could be very helpful, although the small size of effect compared to high impact factors, such as cigarette smoking, will require large data sets from many different countries.

The bottom line, however, remains that when considering trends in air pollution related diseases and the role of air pollution in such trends, all potential confounders and co-exposures must be dealt with before attributing residual effects to air pollution.

References

Aberg N. (1989) *Clin. Exp. Allergy*, **19**, 59–63.

American Thoracic Society. (1995) *Am. J. Respir. Crit. Care Med.* **152**, S77–S120.

Anderson H.R. (1989) *Arch. Dis. Child*, **64**, 172–175.

Anderson H.R. (1989) *Thorax*, **44**, 614–619.

Anderson H.R. *et al.* (1994) *Brit. Med. J.*, **308**, 1600–1604.

Anthonisen N.R. (1989) *Am. Rev. Respir. Dis.* **140**, S95–S99.

Anthonisen N.R. *et al.* (1994) *J. Am. Med. Assoc.* **272**, 1497–1505.

Australian Bureau of Statistics. 1989–90 National Health Survey. (1991) Catalogue No. 4373.0. Canberra, Commonwealth of Australia.

Ayres J.G. (1986) *Thorax*, **41**, 111–116.

Barker D.J.P. & Osmond C. (1986) *Br. Med. J.* **293**, 1271–1275.

Beckwith J.B. (1970) In *Sudden Infant Death Syndrome. Proceedings of the Second International Conference on the Causes of Sudden Death in Infants*, eds. Bergman A.B., Beckwith J.B. & Ray C.G. pp. 14–22, University of Washington Press, Seattle.

Britton W.J. *et al.* (1986) *Int. J. Epidemiol.* **15**, 202–209.

Broder I. *et al.* (1974) *J. Allergy Clin. Immunol.* **54**, 100–110.

Brown R.D. *et al.* (1996) *Stroke*, **27**, 373–380.

Brunekreef B. *et al.* (1989) *Am. Rev. Respir. Dis.* **140**, 1363–1367.

Burney P.G.J. *et al.* (1990) *Br. Med. J.* **300**, 1306–1310.

Butland B. *et al.* (1997) *Br. Med. J.* **315**, 717–721.

Campbell M.J. *et al.* (1997) *Br. Med. J.* **314**, 1439–1441.

Central Health Monitoring Unit. Asthma: an epidemiological overview. London, HMSO, 1995.

Chan-Yeung M. & Malo J.-L. (1995) In *Asthma and rhinitis*, eds. Busse W.W. and Holgate S.T. pp. 44–57, Blackwell Scientific Publications, Boston.

Cockcroft D.W. *et al.* (1992) *J. Allergy Clin. Immunol.* **89**, 23–30.

Committee on the Medical Effects of Air Pollutants (1995) In *Asthma and outdoor air pollution.* pp. 85–129, London, HMSO.

Couriel J.M. (1988) *Arch. Dis. Child*, **63**, 1305–1308.

Cox B.D. (1987) In *The health and lifestyle survey. Preliminary report of a nationwide survey of the physical and mental health, attitudes and lifestyle of a random sample of 9003 British adults.* pp. 17–33, Health Promotion Research Trust, London.

De Boeck K. (1995) *Eur. J. Pediatr.* **154**, 432–436.

Denny F.W. *et al.* (1983) *Pediatrics*, **71**, 871–876.

Dockery D.W. *et al.* (1993) *New Engl. J. Med.* **329**, 1753–1759.

Dodge R.R. & Burrows B. (1980) *Am. Rev. Resp. Dis.* **122**, 567–575.

Douglas A.S. *et al.* (1996) *BMJ*, **312**, 1381–1383.
Emanuel M.B. (1988) *Clin. Allergy*, **18**, 295–304.
Epstein F.H. (1996) *Circulation*, **93**, 1755–1764.
Evans III R. *et al.* (1987) *Chest*, **91**, 65S–74S.
Expert Panel on Air Quality Standards. (1994a) Benzene. HMSO.
Expert Panel on Air Quality Standards. (1994b) 1,3-butadiene. HMSO.
Feinlieb M. *et al.* (1993) *Ann. Epidemiol.* **3**, 458–465.
Fleming D.M. & Crombie D.L. (1987) *Br. Med. J.* **294**, 279–283.
Fletcher C.M. *et al.* (1976) *The natural history of chronic bronchitis and emphysema. An 8-year study of working men in London.* Oxford University Press, Oxford.
Fletcher C. & Peto R. (1977) *Br. Med. J.* **1**, 1645–1648.
Geelhoed G.C. (1996) *Ann. Emerg. Med.* **28**, 621–626.
Gergen P.J. *et al.* (1988) *Pediatrics*, **81**, 1–7.
Gergen P.J. & Weiss K.B. (1995) In *Asthma and rhinitis*, eds. Busse W.W. & Holgate S.T. pp. 15–31, Blackwell Scientific Publications, Boston.
Gillam S.J. *et al.* (1989) *Br. Med. J.* **299**, 953–957.
Gillis C.R. *et al.* (1992) *Brit. Med. J.* **305**, 1331.
Ginter E. (1995) *Eur. J. Epidemiol.* **11**, 199–205.
Gold D. *et al.* (1989) *Am. Rev. Respir. Dis.* **140**, 877–884.
Gortmaker S.L. *et al.* (1982) *Am. J. Public Health*, **72**, 574–579.
Grad R. & Taussig L.M. (1990) In *Kendig's Disorders of the Respiratory Tract in Children*, ed. Chernick V. pp. 336–349, Philadelphia, WB Saunders.
Groves F.D. *et al.* (1993) *Eur. J. Cancer*, **31A**, 941–949.
Haahtela T. *et al.* (1990) *Br. Med. J.* **301**, 266–268.
Halfon N. & Newacheck P.W. (1986) *Am. J. Public Health*, **76**, 1308–1311.
Hanrahan J.P. *et al.* (1992) *Am. Rev. Respir. Dis.* **145**, 1129–1135.
Hogg J.C. (1968) *N. Engl. J. Med.* **278**, 1355–1360.
Holgate S.T. (1998) *Q. J. Med.* **91**, 171–184.
Inman W.H.W. & Adelstein A.M. (1969) *Lancet*, **ii**, 279–285.
Jakovljevic D. *et al.* (1996) *Stroke*, **27**, 1774–1779.
Jee S.H. *et al.* (1998) *Int. J. Epidemiol.* **27**, 365–369.
Lee P.N. *et al.* (1990) *Thorax*, **45**, 657–665.
Lewis S. *et al.* (1994) *Am. J. Respir. Crit. Care Med.* **194**, A574.
Mackay T.W. *et al.* (1992) *Scot. Med. J.* **37**, 5–7.
Mannino D.M. *et al.* (1998) *Int. J. Epidemiol.* **27**, 159–166.
Marine W.M. *et al.* (1988) *Am. Rev. Respir. Dis.* **137**, 106–112.
Marx A. *et al.* (1997) *J. Infect. Dis.* **I176**, 1423–1427.
McWhorter W.P. *et al.* (1989) *Am. Rev. Resp. Dis.* **139**, 721–724.
Medical Research Council Committee On The Aetiology of Chronic Bronchitis. (1965) *Lancet*, **i**, 775–779.

Ministry of Health. (HMSO, London, 1954).
Mitchell E.A. (1983) *NZ Med. J.* **96**, 463–464.
Mitchell E.A. & Becroft D.M.O. (1997) *Acta Paediatr.* **86**, 789–790.
Montgomery Smith J. (1983) In *Allergy: Principles and practice*, eds. Middleton E., Reed C.E., Ellis E.F., Adkinson N.F. & Yunginger J.W. pp. 771–803, CV Mosby, St Louis.
Murray A.B. & Morrison B.J. (1988) *Chest*, **94**, 701–708.
Mygind N. (1985) In *Allergic and vasomotor rhinitis; clinical aspects*, eds. Mygind N. & Weeke B. pp. 15–20, Munksgaard, Copenhagen.
National Institutes Of Health (1995a) In *Global initiatives for asthma.* pp. 9–24, Publication Number 95-3659.
National Institutes Of Health (1995b) *Global initiatives for asthma.* pp. 25–38, Publication Number 95-3659.
Pattemore P.K., Asher M., Harrison A.C., Mitchell E.A., Rea H.H. & Stewart A.W. (1989) *Thorax*, **9**, 168–176.
Pearce N.E. *et al.* (1995) In *Asthma and rhinitis*, eds. Busse W.W. & Holgate S.T. pp. 58–69, Blackwell Scientific Publications, Boston.
Pedersen P.A. & Weeke E.R. (1981) *Allergy*, **36**, 375–379.
Pope C.A. *et al.* (1995) *Am. J. Respir. Crit. Care Med.* **151**, 669–674.
Pride N.B. & Burrows B. (1995) In *Chronic Obstructive Lung Disease*, eds. Calverley P. & Pride N. pp. 69–91, Chapman & Hall, London.
Reid D.D. & Fletcher C.M. (1971) *Brit. Med. Bull.* **27**, 59–64.
Ries L.A.G. *et al.* (1991) Cancer statistics review 1973–1988. National Cancer Institute. *NIH Publications*; No. 91-2789.
Robertson C.F. *et al.* (1991) *Br. Med. J.* **302**, 1116–1118.
Roemer W. *et al.* (1998) *Eur. Respir. Rev.* **8**, 4–11.
Roorda R.J. *et al.* (1994) *J. Allergy Clin. Immunol.* **93**, 575–584.
Rosamond W.D. *et al.* (1998) *NEJM*, **339**, 861–867.
Rothwell P.M. *et al. Lancet*, **347**, 934–936.
Schroder K. *et al.* (1997) *Heart*, **77**, 506–511.
Schwartz J. & Dockery D.W. (1992) *Am. Rev. Respir. Dis.* **145**, 600–604.
Schwartz J. (1997) *Epidemiology*, **8**, 371–377.
Sears M.R. (1991) *Bull. Int. Union Tuberc. Lung Dis.* **66**, 79–83.
Sheffer A.L. (1991) *J. Allergy Clin. Immunol.* **88**, 425–534.
Sibbald B. & Rink E. (1991) *Thorax*, **46**, 378–381.
Sibbald B. & Rink E. (1991) *Thorax*, **46**, 895–901.
Skolnik N.S. (1989) *Am. J. Dis. Child*, **143**, 1045–1049.
Snider G.L. *et al.* (1985) *Am. Rev. Respir. Dis.* **132**, 182–185.
Starr J.M. *et al.* (1996) *Int. J. Epidemiol.* **25**, 276–281.
Statistics Bulletin–Metropolitan Insurance Companies. **74** (1993), 10–18.

Stockwell H.G. *et al.* (1990) *Int. J. Epidemiol.* **19**, S48–S52.
Strachan D.P. (1989) *Br. Med. J.* **299**, 1259–1260.
Strachan D.P. *et al.* (1990) *J. Epidemiol. Community Health*, **44**, 231–236.
Strachan D.P. (1995) In *Chronic Obstructive Lung Disease*, eds. Calverley P. & Pride N. pp. 47–67, Chapman & Hall, London.
Tager I.B. *et al.* (1983) *N. Engl. J. Med.* **309**, 699–703.
Tager I.B. *et al.* (1985) *Am. Rev. Respir. Dis.* **131**, 752–759.
Tager I.B. *et al.* (1988) *Am. Rev. Respir. Dis.* **138**, 837–838.
Tager I.B. *et al.* (1993) *Am. Rev. Respir. Dis.* **147**, 811–817.
Taylor B. & Wadsworth J. (1987) *Arch. Dis. Child*, **62**, 786–791.
Thom T.J. (1989) *Am. Rev. Respir. Dis.* **140**, S27–S34.
Thom T.J. & Epstein F.H. (1994) *Circulation*, **90**, 574–582.
Thorvaldsen P. *et al.* (1997) *Stroke*, **28**, 500–506.
Vege A. & Rognum T.O. (1997) *Acta Paediatr.* **86**, 391–396.
Vineis P. *et al.* (1988) *Int. J. Cancer*, **42**, 851–856.
Watson J. *et al.* (1996) *Eur. Respir. J.* **9**, 2087–2093.
Weiss S.T. *et al.* (1992) *Am. Rev. Respir. Dis.* **145**, 58–64.
Weiss S.T. (1995) *Eur. Respir. Rev.* **5**, 303–309.
Weitzman M. *et al.* (1990) *Pediatrics*, **85**, 505–511.
Weitzman M. *et al.* (1992) *J. Am. Med. Assoc.* **268**, 2673–2677.
Wynder E.L. & Muscat J.E. (1995) *Environ. Health Perspec.* **103** (Suppl. 8), 143–148.
Xu X. *et al.* (1994) *Eur. Resp. J.* **7**, 1056–1061.
Young S. *et al.* (1991) *N. Engl. J. Med.* **324**, 1168–1173.
Yunginger J.W. *et al.* (1992) *Am. Rev. Resp. Dis.* **146**, 888–894.

CHAPTER 3

AN INTRODUCTION TO STATISTICAL ISSUES IN AIR POLLUTION EPIDEMIOLOGY

Fintan Hurley

1. Introduction and Purpose

There has in recent years been an enormous growth in reported epidemiological studies of the possible adverse health effects of ambient air pollution. One feature of this literature is the use of relatively sophisticated statistical methods, for both analysis and presentation of results. These methods are necessary in that air pollution is only one of many factors which may adversely affect health. Indeed, other factors may be much stronger determinants of illness and mortality. In this context, careful statistical analysis is needed both to separate any associations with air pollution from those with other possible causes, and to identify, where feasible, the contribution of pollutants individually. Certainly, the recent advancements in air pollution epidemiology would not have been possible without powerful computing facilities, both for the management of large-scale datasets and for carrying out the intensive computations implied by many modern statistical methods.

At least initially, this gain in analytical power came at a price. The emerging epidemiological literature was of interest to scientists and policy makers from a wide variety of backgrounds. However, the statistical methods employed by that literature were considered difficult by many readers who found themselves unable to make reliable independent judgements about whether the reported conclusions were indeed justified by the data. This difficulty was compounded by what at the time was understood to be the biological implausibility of the findings of adverse effects. Together, these reasons contributed to the question: were the findings of adverse effects simply an artifact, based on mis-application of statistical methods? Or, in its gentler version: how sensitive are the estimated risks to the details of the statistical modelling?

Briefly, it is now widely accepted that the key results of modern air pollution time series studies are indeed derived from the data, become evident only by application of powerful statistical methods and are not artifacts spuriously generated by the use of powerful statistical methods. Nevertheless, these methods can still act as a barrier to transparency for many readers. Similarly, the use of advanced statistical methods in the analysis of cohort studies can cause difficulty. The purpose of this chapter is therefore to help the non-technical reader in reading, and independently assessing, the literature on air pollution epidemiology. I hope that it will at least help some readers to be less intimidated by "the statistics", and so avoid the extremes of either doubting everything, or taking the authors' views entirely on trust. More positively, I hope that it will help those readers less experienced in statistical methods to get a more "hands-on" flavour of the nature of the epidemiological evidence, to exercise some independent judgement about the strengths and weaknesses of particular studies, and to understand how important in practice any limitations might be.

Rather than attempting an abstract discussion of statistical methods, the approach adopted here is to read through two papers, highlighting and discussing methodological issues which might have a bearing on the reliability of conclusions. The papers chosen are Schwartz (1994), as a reasonably early example of the kind of time series study which led to a fundamental re-assessment in recent years of the possible adverse health effects of air pollution; and Pope *et al.* (1995), the largest of a small group of cohort studies which give substantial new evidence that effects may be even greater than the time series studies suggest.

The methodological commentary on these two papers will focus on, but by no means be restricted to, what might usually be understood as "statistical" issues. This is because good statistical analysis in epidemiology involves integrating a capability in statistical methods with an understanding of study design, of the nature of the measurements and their reliability, and of the kinds of relationships which might be expected to obtain between health, air pollution and other factors. We begin, however, with two short reviews of basic concepts. The first concerns regression analysis in statistics because the more modern methods used in air pollution epidemiology are developments of these basic concepts. The second deals with design issues in air pollution epidemiology because design and analysis issues form a unified whole.

2. Statistical Regression Methods: A Short Review of Basic Concepts

Regression analysis is used to explore and describe the relationships (or associations) between a single response (or outcome) variable and a set of explanatory variables. In epidemiology, the response variable is typically some measure of health or disease; explanatory variables might be any of a wide range of possible determinants or predictors of that health measure. As a simple example, think of lung function at a point in time as a response, with age and physique at that time, gender, lifetime smoking habit and exposure to dust as possible explanatory variables. Where several response variables are of interest, it is usual to carry out separate regression analyses of each, and to examine the overall results for *coherence* of findings, i.e. how well they fit together (Bates, 1992).

2.1. The Basic Concepts

The starting point of a regression analysis is that the response variable takes on different values at different observations; and we do not fully understand why. We do however think that, individually and together, the explanatory variables that have been measured may help us understand or "explain" at least some of this variation in response; hence their name. Regression modelling checks this judgement by explicitly constructing one or more equations which are in some sense best, and hopefully useful, at explaining or predicting the response variable. The response actually observed on any occasion can then be represented as the sum of two parts: a part that is predicted or explained by the explanatory variables via the regression equation; and a part that remains unexplained, often called the residual, or chance, or error component.

This partitioning of the response into two components leads to criteria for evaluating when one regression equation is as good as, or better than, another at explaining or predicting the response. One natural measure of *goodness-of-fit* is a high agreement or correlation between the observed values of the response variable and those predicted by the regression equation. Another is based on increasing the amount of variation in the response which the regression succeeds in explaining, while maintaining limits on the number of explanatory variables included. At least in simple cases, these two measures lead to similar conclusions about the usefulness of various equations in explaining the response. In practice, the situation is often simplified in that the focus of comparison is on equations that differ by one or more terms, rather than on a completely open-ended set of comparisons. Also, it can be helpful to base inferences on a series of equations rather than a single one, thereby reducing emphasis on what single equation may be "best".

Formally, a *regression equation* is a sum of several terms, each term consisting of an explanatory variable multiplied by a *regression coefficient.* It is not necessary that the explanatory variables should be represented in their original scale of measurement; the use of variables which have been transformed, say to the log scale or to some power, is commonplace. The regression coefficient then estimates or predicts how the response variable would change on average, for one unit change in the explanatory variable, in an idealised situation where all other explanatory variables in the equation are kept constant. Trivially, a coefficient of zero implies that the relevant explanatory variable makes no contribution at all to the prediction equations because the prediction is the same whatever value that explanatory variable takes on.

As well as estimating a regression coefficient, regression analysis also provides, via the *standard error* (*SE*) of the coefficient, an estimate of how precisely that coefficient has been estimated. The standard error has a similar meaning here to when it is used in describing the distribution of the sample mean around the (true) population mean. In both contexts, the size of the SE depends strongly on the sample size of the dataset. Other factors, for example the range of the explanatory variable, and how it is correlated with other explanatories, also affect the SE of a regression coefficient.

It is conventional to construct a *95% confidence interval* for the coefficient, this interval being the set of values ranging from two SEs below the coefficient to two SEs above it. Its name comes from the fact that, when regression analysis is applied repeatedly to different samples of data, and assuming that no important errors have been made in formulating the regression equation or in measuring the explanatory variables, the true value of the regression coefficient will lie within that confidence interval about 95% of the time.

Occasionally high values of a regression coefficient could arise by chance, even if the relevant explanatory variable was in reality of no use at all in helping explain the response; i.e. even if the "true" regression coefficient were zero. A helpful measure is the probability that a coefficient as extreme (whether in a positive or a negative sense) as that found would arise purely by chance. This probability, especially in large datasets, is related to and can be derived directly from the magnitudes of both the regression coefficient and its standard error. Indeed, the probability will tend to be higher insofar as the regression coefficient is small, and/or as its SE is large; and so is in an inverse relationship to the ratio of coefficient to its SE.

Thus, the greater the observed coefficient, the smaller this probability will be. A regression coefficient is said to be *statistically significant at the 5% level* (written as $P < 0.05$, or abbreviated to "statistically significant") if a value as extreme would arise by chance less than 5% of the time, given the explanatory variables as measured. Equivalently, in large datasets, a coefficient is statistically significant when the 95% confidence interval constructed around that coefficient does not include zero.

It is conventional, when a coefficient is statistically significant, to look for reasons other than chance in understanding why the value is relatively far from zero; though chance can never be excluded with certainty. Similarly, when a coefficient is not significant statistically, regression analysis does not "prove" that the coefficient is a result of chance. It simply denotes that chance cannot be excluded as one plausible explanation. Several other explanations may be plausible too.

There are two final points to be mentioned at this stage. Firstly, there is a difference between the size of a regression coefficient, and its relative size compared with its standard error. The size expresses the variable's *practical significance*, in terms of how a unit change is associated with change in the response; whereas the relative size measures *statistical significance*. These are two relevant but different concepts. It helps, if discussing whether or not a variable is "significant", to be clear about which interpretation is meant. Secondly, both the size and the statistical significance of regression coefficients in epidemiology usually depend on what other explanatory variables are included in the regression equation, a point that will be discussed more fully later. We often speak of size and statistical significance as if they are absolutes, but they are not. Both the coefficient and its statistical significance will change with the addition or removal of any other explanatory variable which is correlated both with the response, and with the explanatory variable of interest; its statistical significance will also change with the addition or removal or any variable which is correlated with the response only.

So, it is essential when describing a coefficient, or its statistical significance, that there is a clear understanding of context, in the sense of what other variables are included, and how these have been represented.

2.2. Some Implications

A few comments on the application of this basic structure may be helpful. First, there are particular difficulties when possible explanatory variables are very highly correlated with one another, as is often the case with different measures of ambient air pollution. There may then be little to choose between various sets of explanatory variables in terms of predictive ability as measured by overall goodness-of-fit. Moreover, there may be no advantage in including more than one of the highly correlated set of variables. This is because two or more highly correlated explanatory variables carry much the same information for prediction of the response as any one of them does individually. An implication is that we may be able to detect an association, but may be unable to attribute it to one variable rather than another. If two or more of a highly correlated set of variables are included in a regression equation, the prediction set can become unstable, being overly sensitive to small peculiarities of the dataset, with little or no gain in overall goodness-of-fit. In these circumstances the estimated coefficients, and their standard errors, may not transfer well to situations where the explanatory variables are correlated with one another in a different way.

A few additional remarks concerning extrapolation beyond the study data may be helpful. The goodness-of-fit of a regression model applies strictly to the dataset on which it was developed (or more exactly to the relationships which gave rise to that dataset). There is therefore some added uncertainty, over and above that captured by the original goodness-of-fit, when the same equations or coefficients are applied, overall or in part, to other circumstances. This uncertainty applies in particular to using the prediction equations in circumstances different from those examined originally. Such extrapolation may be valid; a problem is that the original study cannot confirm whether it is or not.

There is a similar problem with interpreting the regression model in terms of causality. One of the basic tenets of statistical analysis states that an association, say between a health response and an explanatory variable, does not necessarily imply a causal relationship. It may be that in reality the predictive power of one variable derives not from causality in its own right, but from its correlation with one or more other causal factors. Thus, many air pollution studies include the caveat that, even if air pollution is causally related to ill-health, the pollutants actually measured and studied may not be those that are responsible causally. On the other hand, the dictum that association does not imply causality can be overplayed. Statistical associations are informative about causality. For example, if a causal relationship really does exist, then we would expect to see evidence of association in suitably large datasets with well-measured variables. And if a similar association is found in several different datasets and under differing circumstances, then it becomes harder to find a credible alternative to a causal explanation.

The final point is a technical one. It concerns the scale of measurement of the response variable. A key part of statistical regression analysis, not always transparent to the reader, consists of understanding, describing and making use of patterns in the unexplained, chance or residual component of the response variable. What is considered to be the best regression equation depends, among other factors, on what really is the case, and on what is assumed, about the shape of the residuals over the dataset as a whole.

Residuals, though unpredictable individually, as a group tend to follow well-defined patterns described by various families of statistical probability distributions such as normal (Gaussian), log-normal, Poisson and so on. Exactly which of these distributions is appropriate in any situation depends in part on how the outcome variable is represented; and in particular, if it is a categorical or continuous variable. Lung function, for example, is effectively a continuous variable, measured on an interval scale; and there are theoretical and empirical grounds for treating residuals from a well-specified regression model as following a Normal distribution. Regression analysis was first developed in this context of continuous response variables with normal errors; and the associated chi-square, t- and F-tests of statistical significance will be familiar to many readers.

The situation is less straightforward when other kinds of response variables need to be considered. Part of the development of modern statistics has been the extension of regression methods to these more complex situations. The reader may be familiar with logistic regression analysis, which requires a two-valued outcome variable, for example, an individual reporting presence or absence of a respiratory symptom on a given day. Poisson regression analysis, or some extension of it, may be appropriate when the outcome is a count, for example, the number of deaths or hospital admissions on a particular day. Studies based on length of survival present different complications because, though length of survival is a continuous variable, its exact value is known only for those who have died. For those who have survived the study period, we do not know their exact length of survival; only that for each it exceeds the value it took at the end of the study. Extensions of regression analysis to deal with these and some other complexities in air pollution epidemiology are discussed via example below.

2.3. Primary Explanatory Variables

I have described regression analysis as an attempt to estimate a response variable by constructing regression equations which are the best, or nearly the best, in terms of overall goodness-of-fit to the observed data. However, air pollution epidemiology is almost never concerned with finding best overall estimates of health outcomes. Rather, its aim is the more specific one of establishing whether and to what extent air pollutants, both individually and together, are associated with (adverse) health outcomes, by investigating whether they help explain the variation in health responses. This more specific aim is in turn a step in assessing the causal contribution of air pollutants to human health, and so the benefits to health due to pollution reduction.

With this specific purpose, regression analysis is still the core methodology used, because it allows efficient simultaneous assessment of associations with many explanatory variables. Finding the "best" regression equation overall is not however the purpose of regression modelling in this context. Rather, the point is to estimate reliably the regression coefficients associated with various air pollutants, in the context of various *nuisance* variables or *confounders*. This is done by establishing one or more plausible regression models as a context for estimating the coefficients of the air pollutants, and then examining whether the estimated air pollution coefficients are sensitive to other included variables, and if so, in what way.

3. Design Issues in Air Pollution Epidemiology

3.1. The Time Dimension: Acute and Chronic Health Effects

Time periods

The literature of air pollution epidemiology distinguishes between *acute* and *chronic* health effects. These names are not especially informative and may even be misleading. For example, Dockery and Pope (1996) avoid them, and base a distinction on exposure only, using the language *acute exposure* and *chronic exposure* studies.

This is informative but unconventional. For better or worse, the language of acute and chronic health effects of air pollution is now well-established. In understanding it, it is helpful to distinguish conceptually three consecutive time periods in the development of an exposure-related health effect, as follows:

a. *The biologically relevant period of exposure.* This is the time period when an individual is exposed to air pollution which may contribute to the development of the health effect of interest.
b. *The latent period.* There are several, sometimes conflicting, definitions. We define the latent period as the time period between the end of biologically relevant period of exposure, and the time when consequent health effect is diagnosed or identified.
c. *The duration (or transience) of health effect.* This is the time period during which the health effect lasts.

It may be necessary under some circumstances to elaborate on this simple differentiation. The distinctions are sufficient, however, to help show that the difference between acute and chronic effects is not based principally on the transience or persistence of the health effect. Rather, they can best be distinguished by the combined pattern of relevant exposure and latency, as explained below.

Acute effects

In air pollution epidemiology, *acute effects* have a very short biologically relevant period of exposure, and a very short latency; the combined duration of relevant exposure period and latent period is almost never greater than five days. Typically, the exposure period considered is one day, with effects being attributed on the same day, or any of the subsequent days. Acute effects, then, are the more-or-less immediate effects of short-term exposures.

Acute effects vary widely in terms of the transience of the impairment to health. Many small but measurable effects on lung function are indeed reversible, within a matter of days or less, and occurrence of a respiratory symptom on a given day does not necessarily imply recurrence on the following day. On the other hand, admission to hospital can lead to a stay of different durations, measured in days, weeks or months; and at the extreme, there is of course acute mortality, or death in the days immediately following a higher degree of air pollution.

The main point here is that though some acute effects are transient, the terms "acute" and "transient" are certainly not synonymous.

Chronic effects

We will treat chronic health effects as those where the *combined* duration of relevant exposure period and latent period is long: months or years, usually years. In the context of the classical air pollutants (particles, SO_2, NO_x, ozone and CO), it will not be misleading to think of chronic effects as consequences of long-term exposure to air pollution, where damage is incremental over time.

3.2. Studies Designed to Investigate Acute Effects

These are studies designed to examine the more-or-less immediate effects of air pollution on the health of a given population, by examining relationships between short-term (usually daily) variations in ambient pollution and daily measurements of some health outcome. The broad design involves concurrently tracking:

i. the concentrations of ambient pollutants to which the study population is exposed;
ii. the health characteristics in the population; and
iii. other characteristics such as climate which vary from day to day and may confound any associations between air pollution and health.

Studies of acute effects fall into two broad categories according to whether or not the population being studied is identified and followed up at the individual level, or purely at the group level. *Panel studies* track identified individuals who have agreed to take part in the study and whose active co-operation is needed in recording, usually on a daily basis, but in some studies more frequently, outcome measurements such as lung function or occurrence of particular symptoms. It is usual also to record and study other characteristics of the participating individuals, including characteristics which do not vary from day to day such as age, gender, or health status at the start of the study. Typically, panel studies are used when studying outcomes such as symptoms whose risk of occurrence is high for an individual on a given day, irrespective of air pollution levels; or when studying variations in a continuous measurement such as lung function.

Time series studies in air pollution epidemiology involve tracking health events in large populations using systems which do not require the active co-operation of the individuals studied. They are especially useful when studying outcomes such as mortality or hospital admissions of which risk of occurrence for an individual on a given day is on average small, regardless of air pollution levels. It may then be necessary to study several million people for several years, so that daily variations in these relatively rare events can be associated reliably with daily variations in air pollution.

It is impractical to enlist directly the co-operation of such a large study group, for such long periods. Fortunately this is also unnecessary. Time series studies of mortality and hospital admissions are conducted by obtaining, from routinely collected records, the numbers of daily deaths or hospital admissions in some designated region, or catchment area for a group of hospitals. Changes in these numbers can be related to changes in daily levels of pollution, without the need to identify the at-risk population explicitly, on the assumption that the at-risk population in the catchment area is more or less constant from day to day.

In both time series and panel studies, analysis examines the associations, if any, between daily pollution and health on the same or immediately following days; adjusting for season, for other long-term trends in health, and for daily differences in climate. Sometimes it is possible to also adjust for other factors, such as influenza epidemics. Because these studies are of changes *within* a population, it is unnecessary to adjust for factors such as smoking habit, diet and socio-economic status: such characteristics are invariant in populations in the short scale, and changes in them are in any case unlikely to be closely correlated with changes in daily pollution levels. Differences *between* populations in these respects may however contribute to differences between studies in the estimated overall effects of air pollution. It may be possible to investigate this by examining, in time series or panel studies, the effects in particular sub-populations, characterised by age or other factors.

It is very unusual in these studies to examine time-lags of greater than five days between pollution and health outcome. In studies of "acute" mortality, for example, the time-lags studied between increases in ambient PM_{10} concentrations and numbers of deaths, range from same-day effects to effects delayed up to five days, with most studies reporting effects based on intermediate time-lags (US EPA, 1996). The corresponding time-lags for studies of respiratory hospital admissions are generally same-day or one-day lagged effects. This short latency, together with the fact that the pollution changes being studied are on a daily basis (and occasionally combined over successive days), means that time series and panel studies are studies of acute effects only, and thus are uninformative about chronic effects. (Knowledge of acute effects may be indirectly informative about chronic effects; this is an issue on which there is not currently a consensus.)

3.3. Studies Designed to Investigate Chronic Effects

Cross-sectional studies

The defining characteristic of cross-sectional studies is that the health of subjects is examined at one point in time only. Identification of associations between air pollution and health therefore depends on comparisons *between individual subjects* or *between communities* whose amount of exposure to air pollution differs. Whether the focus is on individuals or on groups depends on the frequency of outcome and on the ease of data collection in ways that parallel those described earlier as distinguishing between panel and time series studies. Comparisons between communities typically involve death or hospital usage rates over a defined period (say one year). Comparisons between individuals typically focus on endpoints such as lung function or symptoms, which require fewer subjects than would a corresponding study of one-year death rates.

The reliability of cross-sectional studies to show air pollution effects depends critically on how well the studies can adjust for other factors influencing health differences between subjects or studied communities. Failure to do so appropriately can lead to biased or misleading estimates of air pollution risks. Where the study is individually based, it is usually possible to obtain relevant information (e.g. on smoking habits, on socio-economic status, on occupational exposures) directly from the individuals; and so to adjust reasonably well for possible confounding.

It is however much more difficult to ensure the reliability of cross-sectional studies conducted at the group level. This is because it is often impractical to obtain reliable measures of the community's average experience. Also, it may be that adjustments using community average values of confounding factors will not adjust sufficiently for the aggregate effect of these confounders at the individual level; and that relationships between exposure and health at the group level may not apply to individuals within a group [Treating them as identical is known as the ecologic fallacy: see, e.g. Walter (1991a; 1991b)].

Though it is very difficult to ensure that the biologically relevant exposure period has been studied, and latency allowed for appropriately, the exposure characteristics studied cross-sectionally are measures of (or surrogate measures for) the long-term exposures of individuals or communities. However, cross-sectional studies are not pure studies of chronic effects, because the health responses measured also include any corresponding acute effects operating at the time of measurement.

Studies of morbidity or mortality over time, including cohort studies

Studying changes in morbidity or mortality of communities does not really take us much further forward when these populations are treated as a single entity. Where the focus is on comparing trends in pollution and health over time *within* a population or community, there may be difficulties in adjusting concomitantly for changes in diet, smoking habit, socio-economic status and so on. These changes in non-pollution factors are much more likely to be correlated with, and so distort the estimated effect of, annual average pollution levels than they are to be correlated with the daily variations in pollution examined in time series studies. Where studies involve comparisons *between* populations or communities in different locations, they face similar difficulties to cross-sectional studies in adjusting well for confounding factors.

Longitudinal studies of morbidity involve tracking the health of identified individuals over time; and in this the design is similar to that of panel studies. In practice they differ from panel studies in that the study duration is much longer; health effects are measured on a few occasions only rather than on a daily basis; and annual average or lifetime exposures are examined, not daily variations. Thus, the aim is to study chronic effects principally. As before, the ability to take account of potentially confounding factors at the individual level, including how these change over time, adds to the reliability of these studies compared with a community-level design. Analysis requires some assumptions, which may not be checkable, about how soon changes in pollution may affect changes in outcome. Longitudinal studies of morbidity are as yet rare in air pollution epidemiology.

Just as longitudinal studies of morbidity follow up the experience of study subjects over time, so do *cohort studies of mortality.* These are studies of individuals whose non-pollution characteristics (such as age, smoking habit, education etc.) are determined and adjusted for at the individual level. The cohort studies differ from longitudinal studies of chronic morbidity in the determination of endpoint. For mortality studies, there is in effect continuous monitoring to determine if and when any subject dies and if so, from what cause of death; or else, to establish the individual's survival throughout the study period.

Thus by design, the cohort studies focus on the date of death, or exact age at death, of designated individuals as the primary endpoint of the study; and the analysis uses these dates and various explanatory factors to model how ambient pollution may be related to differences between individuals or cities in their adjusted risks of mortality. In this way, cohort studies take account of the full force of pollution on mortality; i.e. they embody both "acute" effects, which occur in the days more-or-less immediately following higher pollution; and "chronic" effects which develop over the longer term. Indeed, the cohort studies do not differentiate between acute and chronic effects, and do not attempt to estimate them separately. Rather, they estimate an overall effect of air pollution on age- or time-specific mortality.

Cohort mortality studies with individually-based adjustment for confounders are a great advance methodologically compared with the adjustment between communities as used in cross-sectional studies of mortality. For particular cohort studies there still however remain important methodological issues. One is the question of whether adjustment for confounders has been sufficiently comprehensive. Another is the range and distribution of exposures which it has been possible to investigate. More importantly perhaps is the extent to which any cohort study has been able to take account of biologically relevant exposures, and latency. The relative importance of acute effects needs to be considered in interpreting the results. These and other methodological issues will be considered in more detail when Pope *et al.* (1995) is considered in more detail below.

4. A Population-Based Time Series Study: Schwartz (1994), Hospital Admissions in Detroit

4.1. Population

The at-risk population of those aged 65 years or more was about 517 000 persons, from a total population of 4 382 000 (1990 data) in the Detroit metropolitan statistical area ("Detroit"), USA. Study duration was four full years, from 1 January 1986 to 31 December 1989. The study is therefore a typical time-series study in its design.

4.2. Health Effect Measurement

Hospital admission data for all hospitals in Detroit were used to construct daily counts of admissions for pneumonia, for asthma, and for chronic obstructive pulmonary disease (COPD) other than asthma. The hospital admission data are used to obtain reimbursement from the US Medicare health system, and so are standardised and likely to be complete. Diagnosis is recorded at time of discharge and so is likely to be more accurate than diagnosis at time of admission. It was coded according to the current international standard, International Classification for Disease, ninth revision (ICD9).

4.3. Air Pollution Data

Daily air pollution data were obtained from the Environmental Protection Agency (EPA) database. Separately, for PM_{10} and for ozone, daily data (24-hr averages) for the at-risk population were derived by averaging over all relevant monitoring stations in Detroit. The number of PM_{10} stations varied by year. Agreement between the stations is not reported, though based on other studies, we would expect that it was high.

Problems of missing data are described only cursorily. PM_{10} measurements were available for 82% of the study days. The pattern of, or reasons for, missing data are not described; nor does the paper report what was done, either for missing days or when results were unavailable for particular monitors. It is unclear to what extent these were the same as, or different from, those days with missing PM_{10} data. Some information about this, and about how missing days affected the investigation of lagged pollution effects or the construction of multiple-day averages, would have been helpful. We can take it however that with a leading researcher such as Schwartz, there will not be issues on how the missing pollution data were handled such as might have affected the study's conclusions.

4.4. Confounding Factors

Data on daily mean temperature and dew point temperature (an index of humidity) were obtained from one site in Detroit. Daily measurements of these two characteristics were very highly correlated (correlation coefficient = 0.96). Different ways of modelling weather effects have subsequently been investigated in detail in other studies and datasets. Results have shown that more complex modelling neither improves the goodness-of-fit significantly, nor importantly alters the estimated pollution effects.

4.5. Statistical Methods

Schwartz uses Poisson regression analysis, allowing for overdispersion, as the basic framework for statistical modelling. The basic ideas are similar to ordinary regression. In particular, the average or expected value of the response variable is represented as a combination of effects associated with various measured explanatory variables. Specifically, the expected number of hospital admissions among the elderly on any day is represented as the product of coefficients and variables associated with season, climate, air pollution and so on. In practice, analysis is carried out on the log scale where the effects are then additive, strengthening the similarity with "ordinary" regression. Analysis on the log scale affects the meaning of the coefficients; see 4.8 below.

The actual or observed numbers of daily hospital admissions were assumed to vary around these daily averages in a way that is representable by the Poisson distribution, with one difference. In a standard Poisson distribution, the dispersion (variance) is the same as the average or mean value. Schwartz's analysis allowed for the possibility that the variance may be bigger than the mean by a constant factor, to be estimated from the data. Such overdispersion is not uncommon when analysing data as counts. Poisson regression with overdispersion extends "ordinary" regression analysis to provide an appropriate, flexible and now well-established framework for investigating the effects of PM_{10} and other pollutants on daily hospital admissions or deaths, adjusting for confounding factors.

As noted here by Schwartz, and in many other papers also, there may be unexplained associations between observed numbers of hospital admissions on days that are near one another. Analyses which ignore such autocorrelation, when it is present in the residuals, may give spurious precision to the estimated relative risks, by suggesting as statistically significant some coefficients which in reality are not. (The regression coefficients themselves are less sensitive to autocorrelation than are their standard errors.) Allowance for autocorrelation is made using generalised estimating equations (GEE), a relatively new and complex methodology developed to take account of serial correlation in Poisson, logistic and other non-normal regression models.

However, in the present study, no significant serial correlation was found in residuals from the principal Poisson regression models used.

Thus, the reported results are based on Poisson modelling with overdispersion, but without the need to use GEE methods to account for day-to-day carryover. Other time series studies have shown that this result is true more generally. Thus the more complex GEE methods are usually needed for sensitivity analyses only, and not for the modelling on which main conclusions are based. They can be bypassed by those who want to read the main story only.

4.6. Modelling Strategy

In this and in other studies, Schwartz and others safeguard against that danger of attributing to pollution some of the effects of weather or other confounding variables by means of the following modelling strategy. First, adjustments are made for long-term or seasonal patterns and for weather, before seeking to describe associations with air pollutants. The strategy is conservative in the sense that when constructing the baseline model Schwartz risks overfitting climate and seasonal effects, and possibly attributing to them some of the effects of pollution.

Pollution effects are then investigated directly in the context of the baseline model so that, if identified, they are unlikely to be artefacts of the *measured* weather variables. (In principle, they could be artefacts of unmeasured climate characteristics; but, as noted in 4.4 earlier, complex representation of climate effects has not led to significantly changed estimates of air pollution effects.) Finally, various sensitivity analyses are carried out to see if the estimated effects of pollution are robust to the precise formulation of the baseline model.

As a strategy for estimating pollution effects given time-related fluctuations and confounding, this is unarguably sound from the viewpoint of limiting the chances of false positive associations. Limitations, if any, arise not in the core strategy, but in the details of model building. The same strategy would of course not be suitable if the principal research interest were in the longer-term fluctuations or in the effects of weather; and indeed, in the present paper, the author did not report regression coefficients for these characteristics.

4.7. Constructing the Baseline Model

General strategy

In developing a baseline Poisson regression model, Schwartz carried out an iterative process consisting of model fitting, validation through examining residuals, model adaptation and re-fitting, until there was no further apparent association with longer-term time-related variables or with available confounding factors. The modelling attempted not to impose any preconceptions about the shape or structure of longer-term time-related fluctuations and of the effects of temperature, but rather to let these be determined by the data. This is consistent with the strategy of possibly over-fitting climate and seasonal effects.

Time-related factors

Thus, adjustment for time-related patterns over the 4-year study period involved constructing a set of "dummy" or indicator variables to denote in which of the 48 months each day belonged. Typically, a dummy variable takes the value 1 if the day falls within the month in question, and the value 0 otherwise. This way of removing time-related patterns has the effect of removing also any associations between monthly average pollution levels and monthly average hospital admissions. The search for associations between air pollution and hospital admissions is consequently restricted to aggregating, across the 48 months of study, whatever associations are found within individual months.

This approach distinguishes absolutely between months, without presupposing or imposing any relationship between hospital admissions in adjacent months, or in the same month in different years. There is, on the other hand, no differentiation between days within months: exactly the same adjustment is made for all days within the same month and year. The adjustment for time and season, though detailed, was therefore discontinuous. However, linear and quadratic terms of time were also fitted, to give a smooth further adjustment for any systematic longer-term trends in time, and so bringing minor adjustments between days in the same month.

Climate

The shape of relationships with temperature is not well-established, but is likely to be non-linear. Consequently, in this study, Schwartz again used a grouping and dummy variable approach to adjust for temperature effects. This was done separately for daily mean and for dew-point temperatures. The number of groups into which each variable was split was data-dependent but pre-defined to be at least seven. In practice, eight groups per variable were used, residual analyses then showing no remaining patterns linking hospital admissions with either of the two climate characteristics. In principle, over-adjustment for weather might have implied fitting a separate dummy variable for each combination of mean daily and dew-point temperature. In practice, given the extremely high correlation between mean daily and dew-point temperatures, this would have been impractical and almost certainly of no real benefit.

Investigation of alternative representations

Finally, various non-parametric smoothing methods (Hastie & Tibshirani, 1990) were used, both to identify patterns in residuals from the fitted models, and as an alternative method of adjusting for longer-term fluctuations in, and for effects of confounders on, the daily hospital admission counts. These non-parametric smoothing methods are generalisations of weighted running means. They allow a smooth adjustment to be made for the effect of confounding factors, in contrast to the jagged adjustment arising from using grouping and dummy variables. Furthermore, the adjustment is data-driven, in that there are no preconditions that it should follow a regular shape such as linear, quadratic or cyclical. These non-parametric generalised additive modelling methods are a very powerful tool for the integrated task of good data descriptions and modelling; and are increasingly used, to show the shape of relationships between air pollutants and health (with and without adjustment for other factors) as well as being a means of adjusting for confounding factors. They are a particularly good example of how powerful modern computer-intensive methods need not cut the reader off from a sense of the raw data, but can be and are used to show very important features of the data, for example, the shape of any exposure–response relationship, which previously would have remained inaccessible.

Conclusions about adjustment for confounders

In summary, in this paper, Schwartz has made flexible use of a variety of complementary approaches towards eliminating longer-term fluctuations and the effects of confounders. Singly and together, these methods risk overfitting or over-compensating for confounders, in that the methods of adjustment are quite strongly data-driven and so may reflect and adjust for patterns associated with confounders only by chance; and so they justify Schwartz's description of the modelling strategy as conservative with respect to identifying pollution effects.

Any limitations, therefore, should be sought in the choice of confounding factors studied, not in the thoroughness or correctness of the modelling done on those that have been examined. It has been suggested that various other measures of climate, for example night-time temperature, might be relevant; but as noted above, later studies have not shown that estimated pollution effects are sensitive to climate modelling. Similarly, there is no evidence that lagged temperature effects were examined, and these may also be relevant. Or, there may also have been systematic day-of-the-week patterns in the admission data. None of these is, of course, a true confounder unless also related to daily PM_{10} measurements.

Note that in UK studies, where winter pollution is positively associated with cold weather, adjustment for season and for climate usually reduces the apparent association between mortality or hospital admissions and ambient particles. It is interesting that adjustment does not have this effect universally. For example, in the present study, hospital admissions for pneumonia peak in winter, whereas PM_{10} and ozone peak in the summer. Without removal of long-term trends and temperature effects, the naive (unadjusted) association between admissions for pneumonia and PM_{10} or ozone is negative. Clearly, this negative association reflects seasonal patterns which must first be taken into account before valid inferences can be made about hospital admissions and air pollutants. Failure to adjust properly for confounders does not necessarily mean that pollution effects have been overestimated.

4.8. Modelling the Effects of PM_{10} and Ozone

Both 24-hr average PM_{10} and 24-hr average ozone were added, separately and jointly, to a complex baseline model which used dummy variables to distinguish 48 months, eight ranges each of daily mean and of dew-point temperature and which included linear and quadratic terms in time. Both PM_{10} and ozone were included as linear terms in a Poisson regression model analysed on the log scale. This implies an assumption that if there is a percentage increase in average daily admissions associated with PM_{10} or ozone, that percentage increase applies at least approximately at all background levels within the range of ambient pollution implied by this study (e.g. 10th percentile PM_{10} 22 $\mu g/m^3$; 90th percentile PM_{10} 82 $\mu g/m^3$).

Within this framework, positive and statistically significant relationships were found between PM_{10} and, respectively, pneumonia and COPD admissions (excluding asthma), but not with asthma itself. Though described only cursorily, it is clear that the effects of pollution were examined using various lag periods; and that relationships were also investigated using average pollution levels over several previous days. Strongest relationships were found with same-day PM_{10} and previous-day ozone; these are the relationships that are reported.

The PM_{10} coefficient was similar whether or not ozone was included as an explanatory variable; implying that, having adjusted for longer-term fluctuations and weather, daily PM_{10} and previous-day ozone were effectively independent of one another in their association with hospital admissions. This reflects the fact that, unadjusted for confounders, the correlation coefficient between same-day PM_{10} and ozone was not very high, at 0.35. Identifying associations with individual pollutants is not usually so easy; and in particular, it may be difficult to distinguish from one another the effects of particles, SO_2 and NO_x, all of which tend to be more highly correlated with one another than with ozone.

The estimated regression coefficients quantify how hospital admissions in the elderly changed with changes in PM_{10} and ozone. Taking admissions for pneumonia as an example, the estimated Poisson regression coefficient for PM_{10} was 0.00115. This was very highly significant statistically, with a standard error (SE) of 0.00039 implying a *P*-value of 0.0028. The estimated impact of PM_{10} on pneumonia admissions was however quite small, the associated percentage increase, per 10 $\mu g/m^3$ PM_{10}, being 1.16%. (The log-scale coefficient for 10 $\mu g/m^3$ PM_{10} is 0.0115. Transforming back to the original units, exp(0.0115) is equivalent to a factor of 1.011566, implying an increase of 1.157%.)

4.9. *Checking the Robustness of the Estimated Effect*

The final stage of statistical modelling was a series of checks on whether the estimated pollution effects were sensitive to other aspects of the modelling. The estimated percentage increase did not depend on whether adjustment for confounders was by dummy variables or by non-parametric smoothing. Also, it was insensitive to whether high-pollution days (defined as daily mean PM_{10} exceeding 150 $\mu g/m^3$, or 1-hr ozone exceeding 120 ppb; 240 $\mu g/m^3$) were excluded or not from the analysis. Thus, the effect was not attributable to episodes of high pollution, but rather reflected experience across a range of days when pollution levels lay below current US standards. (One of the major changes in understanding in the past 15 years or so, and especially in recent years, has been the realisation based on evidence that the adverse acute health effects associated with air pollution are not restricted to episodes of high air pollution. Rather, they occur across the range of what are considered to be normal pollution levels in various locations. This is an important observation that would not have been possible without large datasets, powerful computing facilities and modern statistical regression methods.)

Finally, to check on the assumption that the percentage increase in expected daily hospital admissions was constant per unit exposure across the range of pollutant measurements studied, both same-day PM_{10} and previous-day ozone measurements were categorised into four groups, consisting of the four quartiles of the distribution. For each pollutant, dummy variables distinguishing the four quartile groups were added to the baseline Poisson regression model. Relative risks (RR) of the four quartile groups were then standardised by arbitrarily fixing the RR of the lowest quartile group as 1.0. Plots of the RR per quartile of PM_{10} against mean PM_{10} of that quartile showed no strong or suggestive evidence of a threshold. Moreover, the RRs between quartiles of PM_{10} were consistent with the assumption of constant percentage change per unit exposure; i.e. assumptions of the earlier modelling are not contradicted by the data.

The agreement between the assumption of constant percentage increase, and the results of quartile modelling which make no assumption about shape between quartiles, is illustrated in the case of pneumonia admissions and PM_{10}. The risk of admissions on days when PM_{10} was in the upper (fourth) quartile was about 1.077 that of days when PM_{10} was in the lowest (first) quartile, having adjusted as before for longer-term patterns and temperature. Now, the mean PM_{10} of the first and of the fourth quartile were about 21 $\mu g/m^3$ and 83 $\mu g/m^3$, respectively, a difference of 62 $\mu g/m^3$. The original log-linear model predicted a corresponding relative risk of $\exp(0.00115 \times 62) = 1.074$. This is equivalent, within rounding error, to the quartile plot RR of 1.077.

A plot of pneumonia residuals versus PM_{10}, with non-parametric smoothing, might have elegantly illustrated the shape of the relationship. This is done in several other papers by Schwartz and other authors and is a good example of how powerful modern statistical methods can help not only in formal modelling, but in presenting results from complex datasets in ways which are accessible to people not specialists in statistics.

5. A Cohort Study: The American Cancer Society Study, Pope *et al.* (1995)

5.1. Study Participants

The study by Pope *et al.* (1995) included more than 500 000 people (56% women, 44% men). All were aged 30 years or more when enrolled in 1982 in a major (1.2 m people) study by the American Cancer Society (ACS). Pope *et al.* selected individuals from within the larger ACS cohort according to availability of relevant data for their analyses of mortality and air pollution. Individuals were assigned to metropolitan areas, based on residence at time of enrolment in 1982 to the ACS study. People from 48 states of the USA were included.

This is a very extensive study, spread widely across the urban or semi-urban areas of the USA. It therefore includes people who live under widely different circumstances from one another, including, for example, in locations that differ from one another in terms of pollution sources, mixtures and concentrations. For example, sulphate concentrations are far higher in Eastern than in Western US, where nitrate, elemental carbon and organic carbon concentrations are higher. These differences partly reflect different sources of particles, with more industrial sources and associated sulphur dioxide in Eastern US, and a greater contribution from traffic, with associated oxides of nitrogen, in the West. Another contributory factor is the greater abundance of ammonium in Western US (RM Harrison, personal communication).

Studying locations with a wide range of particle concentrations should help in identifying effects, if these exist. Studying locations that differ in other respects, including particle composition, should help in generalising results, and in transferring them to other locations, because the diversity reduces the need to extrapolate beyond the range of circumstances studied.

5.2. Air Pollution Data

Summary

Where practicable, two indices of particulate air pollution, one for sulphates and one for fine particles, were derived for these metropolitan areas. Neither sulphates nor fine particles include the relatively coarse inhalable natural dusts which do however contribute to PM_{10}; and so measurements of both indices are strongly dominated by particles from combustion sources.

The most widely available index of particles, available for 151 metropolitan areas (552 138 participants), was derived as the annual average sulphate concentration in one year, 1980, based on all relevant monitoring sites in the US EPA's National Aerometric Data Base. The study mean across the 151 areas was 11.0 $\mu g/m^3$ (sd 3.6, range 3.6–23.5 $\mu g/m^3$). A second index, available for 50 metropolitan areas (295 223 participants), was the median $PM_{2.5}$ concentration 1979–83 as calculated by Lipfert *et al.* (1988) using data from the EPA dichotomous sampler network (study mean 18.2 $\mu g/m^3$, sd 5.1, range 9.0–33.5 $\mu g/m^3$). Note in passing that the 50 metropolitan areas which can be used for investigating the effects of fine particles is almost, but not quite, a subset of the 151 areas available for studying sulphates and mortality.

Both indices were available for 47 metropolitan areas. The two indices of particles were strongly and positively correlated (Pearson coefficient 0.73) within the 47 metropolitan areas where both were available.

Comments on the reliability of the annual average concentrations used by Pope et al. (1995)

It is to be expected that annual average fine particles ($PM_{2.5}$) and sulphates are positively correlated, because both arise principally from the combustion of fossil fuels. It is also to be expected that the correlation between the two indices is imperfect. Even in high-sulphate regions like Eastern USA, sulphates form less than one-half of $PM_{2.5}$;

the fraction in Western US is much lower (RM Harrison, personal communication). Also, the two indices are measured using different samplers with different analytical methods, and they cover different time-periods. The samplers presumably are not necessarily co-located within metropolitan areas.

Lipfert and Wyzga (1995) report that the air quality measurements of Pope *et al.* were based on a single monitoring station in each metropolitan area. However, sulphates and fine particles are well distributed spatially, more so than coarser particles or mixtures such as PM_{10}; and Pope *et al.* are concerned with annual average data and not with short-term fluctuations. For these reasons, one well-sited monitor per location might differentiate adequately between metropolitan areas. We expect that the monitors will have been sited away from particular local sources of pollution. Failure to do so will have added to the random or measurement error within the concentration data. As discussed by COMEAP (1995) and many others, random errors in exposures will usually lead to *under*-estimation of the associated relative risks, though exceptions may occur when errors are present simultaneously in several exposure variables. This possible attenuation of relative risks is noted by Pope *et al.* (1995) who comment that "exposure misclassification is unlikely to result in spurious associations between pollution and mortality".

Aggregation, and biologically relevant individual exposures

It was a major achievement that informative indices of particulate air pollution could be derived for use in such a geographically-dispersed study. From the viewpoint of epidemiology, however, the indices used have two principal limitations. First, the available pollution data refer to metropolitan areas, and not to individuals within them. The experience of individuals is to some extent differentiated through taking account of age in the analysis; but, in other respects, within-city differences in exposures are ignored. The use of city-specific rather than individually-based concentration data contributes to the misclassification error whose effects have been discussed above.

The second issue may be the more important one. The available air pollution data refer to concentrations over a relatively short time-period: one year for sulphates, 4–5 years for fine particles. It is unclear to what extent these relatively recent data are reliable indicators of particle concentrations from earlier time periods which would be biologically relevant to development of disease, if a causal relationship does exist. In fact, there is currently no agreed understanding of what constitutes the biologically-relevant time period. Dockery *et al.* (1993) look at a biologically relevant period of about 15 years before death. Lipfert (1998), citing Rose (1982) and Moolgavkar *et al.* (1989), assumes a latency or "incubation" period of 20–40 years for deaths from cardio-pulmonary disease. The basis for this is not clear; Rose (1982) suggests "10 years or more" between exposure and greatest impact on mortality. It is possible that, if the relationship between mortality and long-term exposure to particles is real, the effect on mortality may result from incremental damage over a long period, and perhaps throughout a lifetime. This implies that for some people, exposure many decades ago may be relevant.

Recent annual concentrations as a surrogate for past concentrations

Pope *et al.* differentiate between the metropolitan areas in terms of their annual average particulate concentrations around 1980. The weak version of the reliability question is: How good is a *ranking* of metropolitan areas based on recent pollution data, as a surrogate for a ranking based on measurements from earlier decades? As discussed above, misclassification of the ranking of metropolitan areas would tend to work against detecting a real relationship. Unless the surrogate is a good one, a real association might remain undetected or obscured.

The strong version of reliability question is: How good quantitatively are the measurements near 1980 as estimates of concentrations in earlier periods? This stronger criterion is relevant to Pope *et al.*'s estimates of relative risk per $\mu g/m^3$ sulphates or $PM_{2.5}$. Briefly, if past concentrations of sulphates and fine particles were much higher than now, the use of only recent concentration data would tend to overestimate real risks when expressed as proportional hazards per $\mu g/m^3$ sulphates or $PM_{2.5}$.

Pope *et al.* address this issue directly. They note that the longer sequence of measurements from the Six-Cities Study (Dockery *et al.* 1993) showed little change in annual sulphate and fine particle concentrations during the 1970s. Separately, Schwartz (1997) provides data showing no marked decline in fine particle measurements throughout the USA over this time-period. He also points to changes in composition of pollution mixtures which would lead to the finer fraction making up a greater proportion of TSP than previously, implying that changes in (say) $PM_{2.5}$ over time will have been less than in PM_{10}. These results suggest that the problem of using relatively recent concentrations as a surrogate for earlier ones may not be a serious one.

More recently, Lipfert (1998) suggests differently. He draws both on emissions and on ambient monitoring data to estimate trends in population exposures to airborne particles in the USA, 1940–90. Very few measurements of fine particles were available before about 1970, but many sources of more recent measurements of $PM_{2.5}$ were found and used. Briefly, the data suggest reductions in annual average concentrations in the order of 4–5% per year, overall, 1970–90. This rate is consistent with the much more limited data from earlier years (1960, 1956 and 1932); though backward extrapolation is complicated by what Lipfert describes as an "emissions plateau" between 1960 and 1970. The estimated rate of decline of total suspended particulates (TSP) was similar, suggesting that the size distribution of ambient particles may have remained fairly constant over time.

Lipfert (1998) takes 1985 as the average year of death for subjects studied by Pope *et al.* Though the issue of biologically relevant exposures is unresolved, concentrations experienced say 30 years prior to death may be a better surrogate of biologically-relevant exposures than recent (e.g. 1980) concentrations. *If* this is true, and *if* the estimated decline of 5% per year is true over that period, then concentrations in 1980 would have been less than 30% of those in 1965; and *if* exposure misclassification did not lead to under-estimation of effects, then Pope *et al.*'s inability to use concentration data from the relevant time-period biologically could have led to a three- to four-fold over-estimation of risks. In practice, any over-estimation will almost certainly have been counter-balanced at least in part by misclassification of exposures, as discussed earlier.

Mixtures

The absence of indices of pollution other than particles implies that Pope *et al.* were unable to investigate whether relationships between mortality and sulphates or fine particles were really (causally) an effect of particles; or whether particles are acting as a surrogate for more complex air pollution mixtures. This disentanglement is complicated by the issue of biologically relevant exposure periods. On the other hand, the diversity of pollution scenarios covered by the study, i.e. the diversity of pollution mixtures within which these fine particles are found, cautions against any easy alternative explanation that covers the dataset as a whole.

5.3. Characteristics of the Participants; Confounders at the Individual Level

As part of the ACS study, information had been gathered by self-administered questionnaire on a range of individual characteristics. Pope *et al.* used values of these characteristics at time of enrolment into the study, including age (mean 56 years), sex, race (94% white), body mass index (BMI, mean 25), drinks per day of alcohol (mean 1), and educational attainment (high school or more: mean 87–88%). People who had ever been regularly exposed occupationally to any of a wide range of industrial pollutants were identified as one group, comprising 20% in total of all participants. Classification by smoking habit in 1982 showed 22% current cigarette smokers, 29% former cigarette smokers,

4% smokers of pipe or cigars only, and by implication, 45% lifelong non-smokers. Information was available, for both current and former cigarette smokers, about number of cigarettes smoked per day (mean 22); and number of years as a smoker (mean 33.5 years for current smokers; 22.3 years for ex-smokers). On average, participants were passively exposed to cigarette smoke for about 3 hours/day.

This is an exceptionally rich set of potentially confounding factors, at the individual level, for a mortality study of air pollution and health. In particular, the information on smoking habit and smoking history is relatively comprehensive; and arguably necessarily so, when smoking is a major risk factor for the causes of death of interest (non-malignant cardio-respiratory causes and lung cancer).

The study's ability to adjust for other potentially confounding influences is much less good. The indicator of regular past occupational exposure is crude; but of course better than none at all. The overall percentage of about 20% exposed occupationally was presumably much higher in men than in women; exact numbers not given. The reporting of alcohol consumption in particular is known to be subject to errors. And educational status is crude as an indicator of socio-economic circumstances; especially when summarised as an indicator available (less than high school education, or not) which identifies only a minority of participants.

It is reasonable to assume that in a study as major as that of the American Cancer Society the basic data collection was done to well-designed protocols. The work was however carried out by volunteers, and it is difficult to maintain high standards of quality with such an arrangement.

In considering these limitations however, it is essential to remember that it was not the purpose of Pope *et al.*'s study to investigate, identify and describe the relationship between mortality and factors such as smoking habit or socio-economic status. Rather, the purpose was to take account of these factors well enough so that any relationship between particulate air pollution and mortality can be identified reliably, and ideally without bias, which can occur only if an omitted or poorly measured variable is related both to mortality and to ambient particles.

5.4. Regional Confounders

Weather variables that identified relatively cold or relatively hot metropolitan areas were included in some analyses. The purpose was to check if the estimated pollution effect was sensitive to their inclusion.

Focusing on the temperature extremes is reasonable in a sensitivity analysis, in that it should identify if temperature did have a distorting effect on the estimated pollution effect.

Mostly in this study, potential confounding factors were adjusted for at the individual level. These climate variables are the only characteristics, other than pollution itself, which were included at the level of metropolitan area rather than individual. It is possible to think of other relevant characteristics which vary with metropolitan area level, for example, health services provision. However, as noted above, such omitted factors would confound the core particle–mortality relationship if, and only if, they were simultaneously related both to mortality and to particulate pollution.

5.5. Mortality Information

Vital status of the participants was determined over an approximately 7-year period (1 Sept 1982–31 Dec 1989), with death certificates obtained for 96% of those identified as dead. Cause of death was coded (9th Revision, ICD) without knowledge of air pollution levels. Three main cause-of-death categories were used: lung cancer, (non-malignant) cardiorespiratory, and all other causes.

This is a very good follow-up rate, especially considering the size of the cohort. It is well-known that there are errors in the routine recording of cause-of-death; but these errors principally involve detailed cause, and not the very broad groupings used in this analysis. Routinely, there is some misclassification between cardiac and respiratory causes as principal cause of death. The nature and extent of such misclassification is age-related; and Pope *et al.* explain that it was to avoid possible consequent biases that the very broad category of non-malignant cardio-pulmonary diseases was not sub-divided.

In summary, it is highly unlikely that any limitations of mortality follow-up or of cause-of-death coding had any material impact on the conclusions of this study.

5.6. Statistical Methods: Time and Adjustment for Age

The usual approach to analysing cohort studies

Central to the analysis and interpretation of cohort studies is the concept of mortality *hazard*, that is, the probability or risk of dying in the next immediate period, assuming survival up to the present. This last assumption is important, in that a phrase like "the risk of dying at age 40" is not well-defined: the risk varies markedly according to the present age of the person in question. Trivially, for someone now aged more than 40 years, the risk is zero. For someone younger than 40, "the risk of dying at age 40" includes the concept of survival to age 40, death before age 40 is a possibility. The concept of hazard is a step towards clarifying this situation, by describing the chances of death at a particular time or age conditionally on having survived up to that time or age.

The hazard is approximated by time- or age-specific death rates.

The modern methods of statistical analysis of cohort studies in occupational and environmental epidemiology draw very heavily on a methodology known as proportional hazards (or 'Cox regression') modelling. This approach, proposed by Cox (1972) and investigated and developed intensively since, is now the standard well-established strategy of analysis.

Proportional hazards regression modelling assumes that the population's basic mortality experience can be represented by one or several underlying hazard functions, giving average age- or time-specific risks of death for the population as a whole, or for specified sub-populations. It then focuses on estimating how these underlying mortality hazards are modified or "customised" according to an individual's experience in terms of smoking habits, occupational exposures, diet, exposure to air pollution and so on. Specifically, the regression modelling represents the effects of these characteristics, individually and jointly, as multiplicative factors affecting the underlying hazard function(s); hence the name *proportional hazards* modelling.

In practical implementations, multiplicative effects on the underlying hazards are estimated as additive effects on the scale of the log hazard, with a zero coefficient implying a multiplicative factor of unity, i.e. no effect on the hazard; a negative coefficient implying a multiplicative factor of less than one, i.e. a reduction in hazard, and so on. There are close connections with the Poisson regression methods described earlier.

The distinctive and outstanding strength of the approach is that the effects of explanatory variables, including air pollution, on underlying hazards can be estimated without making any prior assumptions about the shape of the underlying hazard. The modelling specification focuses only on the proportional way in which explanatory variables affect the hazard(s), which can remain unspecified. Indeed, there are choices in whether the basic hazard is represented in terms of age, or time since first entry into the study (start of follow-up), or some other time-related variable. This gives the methodology great generality.

Developments in recent years allow investigation of alternative ways in which characteristics may affect hazards; i.e. not just in proportional terms. Nevertheless, the proportional hazards approach provides a flexible framework for investigating how environmental factors may influence death rates.

The specific implementation of Pope et al.

Pope *et al.* used proportional hazards regression modelling with time since enrollment as the underlying time variable. The underlying hazards were allowed to vary by combinations of age (5-year groups), gender and race. This was achieved by stratifying the study dataset. Separate supplementary analyses were carried out by smoking habit and gender.

Comments

As noted earlier, the broad methodology is well-established and up-to-date.

With the underlying time dimension being time during the follow-up period, it was essential to take account of how the hazards would vary by other factors. Stratification of the underlying hazard (by age, sex and race) is an effective way of taking account of some of these factors. This is especially appropriate for factors such as age, gender and race whose effect on relative hazards may be complex, but in which the study does not have a direct interest; i.e. it is not the purpose of the study to estimate their effect; but their effect must be accounted for sufficiently so that other relationships can be estimated with minimal bias.

It is questionable if an age stratification in 5-year groups is sufficiently refined to take account of changes in hazard which, at least in older age-groups, may vary rapidly with age. However, lack of sufficiently refined adjustment is a problem only if three circumstances combine: the hazard within 5-year age groups varies importantly; within these 5-year age groups there are important differences in detailed age-distribution between metropolitan areas; and these differences are correlated with differences in the available indices of particulate air pollution. It is possible that some artefacts of this sort will have influenced the results. The paper gives no direct evidence; but it is unlikely that they would be an important contributor to the findings.

Within this framework of different baseline hazards for all combinations of (5-year) age, sex and race, the effect of particulate pollution (i.e. its relative or proportional influence on hazard) was estimated as if this relative effect was independent of age, sex and race. It appears that the same approach was adopted for other factors also; e.g. smoking habit, occupational exposures, etc.

It does not appear that this assumption, that the relative effects (per $\mu g/m^3$) of ambient particles on mortality hazards does not vary by age, was checked against the study data; the paper gives no evidence that it was. Rather, it seems that the reporting of a single relative hazard for all age groups was a simplification imposed *a priori* on the analysis and results. It may be a plausible simplification: hazards vary with age; and so the same multiplicative factor applied to different age-specific hazards implies different effects of pollution on numbers of deaths in each age-group. Nevertheless, it is an approximation; and one which may be wrong. For example, Doll *et al.* (1994), in studying the effect of smoking on mortality, found that the relative risks of smoking on mortality vary by age. Throughout ages 45–74, mortality rates in cigarette smokers were about twice those in non-smokers; with the relative effect reducing to 1.6 at 75–84 years, and 1.3 at ages 85 or more. The effect of cigarette smoking on mortality is much greater than that of air pollution (see below) and so it would be easier to identify age-related patterns in relative risks of smoking compared with air pollution, in studies of comparable size.

On general grounds, and in view of the results on smoking, it would be very helpful if this question (does the relative risk of air pollution vary by age?) could be investigated directly in the dataset studied by Pope *et al.* (1995). Looking from outside the study, it would seem that the useful device of stratifying by age would give a way in to examining this question; though it is unclear how powerful the study is to detect real differences in age-related hazards, if these exist. The US Health Effects Institute (HEI), with the active co-operation of the original authors, is sponsoring an independent re-analysis of this and other cohort studies of air pollution and mortality which presumably will be informative about this and other questions.

5.7. Statistical Methods: Modelling Strategy

The modelling strategy was very similar to that used by Schwartz (1994) and described in some detail above. Briefly, the authors constructed statistical regression models to explain as far as practicable the observed patterns of mortality in terms of non-pollution confounders; then examined whether the further inclusion of particulate air pollution led to an identifiable (i.e. statistically significant) improvement in the goodness-of-fit and associated explanatory power of these models; and finally examined the sensitivity of the estimated pollution effect to changes in the baseline non-pollution model, including by examining whether the estimated effect varied importantly by sub-group defined by gender and/or smoking status.

As noted earlier, this strategy gives a systematic and structured way of investigating the key pollution–mortality relationship in the presence of possible confounding factors; and as such might be viewed as standard good practice in statistical modelling.

5.8. Results

Smoking was the dominant factor in explaining mortality patterns, overall and for each of the cause-of-death groups.

A baseline model of non-pollution explanatory variables was constructed using smoking habit and a very wide range of other characteristics. Some of these characteristics were, on further analysis, shown as not clearly related to mortality. Their inclusion in the baseline model reflects once again the strategy of possibly overfitting the available non-pollution confounding factors.

Having adjusted for these factors, both all cause and cardio-respiratory deaths (but not "all other causes") were related both to sulphates and to $PM_{2.5}$. Results are given in terms of the multiplicative or proportional factors to be applied to the basic hazards. Thus, for fine particles, the estimated effect was a multiplicative factor ("adjusted mortality risk ratio") of 1.17 per 24.5 $\mu g/m^3$ $PM_{2.5}$, or equivalently a proportional hazard of 1.00643 per $\mu g/m^3$ $PM_{2.5}$. As noted earlier, this estimated effect may have been inflated by use of recent and lower concentrations rather than earlier, possibly higher, values which were unavailable. The coefficients both for sulphates and $PM_{2.5}$ were however very highly significant statistically, the reported 95% confidence intervals showing an estimated coefficient of more than four times its standard error.

Sulphates but not $PM_{2.5}$ were also related to lung cancer.

There were no marked differences in the estimated effect of particles in subgroups classified by gender or smoking habit; nor did the inclusion of weather variables influence the observed associations. Plots of adjusted mortality rates against sulphates and $PM_{2.5}$ respectively did not suggest a threshold, though the study is uninformative about conditions below about 4 $\mu g/m^3$ sulphates and 9 $\mu g/m^3$ $PM_{2.5}$.

It is never possible to know if adjustment for confounding has been adequate; and limitations of some of the confounding factors used have been noted earlier. It is however important, as noted by the authors, that the estimated effect of particulate pollution was *not* sensitive to the inclusion or not of many of the main confounders considered. This suggests, though of course does not prove, that a more elaborate adjustment for confounders would not have materially affected the results. Again as noted by the authors, the finding of a similar effect among women and among men, who presumably were differently exposed to dust occupationally, suggests that residual (unaccounted for) confounding by occupational exposure was small: else, it would have shown through as a gender effect.

The absence of an effect on mortality from 'all other causes' adds to the biological plausibility of the positive associations with cardiorespiratory causes.

The similarities of effect within sub-group defined by gender and smoking habit strengthens the case for transferability of results to other populations with a possibly different history of smoking habit, or different gender balance, compared with the study population considered by Pope *et al.*

6. Concluding Remarks

6.1. Similarities and Differences Between Studies

This brief overview of statistical methods in various types of air pollution study will, I hope, have helped the reader who is not technically expert in these aspects to find a way through the studies, and get a better understanding of their strengths and limitations. It is helpful that the broad kind of statistical approach is similar in studies of different design, being based on extensions of ordinary statistical regression analysis; though the nature of these extensions is different in different contexts. Also, it helps that the core modelling strategy (of developing a suitable baseline model, possibly over-fitting confounding factors; then examining the effect of pollutants, singly and jointly; and finally investigating if the apparent effect of pollutants is sensitive to the detailed modelling of other variables) is similar in well-conducted studies of different design. Understanding this framework might justifiably give the reader some confidence in reading and assessing a new study.

6.2. Further Reading

The introductions given here are nevertheless no more than that; an attempt to illustrate that this literature, which I find very interesting scientifically and which is clearly influencing environmental policy in many countries, can indeed be accessible to a broad audience, despite its dependence on relatively modern applied statistical methods. The reader who wishes to follow through the arguments at a greater level of rigour need not be at a loss for helpful resources. Draper and Smith (1998) is one of many classical textbooks which focus principally, but by no means exclusively, on "ordinary" regression analysis. Clayton and Hills (1993) is an excellent overview of statistical modelling, including regression modelling, in epidemiology. Dobson (1990) introduces, and McCullagh and Nelder (1989) is a core reference for, many of the modern extensions of regression methods. Schwartz *et al.* (1996) and Pope and Schwartz (1996) describe the basic ideas of analysing time series studies clearly and with more technical detail than are given here. Dockery and Brunekreef (1996) is a companion and non-mathematical review of panel studies of lung function, including issues of design and analysis. Diggle, Liang and Zeger (1994) is a comprehensive reference for statistical issues in both time series and panel studies, including in particular a description of GEE methods. Its need for rigour implies an exposition that is quite demanding mathematically.

Walter (1991a, b) is a good non-mathematical overview of ecologic studies. There are several good textbooks on regression methods for mortality studies, for example Collett (1994). Breslow and Day (1987) is a very informative book on cohort studies in occupational epidemiology. For a more critical view of the paper by Pope *et al.* (1995), see Gamble (1998). A major series of review articles (Greenland, 1993; Hatch & Thomas, 1993; Morgenstern & Thomas, 1993; Prentice & Thomas, 1993) covers in depth many leading-edge issues touched on in this chapter, especially regarding studies of long-term effects of exposure. Finally, Wilson and Spengler (1996) is a very good introduction to the issues as they arise with respect to particles; Lipfert (1994) is a comprehensive overview of issues at that time, including methodological aspects which remain relevant.

There is a wealth of wisdom about applied regression methods, relevant to air pollution epidemiology, in this material. It is accessible at least in part to those who are less confident with mathematical exposition; though some ease with mathematics is essential in order to gain full benefit. I will consider that this chapter has been a success if it encourages at least some readers, who previously might have steered clear of statistical literature, to browse or read more widely among some of these excellent books and articles; or at least to read papers in air pollution epidemiology with more insight and confidence. Happy reading!

Acknowledgements

I wish to thank my colleague Brian Miller, and Bob Maynard as editor, for helpful comments; Roy Harrison for information about variations in pollution in the USA; other colleagues in COMEAP for many useful insights; and especially Anthony Seaton for discussion of possible mechanisms. Section 4 is an edited version of material first presented in Chapter 7 of COMEAP (1995). Sections 3 and 5 draw heavily on review work funded variously by the Department of Health in London, and by National Power.

References

Bates D.V. (1992) *Environ. Res.* **59**, 336–349.

Breslow N.E. & Day N.E. (1987) *Statistical Methods in Cancer Epidemiology. Vol. II: — The Design and Analysis of Cohort Studies.* IARC Scientific Publications No. 82. Lyon: International Agency for Research on Cancer.

Clayton D. & Hills M. (1993) *Statistical Models in Epidemiology.* Oxford: Oxford University Press.

Collett D. (1994) *Modelling Survival Data in Medical Research.* London: Chapman and Hall.

COMEAP (Department of Health Committee on the Medical Effects of Air Pollutants) (1995) *Non-Biological Particles and Health.* London: HMSO.

Cox D.R. (1972) *Journal of the Royal Statistical Society,* B, **34**, 187–220.

Diggle P.J. *et al.* (1994) *Analysis of Longitudinal Data.* Oxford: Clarendon Press.

Dobson A.J. (1990) *An Introduction to Generalized Linear Models.* London: Chapman and Hall.

Dockery D.W. *et al.* (1993) *N. Engl. J. Med.* **329**(24), 1753–1759.

Dockery D.W. & Brunekreef B. (1996) *Am. J. Respir. Crit. Care Med.* **154**, S250–S256.

Dockery D.W. & Pope C.A. III. (1996) In *Particles in Our Air: Concentrations and Health Effects,* eds. Wilson R. & Spengler J.D. Harvard: Harvard University Press.

Doll R. *et al.* (1994) *Brit. Med. J.* **309**, 901–911.

Draper N.R. & Smith H. (1998) *Applied Regression Analysis* (3rd edn). New York: Wiley.

Gamble J. (1998) *Environ. Hlth. Perspect.* **106**, 535–549.

Greenland S. (1993) *Environ. Hlth. Perspect.* **101**(Suppl 4), 59–66.

Hatch M. & Thomas D. (1993) *Environ. Hlth. Perspect.* **101**(Suppl 4), 49–58.

Lipfert F.W. *et al.* (1988) *A Statistical Study of the Macroepidemiology of Air Pollution and Total Mortality.* Prepared for the Office of Environmental Analysis. US Department of Energy.

Lipfert F.W. (1994) *Air Pollution and Community Health: A Critical Review and Data Sourcebook.* New York: Van Nostrand Reinhold.

Lipfert F.W. & Wyzga R.E. (1995) *J. Air & Waste Management Assoc.* **45**, 949–996.

Lipfert F.W. (1998) *Applied Occupational and Environmental Hygiene,* **13**, 370–384.

McCullagh M. & Nelder J.A. (1989) *Generalised Linear Models* (2nd edn). London: Chapman and Hall.

Moolgavkar S.H. *et al.* (1989) *JNCI,* **81**, 415–420.

Morgenstern H. & Thomas D. (1993) *Environ. Hlth. Perspect.* **101**(Suppl 4), 23–38.

Pope C.A. III *et al.* (1995) *Am. J. Respir. Crit. Care Med.* **151**, 669–674.
Pope C.A. III & Schwartz J. (1996) *Am. J. Respir. Crit. Care Med.* **154**, S229–S233.
Prentice R.L. & Thomas D. (1993) *Environ. Hlth. Perspect.* **101** (Suppl 4), 39–48.
Rose G. (1982) *Br. Med. J.* **284**, 1600–1601.
Schwartz J. (1994) *Am. J. Respir. Crit. Care Med.* **150**, 648–655.
Schwartz J. *et al.* (1996) *J. Epidem. Commun. Hlth.* **50** (Suppl 1), S3–S11.
Schwarz J. (1997) In *Health at the Crossroads: Transport Policy and Urban Health*, eds. Fletcher T. & McMichael A.J., pp. 61–82, Chichester: John Wiley and Sons.
United States Environmental Protection Agency (US EPA) (1996) *Air Quality Criteria for Particulate Matter*, Vol. III. Report No. EPA/600/P-95/001cF. Washington, DC: US EPA.
Walter S.D. (1991a) *Environ. Hlth. Perspect.* **94**, 61–65.
Walter S.D. (1991b) *Environ. Hlth. Perspect.* **94**, 67–73.
Wilson R. & Spengler J.D. (eds) (1996) *Particles in Our Air: Concentrations and Health Effects*. Harvard: Harvard University Press.

CHAPTER 4

CANCER AND AIR POLLUTION

Lesley Rushton

1. Introduction

At the present time, there remains considerable uncertainty and controversy about the extent to which air pollution contributes to the occurrence of cancer. Views range from "risks from many different cancers from pollutants, such as exhaust fumes far outweighing the effects linked with passive smoking, radon or diet" to "there is no demonstrable relationship between air pollution and cancer" (Godlee & Walker, 1992).

This chapter gives an overview of the literature on this topic and assesses the current evidence of the effects of potential carcinogenic urban air pollutants. A general background to the history of the study of cancer and air pollution is given first, followed by a discussion of the potential problems and uncertainties which are encountered in studying this problem. Air pollutants thought to be carcinogenic are identified. More detailed reviews of studies of specific cancer sites are then given. The chapter ends with discussion of prevention opportunities and future research strategies.

2. Background

Air pollution resulting from combustion has been a problem in Britain since the thirteenth century. The first attempt to prevent the use of coal was in 1273, followed by a Royal Proclamation in 1306, a commission of inquiry in 1307, a declaration of annoyance by Queen Elizabeth I and a petition by suffering Londoners (Department of Environment, 1974). Lawley (1994) in his history of current concepts of carcinogenesis, pointed out that chemical sources of cancer in the environment became obvious as a result of the industrial revolution, beginning with the discovery in 1775 by Percival Pott (Pott, 1775) of excess cancers of the scrotum in chimney-sweeps due to prolonged exposure to soot. Lawley (1994) also attributes early studies of the action of soot as among the first to highlight (i) genetic–environment interaction (Pott's grandson Earle suggested that a "constitutional predisposition" was required (Earle, 1823)), and (ii) the existence of latent periods in chemical carcinogenesis (Curling (1856) noted a case 20 years after early exposure to soot). Experimental confirmation of the production of skin cancer after coal tar exposure followed in 1915 (Yamagiwa & Ichikawa, 1918) culminating with the identification of benzo(α)pyrene in coal tar in 1933 (Cook *et al.*, 1933). Swanson (1988) describes the history of studies of exposure to arsenic from harmful effects, including cancer, in occupational groups beginning with Agricola in the 16th century (Bauer, 1556), through to studies of arsenic in ambient air from a copper smelter (Greenberg, 1984).

Epidemiological studies of temporal and geographical variations in cancer incidence and mortality have consistently identified increased rates in urban populations compared with rural populations (Pershagen & Simonato, 1990; Greenburg *et al.*, 1967; Tornqvist & Ehrenberg, 1992; Moller, 1993). The earliest experimental evidence that air pollution was carcinogenic was reported in 1942 (Leiter *et al.*, 1942). Of nearly 3,000 chemicals identified as air pollutants, only approximately 10% have been studied in experimental bioassays, and nearly 60% of those compounds bioassayed are hydrocarbons, nitrogen-containing organics or halogenated organics (Lewtas, 1993). The greatest human exposure and risk in urban air pollution appears to be associated with mixtures of polycyclic aromatic compounds (PACs) such as polycyclic aromatic hydrocarbons (PAHs), derived from incomplete combustion. The evaluation of the carcinogenicity of these compounds to humans has been based mainly on studies of occupational exposure (Lewtas & Gallagher, 1990). However, vehicular emissions are a major source of urban PAC pollution and quantitative estimates of the possible contribution to human cancer risk in the United States (US Environmental Protection Agency, 1990) have suggested that approximately half of the estimated cancer risk from outdoor air pollution may result from exposure to motor vehicle emissions.

3. Challenges

Researchers are faced with the challenges of determining if and to what extent (i) air pollution has an effect on the occurrence of cancer, (ii) whether certain members of the population are more sensitive than others, and (iii) whether current air quality guidelines protect the population from these effects. The International Agency for Research into Cancer (IARC) classifies substances into four groups according to the evidence as shown in Table 1 (International Agency for Research on Cancer, 1989). For human data, sufficient evidence is defined as the establishment of a causal relationship between exposure to the agent and human cancer. Limited evidence is defined as the observation of a positive association between exposure to the agent and human cancer, for which a causal interpretation is considered credible, but chance, bias or confounding, could not be ruled out with reasonable confidence. Similar definitions relate to the evidence from experimental data [see the IARC Monographs on the evaluation of carcinogenic risks to humans for a full description (International Agency for Research on Cancer, 1989)].

Different kinds of research contribute different relevant information. Toxicological studies conducted on experimental animals can rigorously assess the biological response to a range of levels of exposure to a given

Table 1. Carcinogenicity defined by the International Agency for Research into Cancer.

Group	Definition	Used when
1	Carcinogenic to humans	Evidence is sufficient
2A	Probably carcinogenic to humans	Limited evidence in humans, and sufficient evidence in experimental animals
2B	Possibly carcinogenic to humans	Limited evidence in humans, and absence of sufficient evidence in experimental animals *or* inadequate evidence in humans or human data non-existent and sufficient evidence in experimental animals
3	Not classifiable as to carcinogenicity to humans	Not classifiable to any other group
4	Probably not carcinogenic to humans	Evidence suggests a lack of carcinogenicity in humans and in experimental animals

substance or combination of substances. This has the advantage of specificity but presents a problem when extrapolating to humans. The levels and mode of exposure are often very different in humans (Read, 1991; Davis & Muir, 1995). Swanson (1988) identifies other restrictions in quantifying human risk from experimental data, including inadequate knowledge regarding the mechanisms and progression of carcinogenesis, the inability to specify the relationship between cancer, ageing and lifespan, and the need, often, to extrapolate from benign animal tumours to human malignancy.

Clinical studies employing human volunteers have limited use as these can only test response to a narrow range of exposure levels and exposure to carinogenic compounds is likely to be regarded as unethical. Epidemiological methods have, thus, been widely utilised to test causal hypotheses in human cancer, both in specific groups defined, for example, by occupation, or in the more general population. Epidemiological studies may be more relevant to public health than toxicological and clinical research because they reflect the real world rather than artificial experimental environments. However, as Higginson (1988) points out, such studies have limitations which include (i) a lack of sensitivity to detect very low risks, (ii) difficulty in discriminating between several plausible risk factors in complex situations, (iii) the inability to evaluate the impact of recent exposures, and (iv) uncertainty in interpreting "negative" studies or inverse relationships. Tornqvist and Ehrenberg (1994) suggest that, because of the above problems, epidemiological studies are best applied in situations with a considerably raised exposure, such as a work environment, and that they are of limited use in quantitative risk assessment. Well-established relationships between such lifestyle variables as smoking habits and socio-economic status (low socio-economic status areas also tend to have higher air pollution levels) and some cancers emphasise the importance of evaluating the influence of these factors simultaneously with exposure to hazardous substances. Problems lie, not only in assessing the role of possible confounders, but in establishing the nature of the dose-response curve, particularly at the lower exposure concentrations generally experienced in ambient air. Kaplan and Morgan (1981) also point out that in studying the effects of air pollution, migration patterns may be important. The implicit assumption in studies of geographically based populations is often that the people have lived there all their lives, whereas in reality, considerable migration into and out of the area may have taken place. Humans are also exposed to a complex mixture of potential carcinogens in ambient air. Data from epidemiological studies suggest that exposure to more than one known carcinogen, for example, cigarette smoking and inhalation of radon gas, may lead to synergistic interactions that amplify risk (Witschi, 1994).

Whatever the type of study design, the results are ultimately dependent on the quality of the data. As Higginson (1993) points out, global cancer statistics of good quality have now become available (see Muir *et al.*, 1987). The identification of populations exposed to outdoor contaminants may be difficult due to inadequate identification of specific carcinogens and the distribution of contaminants in the ambient air, and knowledge of the duration and concentration of specific carcinogens over many years of exposure. It is also important to be able to identify sensitive population subgroups i.e., those who are particularly susceptible to the effects of a pollutant, and to be able to evaluate the variation of both individual susceptibility and individual dose.

Routes of exposure for humans may be complex, involving not only direct exposure through inhalation and dermal contact, but from indirect sources, for example, ingestion via soil deposition and water contamination. The isolation of the role of air pollution may be difficult or even unrealistic if other major sources of pollution through the work and home environment also exist. It may be that the estimation of total dose is more important than evaluation of the contribution of a single component.

Measurements of exposure, duration and concentration may be difficult to obtain and there are many potential areas of uncertainty in assessing exposure. Individual exposure can be measured using a suitable personal monitoring device. However, the most common method of estimating outdoor exposure in large populations is the use of fixed-point monitoring. This may be useful in longitudinal epidemiological studies where the absolute values of the concentrations are less important than their variation with time over the period of the study. Often, however, the number of monitoring sites is, inevitably, limited. Colls (1997) describes several networks of airborne concentration measurement stations in the UK. He points out that the earlier networks were not representative of the country as a whole, with pollutant information been confined to areas of high population density, and with sites being taken out of service if concentrations were below set values for three consecutive years, causing a decline in the number of monitoring sites. The issue also arises as to which pollutant statistics derived from these data are most appropriate. Possibilities include the long-term average, an estimate of short-term peaks, and measures of "excess" concentration, i.e. above a certain threshold. To address these inadequacies, there has been increasing use of complex modelling methods, particularly for assessing ranges and variability of exposures.

4. Airborne Carcinogens

Ambient urban air is a complex mixture containing a variety of substances, some of which are known to be carcinogenic. Table 2 lists some of those thought to be of most importance.

A person inhales about 20 000 litres of air per day, so even modest contamination of the atmosphere can result in inhalation of appreciable doses of a pollutant. Urban air contains thousands of chemicals and, as has been previously noted, only 10% have published data on either genotoxicity or carcinogenicity (Lewtas *et al.*, 1992). In 1990, the US Environment Protection Agency (EPA) published estimations of the relative contribution by various air pollutants to total estimated cancer cases in the US per year, using current outdoor air exposure data bases in combination with cancer unit risk values. The percent contribution

Table 2. Airborne carcinogens of particular importance.

Substance	IARC evaluation of carcinogenicity
Arsenic	1
Asbestos	1
Beryllium	2A
Cadmium	2A
Chromium — hexavalent compounds	1
Coal gasification	1
Coke production	1
Dioxane 1,4	2B
Formaldehyde	2B
Lead — inorganic	2B
Polycyclic aromatic hydrocarbons	
Benzene	1
Benzo(α)pyrene	2A
1,3-butadiene	2B
Soots	1
Whole diesel engine exhaust	2A
Whole petrol engine exhaust	2B
Vinyl chloride	3

ranged from 35% for products of incomplete combustion (PIC), 12% for 1,3-butadiene, 8% for benzene, 4% for asbestos down to 1% for vinyl chloride.

Lewtas (1990) lists 47 chemicals identified in air which, using the IARC definition, have been defined as Group 1, 2A or 2B. Benzene (Group 1) and formaldehyde (Group 2A) have well-characterised human exposure. Of others in Groups 2A and 2B, 1,3-butadiene, chloroform, ethylene dibromide and carbon tetrachloride have been identified as potentially important contributors to human cancer risk. Other components of outdoor air which are of concern include nitrogen dioxide, ozone, sulphur dioxide and sulphuric acid in aerosol form. Several contaminants of the outdoor air have been regulated in the USA (Swanson, 1988; Robinson & Paxman, 1992) and in Europe (World Health Organisation Regional Office for Europe, 1987). For example, in the USA, sulphur dioxide, ozone, nitrogen dioxide, carbon monoxide and lead are regulated through the Ambient Air Quality Standards and asbestos, beryllium, nitrogen dioxide and vinyl chloride are regulated by Hazardous Pollutant Standards (Swanson, 1988). Robinson and Paxman (1992) present cancer risks for 20 IARC-designated carcinogens whose acceptable ambient air level guidelines are based on threshold limit values (TLVs), but points out that caution is required in using TLVs to represent no observed effect levels (NOEL) for regulatory purposes.

In reality, humans are generally not exposed to individual chemicals in urban air but to a complex mixture of substances, the most important being PACs. Lewtas (1990) lists 13 mixtures for which the weight of evidence of human cancer risk has resulted in an IARC evaluation of Group 1, including soot, coke production, iron and steel founding, and aluminium production. Diesel exhaust (Group 2A) and petrol exhaust (Group 2B) are also important. The most widely distributed outdoor air pollutant is soot which may contain PACs, such as PAHs. Benzo(α)pyrene is often used as an indicator of PAHs. Although the first soot to be recognised as a human carcinogen was the soot from coal combustion (International Agency for Research on Cancer, 1985), there is increasing evidence that soot from the combustion of other fuels, such as petroleum and wood, are similar in their composition, genetic activity and induction of tumours in animals (Hoffman & Wynder, 1976; Lewtas, 1985). The IARC evaluation of diesel and petrol engine exhausts relied heavily on the animal evidence of carcinogenicity (International Agency for Research on Cancer, 1989).

5. Specific Cancers

5.1. Lung Cancer

A peculiar pulmonary disease — *mala metallarium* — was first associated with the occupation of mining as far back as the sixteenth century (Higginson & Jensen, 1977), and identified as lung cancer in miners producing nickel, magnesium, bismuth and radium by Harting and Hesse in 1879. Reasons for believing that air pollution might be an important factor in the development of lung cancer were the presence of known carcinogens, such as benzo(α)pyrene in ambient air (Waller, 1952) and distinct urban/rural gradients in mortality from lung cancer. There have been many reviews of research into the relationship between air pollution and lung cancer (Shy, 1976; Speizer, 1983; Higgins, 1984; Matanoski *et al.*, 1986; Pershagen, 1990; Pershagen & Simonato, 1990; Tomatis, 1991; Hemminki & Pershagen, 1994; Pan *et al.*, 1994; Cohen & Pope, 1995). Epidemiological studies have included descriptive studies of geographical populations, migrant studies, case-control and cohort studies, and studies of specific occupational groups.

A considerable number of studies have compared lung cancer rates in urban and rural residents of the same country, or lung cancer rates for urban areas stratified according to population size (Prindle, 1959; Levin *et al.*, 1960; Stocks, 1960; Stocks, 1966; Winkelstein *et al.*, 1967; Zeidberg *et al.*, 1967; Ashley, 1969; Lave & Seskin, 1970; Carnow & Meier, 1973; Menck *et al.*, 1974; Doll, 1978; Weiss, 1978; Kanarek *et al.*, 1979; Ford & Bialik, 1980; Lyon *et al.*, 1980; Borch-Johnsen, 1982; Walker *et al.*, 1982; Weinberg *et al.*, 1982; Jedrychowski, 1983; Rantanen, 1983; Epstein *et al.*, 1984; Xiao & Xu, 1985; Jedrychowski, 1995). In most countries for which data were available in the 1950s and 1960s, the mortality incidence rates for lung cancer were 2 to 3 times higher in urban areas than rural areas (Stocks, 1947; Curwen *et al.*, 1954; Hoffman & Gilliam, 1954; Haenszel *et al.*, 1956; Stocks, 1958; IARC, 1976). Whilst most urban–rural studies did not consider cigarette smoking, the study by Prindle (1959) found 49% of adults in urban areas smoked cigarettes compared to 43% in rural areas, a difference which did not explain the doubling of lung cancer mortality he found in large cities as compared with rural areas.

Some case-control or cohort studies (see below) have collected data on cigarette smoking for the populations studied and investigated the relationship with urban/rural residence (Stocks & Campbell, 1955; Hammond & Horn, 1958; Haenszel *et al.*, 1962; Dean, 1966; Buell *et al.*, 1967). In addition to the expected differences between smokers and non-smokers, a consistent finding in these studies is that the overall urban-rural gradient and the gradient with increasing urbanisation are larger for non-smokers than for smokers. A detailed review of many of these studies is given by Shy (1976). In some geographical studies, the rates were correlated with ambient air concentration of specific pollutants e.g. benzo(α)pyrene, carcinogenic metals and particulate matter (Stocks, 1959; Winkelstein *et al.*, 1967; Zeidberg *et al.*, 1967; Boucot *et al.*, 1972; Hammond & Garfinkel, 1980; Matanoski *et al.*, 1981; Brown *et al.*, 1984). In spite of an apparent reduction in urban/rural gradients in lung cancer risk since the 1970s (Lawther & Waller, 1978), Table 3 shows that it still exists. Tomatis (1991) suggests that this reduction can be partly explained by an actual improvement in urban air quality, but is probably due to the progressive extension of areas with urban characteristics, the growing proportion of the population living in urban and suburban areas, and the consequent contraction of areas with truly rural characteristics. He also suggests that the greater prevalence and earlier beginning of smoking and higher levels of occupational exposure to carcinogens contribute to the "urban factor". Speizer (1983) supports this latter view and reviews how lack of details of smoking habits may interfere with an assessment of either an occupational

Table 3. Age-standardised rates per 100 000 population for mortality from carcinoma of the bronchus and trachea in urban and rural areas*.

Registry	Males			Females		
	Urban	Rural	Ratio	Urban	Rural	Ratio
Japan, Miyagi	30.9	28.4	1.1	9.2	8.1	1.1
Czechoslovakia, Slovakia	68.2	70.5	1.0	9.4	6.5	1.4
FRG, Saarland	77.7	63.0	1.2	7.7	6.0	1.3
France, Calvados	46.1	39.6	1.2	3.4	2.9	1.2
France, Doubs	56.9	40.1	1.4	3.3	2.0	1.7
Hungary, Szabolcs	61.8	50.9	1.2	10.3	6.2	1.7
Norway	39.4	24.5	1.6	9.6	5.2	1.9
Romania, Cluj County	35.2	35.3	1.0	6.7	4.7	1.4
Switzerland, Vaud	63.8	56.6	1.1	8.7	5.6	1.6
UK, England and Wales	74.8	56.2	1.3	19.7	15.1	1.3
Australia, New South Wales	55.5	46.8	1.2	12.2	8.3	1.5

*From *IARC Cancer Incidence in Five Continents* Vol. V, Muir C. *et al.* (eds), Publ. No. 88, International Agency for Research on Cancer.

exposure or a general environmental exposure effect on the observed rates of lung cancer. In particular, he shows how relatively minor differences in smoking rates for different populations may explain a difference of up to 20% in lung cancer rates, and draws attention to the fact that mis-classification of smokers by not knowing the age they started smoking (and thus their duration of smoking) may swamp any attempt to find an association of a general environmental exposure and lung cancer.

The relationship of lung cancer to urban pollution has been further studied in a series of studies on changes in lung cancer experience of immigrants to various countries. Several investigations on emigrants from the United Kingdom to New Zealand (Eastcott, 1956), South Africa (Dean, 1964; Dean, 1959) and the USA (Reid, 1966) showed that, in general, the lung cancer rates of emigrants were lower than that of residents in Britain but higher than those born in the new country.

Table 4. Summary of cohort studies on lung cancer in urban areas.

Population	Area	Summary of findings on lung cancer	Standardisation variables (in addition to age)	References
187 783 white males followed 1952 to 1955	United States (9 states)	Relative death rate = 1.33 in cities with over 50 000 population in comparison with rural areas.	Smoking	Hammond & Horn, 1958
69 868 men followed 1958 1962	California, United States	The two major metropolitan areas, Los Angeles and San Francisco Bay area plus San Diego had higher lung cancer mortality rates than remaining mixed rural and urban counties. These differences were relatively greater among non-smokers.	Smoking	Buell *et al.*, 1967 to
About 500 000 men followed 1959 to 1965	United States (25 states)	Relative death rates = 1.23, 1.14 and 0.98 in metropolitan areas with more than 1 million inhabitants, less than 1 million and nonmetropolitan areas, respectively, among men with occupational exposure to dust, fumes, gases or X-rays. Corresponding rates for men without such occupational exposures were 0.98, 0.97 and 0.92.	Smoking	Hammond, 1972
25 444 men and 26 467 women followed 1963 to 1972	Sweden	Relative death rate about 1.6 and 1.2 in male smokers of cities and towns, respectively, in comparison with smokers in rural areas. Similar trend in women (based on small numbers).	Smoking	Cederlof *et al.*, 1975
34 440 doctors followed 1951 1971	United Kingdom	No increase in relative death rate in "conurbations, large towns, or small towns" in comparison with rural areas.	Smoking	Doll Peto, 1981 to
About 7.5 million men and women followed 1961 to 1973	Sweden	About 40% and 20% of male and female lung cancer incidences, respectively, statistically explainable in terms of urbanisation variables after subtraction of the effects of diagnostic intensity and smoking.	Diagnostic intensity, smoking (only available for 1% of cohort)	Ehrenberg, 1985
4475 men followed 1964 to 1980 areas)	Finland (3 urban and 3 rural	Increased incidence in urbanised (born rural then moved to urban) in relation to married smokers in rural areas.	Marital status, smoking	Tenkanen & Teppo, 1987

Table 4 (*Continued*)

Population	Area	Summary of findings on lung cancer	Standardisation variables (in addition to age)	References
6340 Seventh-Day Adventist non-smokers followed 1977 to 1982	California, United States	Risk of malignant neoplasms in females increased concurrently with annual exceedance frequencies for all cut offs chosen for total suspended particles. Two fold increase for respiratory cancer.	Years lived with smoker, years worked with a smoker, employment in an occupation with high exposures to airborne contaminant (males)	Mills *et al.*, 1991
8111 adults followed	6 US cities	Associations between mortality and air pollution were strongest for respirable particles and sulphates. The adjusted mortality rate ratio for the most polluted city compared with the least polluted was 1.37.	Sex, cigarette smoking, body mass index, education	Dockery *et al.*, 1993
552 138 adults from 1982 to 1989	151 US cities	Lung cancer mortality was associated with combustion-source air pollution when sulphates was used as the index (adjusted mortality risk ratio = 1.36) but not when fine particles was used as the index (adjusted mortality risk ratio = 1.03).	Sex, race, cigarette smoking exposure to passive smoking, occupational exposure, education, BMI, alcohol use	Pope, 1995

It has been suggested that this reflects a lasting effect of early environment on lung cancer mortality later in life. Comparison of smoking habits did not account for the differences and, generally, data on potential confounding variables were lacking.

A series of papers by Pershagen and colleagues (Pershagen, 1990; Pershagen & Simonato, 1990; Hemminki & Pershagen, 1994) summarise the findings from the many cohort and case-control studies which have been carried out to investigate the relationship between lung cancer and air pollution. Tables 4 and 5 are adapted from Hemminki and Pershagen, amended and updated with several more recent studies, and summarise the findings of those papers. There is considerable

Table 5. Summary of case-control studies on lung cancer in urban areas.

Study population	Area	Summary of findings on lung cancer	Standardisation variables (in addition to age)	References
725 male lung cancer cases and about 12 000 pital controls without cancer identified 1952 to 1954	North Wales and Liverpool, England, United Kingdom	Relative risks ranging from 1.1 to 3.4 in different groups of smokers comparing urban and rural areas. Additivity of effects from urban residence and smoking suggested.	Smoking	Stocks & Campbell, 1955 hos-
2381 white male lung cancer deaths and 31 516 population controls identified 1958	United States	Overall SMR[a] of 1.43 comparing urban and rural areas, with positive trend in relation to duration of residence. Joint effect of urban residence and smoking was far greater than those expected on the assumption of additivity of the separate effects.	Smoking	Haenszel *et al.*, 1962
749 white female lung cancer deaths and 34 339 population controls identified 1958 to 1959	United States	Overall SMR[a] of 1.27 comparing urban and rural areas with positive trend in relation to duration of residence. Additivity of effects from urban residence and smoking suggested.	Smoking	Haenszel & Tauber, 1964
2873 male and 167 female lung cancer deaths; an equal number of deceased controls with non-respiratory illness identified 1960 to 1962	Northern Ireland, United Kingdom	Mortality rate ratios of about 1.5 to 5 in men and women of different smoking groups comparing urban and rural areas. Joint effects of urban residence and smoking appear to exceed additivity.	Smoking	Dean, 1966
180 male and 79 female lung cancer deaths; 2241 male and 2475 female population controls identified 1960 to 1966	Two cities near Osaka, Japan	Relative risks range from 1.2 to 1.8 in men and women comparing areas having high or intermediate levels or air pollution with those having low levels. Effects primarily seen in smokers.	Smoking	Hitosugi, 1968

Table 5 (*Continued*)

Study population	Area	Summary of findings on lung cancer	Standardisation variables (in addition to age)	References
780 male and 199 female lung cer deaths; 2563 male and 2958 female population controls identified 1969 to 1972.	Northeast England, United Kingdom	Relative risks of 1.6 and 1.7 for women and men, respectively, in areas having high air pollution levels compared with areas having intermediate levels. Relative risk increases lower when compared with areas having low pollution levels.	Smoking	Dean *et al.*, 1977 can-
785 male and 138 female lung cer cases identified 1963 to 1972; 1371 male and 1571 female population controls	Northeast England, United Kingdom	Relative risks of 1.4 and 2.3 for women and men, respectively, in areas having high air pollution levels in comparison with areas having intermediate levels.	Smoking	Dean *et al.*, 1978 can-
1425 white male and 576 female cases diagnosed 1972 to 1975; 445 male and 186 female population controls	Los Angeles County, California, United States	No association with long-term residence in high air pollution areas.	Smoking	Pike *et al.*, 1979
417 white male lung cancer cases and 752 hospital controls with non-neoplastic disease identified 1957 to 1965	Erie County, New York, United States	Relative risks of about 1.5, associated with 50 years or more of residence in areas with high or medium levels of air pollution. Exposure to air pollution alone did not significantly increase the risk of lung cancer and was only significant in combination with smoking and/or occupation.	Smoking, occupation	Vena, 1982
283 male and 139 female lung cancer cases; 475 population controls identified 1980 to 1982	New Mexico, United States	No consistent association between residence in urban areas and cancer risk.	Smoking, occupation, ethnic group	Samet *et al.*, 1987

Table 5 (*Continued*)

Study population	Area	Summary of findings on lung cancer	Standardisation variables (in addition to age)	References
729 male and 520 female cancer cases diagnosed 1985 to 1987; 788 male and 577 femela population controls	Shenyang, China	Relative risks of 1.5 and 1.4 for males and females, respectively, residing in somewhat/slightly smoky areas compared with subjects in not smoky areas. Corresponding relative risks for subjects in smoky areas 2.3 and 2.5.	Smoking, education, indoor air pollution	Xu *et al.*, 1989
901 male and 198 female lung cancer deaths; 875 males and 198 female deceased controls with non-respiratory diseases identified 1980 to 1985	Cracow, Poland	Relative risks of 1.48 for men in areas having high air pollution levels and 1.17 in women in areas having medium or high levels compared with subjects in areas with low levels of air pollution. Multiplicative interaction suggested between smoking, occupational exposure, and air pollution.	Smoking, occupation	Jedrychowski *et al.*, 1990
101 female lung cancer cases diagnosed 1987 to 1989 and 89 hospital controls with orthopaedic disease	Athens, Greece	Relative risks of 0.81 and 1.35 and 2.23 comparing highest and lowest quartiles of air pollution exposure in non-smokers and smokers of 15 and 30 years duration, respectively.	Smoking, education, interviewer	Katsouyanni *et al.*, 1991
194 lung cancer cases, 194 hospital controls, 194 population controls	5 German cities	Subset of 80 cases, 160 controls within the cities (excluding rural areas around the cities) OR[b] of 1.01 for an emission index for sulphur dioxide and OR[b] of 1.16 for semi-quantitative sulphur dioxide and OR[b] of 1.16 for semi-quantitative index using information on benzo(α)-pyrene, particles, sulphur dioxide, energy consumption, use of coal, degree of industrialisation.	Smoking, occupational exposures	Jockel *et al.*, 1992
755 lung cancer autopsy cases, 755 controls	Trieste, Italy	Risk of lung cancer increased for all types of lung cancer with increasing level of air pollution. OR[b] = 0.6 for rural residents, OR[b] = 1.5 for centre of city, OR[b] = 1.4 for residents of the social group industrial area compared with the residential area.	Smoking, occupational carcinogens,	Barbone *et al.*, 1995

[a]SMR, standardised mortality ratio
[b]OR, odds ratio

consistency in the results with many of the odds ratios or relative risks being increased to about 1.5 for urban, more densely populated or industrialised areas compared to rural, less populated or industrialised areas, and for higher exposures to particular components of exposure. The more recent studies tend to be more sophisticated in their study design, the number of confounding variables used for adjustment and the statistical methods utilised, and focus on exposure measurements for specific compounds of air pollution. The study by Pope in 1955 suggested, for example, that lung cancer is more closely related to sulphates as an index of pollution than fine particles. Exposure to sulphuric acid, in aerosol form, has been associated with the development of lung cancer in some studies of occupational groups (Beaumont *et al.*, 1987; Forastiere *et al.*, 1987). These and others are reviewed in detail by Saskolne *et al.* (1989). Animal studies also suggest that chronic exposure to sulphuric acid can cause irritations to the upper respiratory tract (Soskolne *et al.*, 1989).

These types of studies and the geographical studies often have problems of imprecise air pollution data, lack of reliable information on potential confounding factors (a particular problem in case-control studies is when the lung cancer case is often dead, so data collection is reliant on a proxy informant) and non-assessment of indoor or work exposures.

A different approach has been taken in a number of studies by examining populations living near specific industries, although many of the above caveats are also relevant. High lung cancer mortality has been found in areas around such industries as chemical, paper and pulp, and petroleum (Blot & Fraumeni, 1976); ferro-chromium alloy (Axelsson & Rylander, 1980); arsenical pesticide production (Matanoski *et al.*, 1986); iron and steel foundries (Lloyd, 1978; IARC, 1984; Lloyd *et al.*, 1985; Smith *et al.*, 1987); non-ferrous smelters (Blot & Fraumeni Jr., 1975; Newman *et al.*, 1975; Pershagen *et al.*, 1977; Cordier *et al.*, 1983; Xiao & Xu, 1985).

The hazards of exposure to asbestos are well documented (McDonald, 1985) and it is known that lung cancer is associated with having prolonged occupational exposure to asbestos (Churg, 1993). Recent findings

(Browne, 1986a; 1986b; 1986c) suggested that the risk of lung cancer only occurs at levels of asbestos exposure high enough to develop asbestosis. Commins (1989) estimates that the level of environmental lifetime risk from exposure to airborne asbestos appears to be about 1 in 100 000 or even lower.

Studies in a variety of occupational settings have also demonstrated excess lung cancers associated with some of the above industries, particularly with exposure to combustion products, e.g., studies of coal gas workers (Doll *et al.*, 1965; 1972) coke oven workers (Lloyd, 1971; Redmond *et al.*, 1976) and roofers (Hammond *et al.*, 1976). Polycyclic hydrocarbon exposure from traffic exhaust, in particular diesel, is also of increasing concern and a number of studies have been carried out to investigate the effect of diesel and gasoline emissions on lung cancer (Kaplan, 1959; Waller, 1981; Howe *et al.*, 1983; Garshick *et al.*, 1988). A detailed review of both animal and human studies is given in the 1989 IARC Monograph (International Agency for Research on Cancer, 1989). The risk of lung cancer increased in many of the studies with increasing exposure to diesel engine exhaust, although not all increases were statistically significant. There were fewer studies which reliably measured gasoline engine exhaust.

It is difficult to draw firm conclusions from the epidemiological evidence on ambient air pollution and lung cancer. Data relating to the measurements of air pollutants in many of the studies varied in detail, precision and quality, making comparison across studies difficult. Most exposure information referred to recent measurements, whilst the relevant exposures may have occurred many years before, and few studies attempted to assess total exposure and to include that from occupation or the home environment. The relationship between indoor and outdoor air quality is of particular concern. The simultaneous evaluation of the effect of confounding factors, such as smoking and dietary habits, was not consistently addressed.

In 1990, Pershagen and Simanto concluded from their review of the epidemiological literature that smoking-standardised excess relative risks of lung cancer in urban areas are generally of the order of 50%

Table 6. Levels of attributable risk of lung cancer to air pollution.

Year of publication	Comment	Author(s)
1955	Urban air (Liverpool) adds approximately 100 deaths to the death rate per 100 000 for lung cancer, regardless of smoking habits	Stocks & Campbell
1962, 1964	5% of all lung cancer	Haenzel *et al.*
1973	5% increase of pulmonary cancer for each increase of 1 μg/1000 m^3 of benzo(α)pyrene	Carnow & Meier
1976	Possibly a tenth of the effect of cigarette smoking	Higgins
1976	0.4 deaths/100 000 per μg/1000 m^3 of benzo(α)pyrene in smokers; 1.4 deaths/100 000 per μg/1000 m^3 benzo(α)pyrene in smokers. In US 1 cigarette/day is equivalent to 10 μg/1000 m^3 of benzo(α)pyrene	Pike *et al.*
1978	5–10 cases/100 000 persons acting together with cigarette smoking	Task Group
1981	1–2% of lung cancer. Less than 1% of all cancers in the future	Doll & Peto
1981	12% of 1980 lung cancer. 10 to 19% of future lung cancer based on 1980 levels of total suspended particles	Karch & Schreiderman
1987	Lung cancer risk of 9×10^{-2} from lifetime (70 years) exposure to concentration of benzo(α)pyrene of 1 μg/m^3 in air	WHO
1990	$<$ 1% at current levels of pollution	US EPA

or less. This would imply that up to one-third of lung cancer cases in urban areas are related to living in such areas. However, the proportion of the 50% excess that could be attributed to air pollution would be much less (Tomatis, 1991). There have been many estimates over the years of the attributable risk of lung cancer to air pollution. Table 6 [adapted from Speizer (1983) and Cohen & Pope (1995)] summarises some of the estimates.

The differences between these estimates are partly due to the different methods used to calculate them. For example, Doll and Peto (1981) used estimates of benzo(α)pyrene in urban air and extrapolated from an occupational study of PAH-exposed workers. The EPA estimate was obtained from applying unit risks, derived either from animal experiments or extrapolation from occupational studies, for over 20 known or suspected human carcinogenics found in outdoor air to estimates of the ambient concentrations and numbers of individuals potentially exposed. The steady reduction in the risk estimates assumes that a decline in air pollution will persist. Future research strategies to evaluate this and to investigate the carcinogenic effect of air pollution, in particular for lung cancer, are addressed in the final section of this chapter.

5.2. Lymphatic and Haematopoietic Cancers

The component of urban air which has been most strongly associated with the development of these cancers is benzene, the main source of which is from gasoline. There have been many studies reporting the carcinogenicity and haematotoxicity of benzene at high exposures, both in animal and epidemiological studies (Linet *et al.*, 1987; Aksoy, 1988). Benzene exposures in excess of 25 parts per million (ppm) have been associated with leukopaenia, thrombocytopaenia, and aplastic anaemia (Vigliani & Saita, 1964; Aksoy *et al.*, 1972; Vigliani & Forni, 1976). The association of leukaemia with high exposures to benzene has also been reported, with acute myeloid leukaemia being the most commonly described (Vigliani & Saita, 1964; Aksoy *et al.*, 1976; Vigliani,

1976; Brandt *et al.*, 1978; Doll & Peto, 1981). Evidence from human studies linking benzene to leukaemias other than acute myeloid leukaemia is contradictory and inconsistent, with, for example, chronic lymphocytic leukaemia comprising only a small percentage (of the order of 2%) of the leukaemias found (McMichael *et al.*, 1975; Aksoy *et al.*, 1976; Checkoway *et al.* 1984; Infante *et al.*, 1990). Lymphatic and haematopoietic cancers other than leukaemias have also been associated with benzene exposure, including lymphoma (Young, 1989) and multiple myeloma (Goldstein, 1990; Bisby, 1993).

Almost all of the above effects have been observed in studies of occupational groups. Several ecological studies have examined adverse health effects in populations around point sources of benzene, such as the petrochemical industry (Blot *et al.*, 1977; Hearey *et al.*, 1980; Kaldor *et al.*, 1984; Knox, 1994), although only the one by Knox implicated the use of fossil fuels.

More recently, there have been attempts to measure exposure and risk directly for acute myeloid leukaemia in order to derive dose-response relations (Bond *et al.*, 1986; Rinsky *et al.*, 1987; Wong, 1987; Paxton *et al.*, 1994; Schnatter *et al.*, 1996; Rushton & Romaniuk, 1997). Considerable uncertainty still remains regarding risks in the low dose range (below 10–20 ppm) because of a critical lack of data and controversy about the exact shape of the dose-response relationship. A risk characterisation carried out for the European Union (Exxon Biomedical Sciences, 1996) concluded that excess acute myeloid leukaemia was unlikely to occur at concentrations of less than 1 ppm.

Benzene has been monitored in ambient outdoor air both by passive and automatic monitoring. Typical mean outdoor benzene concentrations at roadside sites in urban areas range from 0.4–10.9 parts per billion (Broughton *et al.*, 1998). This suggests that the risk of leukaemia associated with environmental levels of benzene must be extremely low.

5.3. Other Cancers

The evaluation of the role of air pollution in the aetiology of other cancers is even more difficult to resolve than for cancer of the lung.

Urban/rural differences in cancer rates have been found in several studies for sites other than the respiratory tract. In the first of a series of studies undertaken by Mills on air pollution and cancer in Cincinnati, mortality from the digestive and gastrointestinal tracts were substantially higher in 15 census areas directly downwind from the industrial area of the city, compared to other parts of the city (Mills, 1943). In a comparison of cancer incidence in Connecticut and Iowa, Levin *et al.* (1960) found some large urban/rural ratios for cancers of the oesophagus, intestine and rectum (males). Within New York State, the urban/rural differences persisted when data for metropolitan and non-metropolitan counties were examined separately.

Positive associations between stomach cancer and indices of particulate pollution were found in two studies of Nashville, Tennessee and Buffalo, New York (Winkelstein, Jr. & Kantor, 1969a; Hagstrom, *et al.* 1998a). In the latter study, after adjusting for age, economic level and ethnic background, mortality was found to be more than three times higher in the most polluted area than in the least polluted. Both studies also showed positive associations between particulate pollution and cancers of the prostate (Winkelstein, Jr. & Kantor, 1969b) and oesophagus.

In a study based on the Utah cancer registry (Lyon *et al.*, 1980), increased mortality ratios in urban areas were found for cancers of the oral cavity and pharynx, oesophagus, bladder, colon, and prostate among non-Mormon males. For Mormons, the only increase in urban areas was seen for cancer of the colon. Cederlof *et al.* (1975) found positive trends in incidence of bladder cancer in men and cancer of the uterine cervix in women with increasing urbanisation, among both non-smokers and smokers.

Analysis of inter-district variations in the incidence of head and neck squamous carcinoma in the period 1978–87 in the West Midlands found higher urban rates compared with neighbouring rural communities. Mean sulphur dioxide and smoke concentrations were positively correlated with squamous cancer of the larynx and pharynx (Wake, 1993).

Many of the above cancers have been shown to relate quite strongly to smoking. Even if smoking habits were controlled for in some of the studies, residual confounding factors from smoking may occur and other factors, such as occupational exposures, alcohol consumption and diet may have also contributed.

The association between engine exhaust and combustion products and non-respiratory cancers has been shown in several studies. Siemiatyci *et al.* (1988), in a population-based case-control study in Canada, studied the associations between 10 different types of engine exhaust and 12 different cancer sites. Marginally elevated odds ratios were seen for colon cancer and exposure to diesel engine exhaust, for cancers of the rectum and kidney with long-term high-level exposure to gasoline engine exhaust and for rectal cancer in bus, truck and taxi drivers. An elevated risk for kidney cancer was also found in a study of workers in gasoline self-service stations in the Nordic countries (Lynge *et al.*, 1997). Increased risk for cancer of the gastrointestinal tract was found in a cohort of professional drivers in Geneva (Guberan *et al.*, 1992).

The association between bladder cancer and exposure to diesel exhaust has been shown in several case-control studies, both population-based (Howe *et al.*, 1980; Hoar & Hoover, 1985), hospital-based (Vineis & Magnani, 1985; Iscovitch *et al.*, 1987) and death certificate-based (Coggon *et al.*, 1984; Steenland *et al.*, 1987). Bladder cancer was also found to be significantly increased among car drivers in a cohort study of urban policeman in Rome (Forastiere *et al.*, 1994).

In addition to the well-recognised risk factors for breast cancer, researchers are also now questioning the role of environmental chemical contaminants, particularly polycyclic aromatic hydrocarbons (PAHs), total suspended particulates (TSPs) and acid haze pollution. PAHs have been shown to be mammary carcinogens in rodents (el-Bayoumy, 1992; Snedeker & Diaugustine, 1996). A study of cancer incidence and mortality in 6000 Seventh-Day Adventists (Mills *et al.*, 1991) found an elevated (non-significant) risk of breast cancer associated with an exceedance frequency for a cut-off of 200 $\mu g/m^3$ TSPs. In a study of 20 Canadian cities, a statistically significant positive association was found between sulphur dioxide concentration as a measure of air pollution and breast cancer (Gorham *et al.*, 1989).

6. Prevention Opportunities and Future Research Strategies

Although cancer is only one of many potential adverse consequences of outdoor air pollution, the risk of cancer may spur decision-makers to explore strategies for reducing air pollution (Graham, 1990). Davis and Muir (1995) advocate the use of the "precautionary principle" i.e. that society should take care not to engage in activities which appear likely to increase risks, eventhough uncertainty exists about the size and extent of those risks. They suggest that even though the cellular and genetic mechanisms of cancer are not fully understood, basic improvements in lifestyle, such as diet, smoking (and other drug) habits and exercise, and in our chemical-physical environment, such as reduced toxic emissions, may have a beneficial effect on general public health. This will be similar to the improvements observed in infectious disease occurrence in the nineteenth century. Multiple exposures or combinations of low levels of commonly occurring carcinogens could be part of the explanation for persisting patterns of cancer that are, otherwise, inexplicable. Ames *et al.* (1987) highlight the fact that we are largely ignorant of the carcinogenic potential of the enormous background of natural chemicals in the world.

Higginson (1993) draws attention to challenges faced by scientists and politicians in addressing the genuine concerns of the public regarding the health effects (including cancers) of what are often, very small or trivial exposures. The explanation of occurrences of so-called clusters of cancers, sometimes around a particular source of pollution, and sometimes without a putative source, are also a common problem faced by public health physicians. In the UK, the much studied occurrence of childhood leukaemia around the nuclear reprocessing plant at Sellafield led to the establishment of the Small Area Health Statistics Unit (SAHSU) to give the ability to respond quickly to reports of unusual clusters of disease and to foster research into the methodology for studying this topic.

Exposures to many carcinogenic chemicals identified in animal bioassays usually occur at levels below that at which a carcinogenic effect can be detected by classical analytical epidemiological methods (Higginson & Muir, 1979; Doll & Peto, 1981; Hemminki & Vainio, 1984) which are often insufficiently sensitive to detect cancer increases or decreases below 1 : 1000, except for certain rare tumours (Higginson, 1988). Graham (1990) discusses how scientific data are currently used by the US Environmental Protection Agency to make quantitative estimates of cancer risk due to exposure to outdoor air pollutants, illustrating this for benzene and formaldehyde. The former was based mainly on occupational epidemiology and the latter from long-term rodent bioassays. Criticisms directed at these risk assessments include (i) small number of cases, uncertain exposure assessment and inappropriate choice of an exposure metric in epidemiological studies and (ii) inaccurate extrapolation of risk across species and inappropriate use of a linear relationship between delivered and administered dose and between exposure and cancer incidence, in animal studies.

Although recognising public concern, Ames *et al.* (1987) suggest that, when setting research priorities, it is important not to divert society's attention away from the few known really serious hazards, such as tobacco, in the pursuit of hundreds of minor or non-existent hazards. The lengthy debate about the role of cigarette smoking on public health over the last 50 years, and the complexity of achieving effective action to discourage smoking does not provide an encouraging model for the control of other, less well-established hazards.

Epidemiological and toxicological research have yet to demonstrate definitive relationships between increased cancer risk and exposure to airborne carcinogens. Priorities include:

1. the investigation of the relative roles of outdoor air pollution, indoor air contaminants and other sources of exposure, i.e., total exposure estimation.
2. the development of epidemiological methods for simultaneous evaluation of the effect of potential confounding factors.
3. the development of non-invasive measures of biological exposure.

4. the implementation of emission control policies and alternative methods of generating energy, and the monitoring and evaluation of subsequent health outcomes.
5. the identification of sensitive subpopulations and the evaluation of the role of individual susceptibility.

Huff (1993) argues strongly for the need for more long-term experimental (animal) studies to establish carcinogenicity of substances. However, Robinson and Paxman (1992) draw attention to the considerable controversy which results when cancer risk assessments are made using only evidence from animal experiments. As has been discussed earlier, epidemiological methods also have severe limitations. Speizer (1983) suggests that studies of general populations to assess an effect with a maximum attributable risk of less than 2%, with poor ability to control for confounding factors or assess exposure misclassification are a fruitless exercise. For epidemiology to be useful in future, both Speizer (1983) and Cohen and Pope (1995) advocate the development of large-scale multidisciplinary collaborative studies, involving epidemiology, industrial hygiene and toxicology. They acknowledge the need for large populations in order to detect the relatively small risk excesses from ambient air pollution. The development of meta analysis techniques may help overcome the problem of achieving an adequate study size.

Methods of estimating exposures accurately and the use of biological and chemical biomarkers provide another essential area of research development. A workshop held in 1994 debated the uncertainties inherent in predicting the health effects of exposure to air pollutants (Williams 1994). The recommendations on future exposure assessment included: (i) better use of existing data and of modelling techniques to identify areas of high concentrations, (ii) the development of improved monitoring methods, both passive and personal, (iii) increased collection of data on specific compounds of concern such as volatile organic compounds and PAHs.

The establishment of a causal relationship between air pollution and cancer continues to be a major challenge to public health researchers. However, as Lawther wrote thirty years ago (Lawther, 1966), "... it behoves us to see that combustion is as complete as possible and that pollution of the air at ground level is abated, for whatever the complexities of the case against air pollution it can be truly said that it can be lethal, it never did anyone any good and is always disgusting".

References

Aksoy M. *et al.* (1972) *Br. J. Ind. Med.* **29**, 56–64.

Aksoy M. *et al.* (1976) *Acta Haematologica,* **55**, 65–72.

Aksoy M. (1988) *Benzene Carcinogenicity.* CRC Press Inc.

Ames B.N. *et al.* (1987) *Science,* **230**, 271–280.

Ashley D.J. (1969) *British Journal of Preventive & Social Medicine,* **23**(3), 187–189.

Axelsson G. & Rylander R. (1980) *Environ. Res.* **23**, 469–476.

Barbone F. *et al.* (1995) *American Journal of Epidemiology,* **141**(12), 1161–1169.

Bauer G. (1556) *De Re Metallica Libri XII.* Basel Folio, Basel.

Beaumont J.J. *et al.* (1987) *J. Natl. Cancer Inst.* **79**, 911–921.

Bisby J.A. (1993) *Health Watch, the Australian Institute of Petroleum Health Surveillance Program. Ninth Report.* Department of Public Health and Community Medicine, University of Melbourne, Melbourne.

Blot W.J. *et al.* (1977) *Science,* **198**, 51–53.

Blot W.J. & Fraumeni J.F., Jr. (1975), *Lancet,* **2**(7926), 142–144.

Blot W.J. & Fraumeni J.F., Jr. (1976) *American Journal of Epidemiology,* **103**, 539–550.

Bond G.G. *et al.* (1986) *Br. J. Ind. Med.* **43**, 685–691.

Borch-Johnsen K. (1982) *Ugeskrift Laeger,* **144**, 1713–1718.

Boucot K.R. *et al.* (1972) *American Journal of Epidemiology,* **95**, 4–16.

Brandt L. *et al.* (1978) *Br. Med. J.* **March 4**, 553.

Broughton G.F.K. *et al.* (1998) *Air Pollution in the UK: 1996.* AEA Technology Plc, Abingdon.

Brown L.M. *et al.* (1984) *Environ. Res.* **34**(2), 250–261.

Browne K. (1986a) In *Proceedings, Third International Conference on Environmental Lung Disease.* American College of Chest Physicians, Montreal.

Browne K. (1986b) *Br. J. Ind. Med.* **43**, 556–558.

Browne K. (1986c) *Br. J. Ind. Med.* **43**, 145–149.

Buell P. *et al.* (1967) *Cancer,* **20**, 2139–2147.

Carnow B. & Meier P. (1973) *Archives of Environmental Health,* **27**, 207–218.

Cederlof R. *et al.* (1975) *The Relationship of Smoking and Some Social Covariables to Mortality and Cancer Morbidity.* Department of Environmental Hygiene, the Karolinska Institute, Stockholm.

Checkoway H. *et al.* (1984) *Am. J. Ind. Med.* **5**, 239–249.
Churg A. (1993) *Monographs in Pathology*, **36**, 54–77.
Coggon D. *et al.* (1984) *J. Natl. Cancer Inst.* **72**, 61–65.
Cohen A.J. & Pope C.A., 3rd. (1995) *Environ. Hlth. Perspect.* **103**(Suppl 8), 219–224.
Colls J. (1997) *Air Pollution. An Introduction.* E & FN SPON, London.
Commins B.T. (1989) *IARC Scientific Publications*, **90**, 476–485.
Cook J.W. *et al.* (1933) *J. Chem. Soc.* **1**, 395–405.
Cordier S. *et al.* (1983) *Environ. Res.* **31**, 311–322.
Curling T.B. (1856) *A Practical Treatise on the Diseases of the Testis and of the Spermatic Cord and Scrotum.* Blanchard and Lea, Philadelphia.
Curwen M.P. *et al.* (1954) *British Journal of Cancer*, **8**, 181–198.
Davis D.L. & Muir C. (1995) *Environ. Hlth. Perspect.* **103**(Suppl 8), 301–306.
Dean G. (1959) *Br. Med. J.* **2**, 852.
Dean G. (1964) *Proceedings of the Royal Society of Medicine*, **57**, 984.
Dean G. (1966) *Br. Med. J.* **1**, 1506–1514.
Dean G. *et al.* (1977) *NLM Report on a Second Retrospective Mortality Study in North-East England Part 1.* Tobacco Research Council, London.
Dean G. *et al.* (1978) *NLM Report on the Second Retrospective Mortality Study in North-East England.* Part 2. Tobacco Research Council, London.
Department of Environment. (1974) In *Clean Air Today.* HMSO, London.
Dockery D.W. *et al.* (1993) *N. Engl. J. Med.* **329**(24), 1753–1759.
Doll R. *et al.* (1965) *Br. J. Ind. Med.* **22**, 1–12.
Doll R. *et al.* (1972) *Br. J. Ind. Med.* **29**, 394–406.
Doll R. (1978) *Environ. Hlth. Perspect.* **22**, 23–31.
Doll R. & Peto R. (1981) *The Causes of Cancer: Quantitative Estimates of Avoidable Risks of Cancer in the United States Today.* Oxford University Press, London.
Earle H. (1823) *Medico-Chirurgical Trans.* **12**(ii), 296–307.
Eastcott D.F. (1956) *Lancet*, **1**, 37.
Ehrenberg L. *et al.* (1985) *Environment International*, **11**, 393–399.
el-Bayoumy K. (1992) *Chemical Research in Toxicology*, **5**(5), 585–590.
Epstein L.M. *et al.* (1984) *Israel Journal of Medical Science*, **20**, 27–32.
Exxon Biomedical Sciences (1996) Benzene Risk Characterisation Report prepared for EUROPIA, CONCAWE and CEFIC. Exxon Biomedical Sciences, New Jersey.
Forastiere F. *et al.* (1987) *Scand. J. Work. Environ. Hlth.* **13**, 258–260.
Forastiere F. *et al.* (1994) *Am. J. Ind. Med.* **26**(6), 785–798.
Ford A.B. & Bialik O. (1980) *Archives of Environmental Health*, **35**(6), 350–359.
Garshick E. *et al.* (1988) *American Review of Respiratory Disease*, **137**, 820–825.
Godlee F. & Walker A. (1992) *Health and the Environment.* British Medical Journal, London.

Goldstein B.D. (1990) *Annals of the New York Academy of Sciences*, **609**, 225–230.
Gorham E.D. *et al.* (1989) *Canadian Journal of Public Health*, **Revue Canadienne de**(2), 96–100.
Graham J.D. (1990) In *Air Pollution and Human Cancer*, ed. Tomatis L. pp. 75–84, Springer Verlag, Berlin.
Greenberg D.S. (1984) *National Wildlife*, **22**, 29–32.
Greenburg L. *et al.* (1967) *Archives of Environmental Health*, **15**(3), 356–361.
Guberan E. *et al.* (1992) *Br. J. Ind. Med.* **49**(5), 337–344.
Haenszel W. *et al.* (1956) *Cancer Morbidity in Urban and Rural Iowa. Public Health Monograph No. 37.* US Government Printing Office.
Haenszel W. *et al.* (1962) *J. Natl. Cancer Inst.* **28**, 947–1001.
Haenszel W. & Tauber K.E. (1964) *J. Natl. Cancer. Inst.* **28**, 803–838.
Hagstrom R.M. *et al.* (1998) *Archives of Environmental Health*, **15**, 237–248.
Hammond E.C. (1972) In *Environment Factors in Respiratory Disease*, ed. Lee D. pp. 177–198, Academic Press, New York.
Hammond E.C. *et al.* (1976) *Annals of the New York Academy of Science*, **271**, 116–124.
Hammond E.C. & Garfinkel L. (1980) *Preventive Medicine*, **9**, 206–211.
Hammond E.C. & Horn D. (1958) *Journal of the American Medical Association*, **166**, 1294–1308.
Harting F.H. & Hesse W. (1879) *Viertjhrs. ger. Med. Off. Sanit.* **30**, 296–301.
Hearey C.D. *et al.* (1980) *J. Natl. Cancer Inst.* **64**(6), 1295–1299.
Hemminki K. & Pershagen G. (1994) *Environ. Hlth. Perspect.* **102**(Suppl 4), 187–192.
Hemminki K. & Vainio H. (1984) In *Monitoring Human Exposure to Carcinogenic and Mutagenic Agents. IARC Scientific Publications No. 59*, eds. Berlin A., Draper M., Hemminki K. *et al.* pp. 37–45, International Agency for Research on Cancer, Lyon.
Higgins I.T. 1976 *IARC Scientific Publications*, **13**, 41–52.
Higgins I.T. 1984 *Preventive Medicine*, **13**(2), 207–218.
Higginson H. & Muir C.S. (1979) *J. Natl. Cancer Inst.* **58**, 825–832.
Higginson J. (1988) *Cancer Research*, **48**(6), 1381–1389.
Higginson J. (1993) *Cancer Suppl.*, **72**(3), 971–977.
Higginson J. & Jensen O.M. (1977) *IARC Scientific Publications*, **16**, 169–189.
Hitosugi M. (1968) *Bulletin of the Institute of Public Health*, **17**, 236–255.
Hoar S.K. & Hoover R. (1985) *J. Natl. Cancer Inst.* **74**, 771–774.
Hoffman D. & Wynder E.L. (1976) In *Chemical Carcinogens. ACS Monograph 173*, ed. Searle C.E. pp. 324–365, American Chemical Society, Washington D.C.
Hoffman E.F. & Gilliam A.G. (1954) *Public Health Report*, **69**, 1033–1042.
Howe G.R. *et al.* (1980) *J. Natl. Cancer Inst.* **64**, 701–713.

Howe G.R. *et al.* (1983) *J. Natl. Cancer. Inst.* **70**, 1015–1019.
Huff J. (1993) *Pharmacology & Toxicology*, **72**(Suppl 1), 12–27.
IARC (1976) Waterhouse J, ed. *Cancer Incidence in Five Continents* Vol III. International Agency for Research on Cancer, Lyon.
IARC (1984) *Monographs on the Evaluation of Carinogenic Risk of Chemicals to Humans Vol 34.* pp. 133–190, International Agency for Research on Cancer, Lyon.
IARC (1985) *IARC Monographs on the Evaluation of Carcinogenic Risks to Humans. Polynuclear Aromatic Compounds, Bitumens, Coal-Tar and Derived Products, Shale Oils and Soots, Volume 35.* International Agency for Research on Cancer, Lyon.
IARC (1989) *IARC Monographs on the Evaluation of Carcinogenic Risks to Humans*, **46**, 1–458.
Infante P.F. *et al.* (1990) *Lancet*, **336**, 814–815.
Iscovitch J. *et al.* (1987) *International Journal of Cancer*, **40**, 734–740.
Jedrychowski W. (1983) *Neoplasma*, **30**, 603–609.
Jedrychowski W. *et al.* (1990) *Journal of Epidemiology & Community Health*, **44**, 114–120.
Jedrychowski W. (1995) *Environ. Health. Perspect.* **103**(Suppl 2), 15–21.
Jockel K.H. *et al.* (1992) *International Journal of Epidemiology*, **21**(2), 202–213.
Kaldor J. *et al.* (1984) *Environ. Health. Perspect.* **54**, 319–332.
Kanarek H.C. *et al.* (1979) *Journal of Environmental Science and Health*, **14**, 641–681.
Kaplan I. (1959) *Journal of the American Medical Association*, **171**, 97–101.
Kaplan S.D. & Morgan R.W. (1981) *Reviews on Environmental Health*, **3**(4), 329–368.
Karch N.J. & Schneiderman M.A. (1981) *Explaining the Urban Factor in Lung Cancer Mortality: A Report to the Natural Resources Defense Council.* Clement Associates, Washington.
Katsoyanni K. *et al.* (1992) *Preventative Medicine.*
Knox E.G. (1994) *J. Epidemiol. Community. Health.* **48**, 369–376.
Lave L.B. & Seskin E.P. (1970) *Science*, **169**, 723–733.
Lawley P.D. (1994) *Advances in Cancer Research*, **65**, 17–111.
Lawther P.J. (1966) *Postgraduate Medical Journal*, **42**(493), 703–708.
Lawther P.J. & Waller R.E. (1978) *Environ. Hlth. Perspect.* **22**, 71–73.
Leiter J. *et al.* (1942) *J. Natl. Cancer. Inst.* **3**, 155–165.
Levin M.L. *et al.* (1960) *J. Natl. Cancer. Inst.* **24**, 1243–1257.
Lewtas J. (1985) In *Carcinogens and Mutagens in the Environment, Volume V. The Workplace*, ed. Stich H.F. pp. 59–74, CRC Press, Boca Raton.
Lewtas J. *et al.* (1992) *Pharmacogenetics*, **2**(6), 288–296.

Lewtas J. (1993) *Pharmacology & Toxicology,* **72**(Suppl 1), 55–63.
Lewtas J. & Gallagher J. (1990) *IARC Scientific Publications,* **104**, 252–260.
Lewtas L. (1990) In *Air Pollution and Human Cancer,* ed. Tomatis L. pp. 49–61, Springer-Verlag, Berlin.
Linet M.S. *et al.* (1987) *J. Occup. Med.* **29**, 136–141.
Lloyd J.W. (1971) *J. Occup. Med.* **13**, 53–68.
Lloyd O.L. (1978) *Lancet,* **1**(8059), 318–320.
Lloyd O.L. *et al.* (1985) *Br. J. Ind. Med.* **42**, 475–480.
Lynge E. *et al.* (1997) *American Journal of Epidemiology,* **145**(5), 449–458.
Lyon J.L. *et al.* (1980) In *Cancer Incidence in Defined Populations (Bancbury Report No. 4),* eds. Cairns J., Lyon J.L. & Skolnik M. pp. 3–30. CSH Press, Cold Spring Harbour, NY.
Matanoski G. *et al.* (1986) *Environ. Health. Perspect.* **70**, 37–49.
Matanoski G.M. *et al.* (1981) *Environ. Res.* **25**, 8–28.
McDonald J.C. (1985) *Environ. Hlth. Perspect.* **62**, 319–328.
McMichael A.J. *et al.* (1975) *J. Occup. Med.* **17**(4), 234–239.
Menck H.R. *et al.* (1974) *Science,* **183**(121), 210–212.
Mills C.A. (1943) *American Journal of Hygiene,* **37**, 131–141.
Mills P.K. *et al.* (1991) *Archives of Environmental Health,* **46**(5), 271–280.
Moller H. (1993) *Pharmacology & Toxicology,* **72**(Suppl 1), 39–45.
Muir C.S. *et al.* (1987) *Cancer Incidence in Five Continents. International Agency for Research on Cancer (IARC) Scientific Publications No. 88.* IARC, Lyon.
Newman J.A. *et al.* (1975) *Annals of the New York Academy of Science,* **271**, 250–268.
Pan B.J. *et al.* (1994) *Journal of Toxicology & Environmental Health,* **43**(1), 117–129.
Paxton M.B. *et al.* (1994) *Risk. Anal.* **14**, 155–161.
Pershagen G. *et al.* (1977) *Environ. Hlth. Perspect.* **19**, 133–137.
Pershagen G. (1990) *IARC Scientific Publications,* **104**, 240–251.
Pershagen G. & Simonato L. (1990) In *Air Pollution and Human Cancer,* ed. Tomatis L. pp. 63–74, Pringer-Verlag, New York.
Pike M.C. *et al.* (1975) In *Persons at High Risk of Cancer,* ed. Fraumeni J.F. pp. 225–239, Academic Press, New York.
Pike M.C. *et al.* (1979) In *Energy and Health,* eds. Breslow L. & Shittermore A. pp. 3–16, Philadelphia.
Pope C.A., 3rd *et al.* (1995) *American Journal of Respiratory & Critical Care Medicine,* **151**(3 Pt 1), 669–674.
Pott P. (1775) *Chirurgical Observatons Relative to the Cataract, the Polypus of the Nose, the Cancer of the Scrotum, the Different Kinds of Ruptures and the Mortification of the Toes and Feet.* Hawes, Clarke and Collins, London.

Prindle R.A. (1959) *Journal of Air Pollution Control Association*, **9**, 12–18.
Rantanen J. (1983) *Environ. Hlth. Perspect.* **47**, 325–332.
Read C. (1991) *Air Pollution and Child Health.* Greenpeace UK, London.
Redmond C.K. *et al.* (1976) *Annals of the New York Academy of Science*, **271**, 102–115.
Reid D.C. (1966) *National Cancer Institute Monograph*, **19**, 321.
Report of the Task Group. (1978) *Environ. Hlth. Perspect.* **22**, 1–12.
Rinsky R.A. *et al.* (1987) *N. Engl. J. Med.* **316**, 1044–1050.
Robinson J.C. & Paxman D.G. (1992) *Am. J. Ind. Med.* **21**(3), 383–396.
Rushton L. & Romaniuk H.M. (1997) *Occ. Env. Med.* **54**(3), 152–166.
Samet J.M. *et al.* (1987) *American Journal of Epidemiology*, **125**, 800–811.
Schnatter A.R. *et al.* (1996) *Occ. Env. Med.* **53**, 773–781.
Shy C.M. (1976) In *Clinical Implications of Air Pollution Research*, eds. Finkel A.J. & Duel W.C. pp. 3–38, Publishing Sciences Group Inc. Acton, Massachusetts.
Siemiatycki J. *et al.* (1988) *Scand. J. Work. Environ. Hlth.* **14**, 79–90.
Smith G.H. *et al.* (1987) *Br. J. Ind. Med.* **44**(12), 795–802.
Snedeker S.M. & Diaugustine R.P. (1996) *Progress in Clinical & Biological Research*, **394**, 211–253.
Soskolne C.L. *et al.* (1989) *Archives of Environmental Health*, **44**(3), 180–191.
Speizer F.E. (1983) *Environ. Hlth. Perspect.* **47**, 33–42.
Steenland K. *et al.* (1987) *American Journal of Epidemiology*, **126**, 247–257.
Stocks P. (1947) *Studies on Medical and Population Subjects. Regional and Local Differences in Cancer Deaths Rates. No. 1.* HMSO, London.
Stocks P. (1958) In *British Empire Cancer Campaign 35th Annual Report 1957. Supplement to Part II.* British Empire Cancer Campaign, London.
Stocks P. (1959) *Br. Med. J.* **1**, 74–79.
Stocks P. (1960) *British Journal of Cancer*, **14**, 397–418.
Stocks P. (1966) *British Journal of Cancer*, **20**, 595–623.
Stocks P. & Campbell J. (1955) *Br. Med. J.* **2**, 923–929.
Swanson G.M. (1988) *Cancer*, **62**(8 Suppl), 1725–1746.
Tenkanen L. & Teppo L. (1987) *Scandinavian Journal of Social Medicine*, **15**, 67–72.
Tomatis L. (1991) *Annals of Oncology*, **2**(4), 265–267.
Tornqvist M. & Ehrenberg L. (1992) *Pharmacogenetics*, **2**(6), 297–303.
Tornqvist M. & Ehrenberg L. (1994) *Environ. Hlth. Perspect.* **102**(Suppl 4), 173–82.
US Environmental Protection Agency. (1990) *Cancer Risk from Outdoor Exposure to Air Toxics. Volume 1 EPA-450/1-90-004.* US Environmental Protection Agency, Office of Air Quality Planning and Standards, Research Triangle Park.

Vena J.E. (1982) *American Journal of Epidemiology,* **116**(1), 42–56.
Vigliani E.C. (1976) *Ann. N. Y. Acad. Sci.* **271**, 143–151.
Vigliani E.C. & Forni A. (1976) *Environ. Res.* **II**(1), 122–127.
Vigliani E.C. & Saita G. (1964) *N. Engl. J. Med.* **271**, 872–876.
Vineis P. & Magnani C. (1985) *International Journal of Cancer,* **35**, 599–606.
Wake M. (1993) *Clinical Otolaryngology,* **18**(4), 298–302.
Walker R.D. *et al.* (1982) *Environ. Res.* **28**, 303–312.
Waller R.E. (1952) *British Journal of Cancer,* **6**, 8–21.
Waller R.E. (1981) *Environment International,* **5**, 479–483.
Weinberg G.B. *et al.* (1982) *American Journal of Epidemiology,* **115**, 40–58.
Weiss W. (1978) *American Journal of Public Health,* **68**(8), 773–775.
Williams M. (1994) In *Report on Air Pollution and Health. Understanding the Uncertainties.* Institute for Environment and Health, Leicester.
Winkelstein W. *et al.* (1967) *Archives of Environmental Health,* **14**, 162–171.
Winkelstein W., Jr. & Kantor S. (1969a) *Archives of Environmental Health,* **18**(4), 544–547.
Winkelstein W., Jr. & Kantor S. (1969b) *American Journal of Public Health & the Nations Health,* **59**(7), 1134–1138.
Witschi W. (1994) In *Report on Air Pollution and Health. Understanding the Uncertainties.* Institute for Environment and Health, Leicester.
Wong O. (1987) *Br. J. Ind. Med.* **44**, 365–381.
World Health Organization Regional Office for Europe. (1987) *Air Quality Guidelines for Europe.* WHO Regional Publications, Copenhagen.
Xiao H.P. & Xu Z.Y. (1985) *National Cancer Institute Monograph,* **69**, 53–58.
Xu Z.Y. *et al.* (1989) *J. Natl. Cancer. Inst.* **8**, 1800–1806.
Yamagiwa K. & Ichikawa K. (1918) *Journal of Cancer Research,* **3**, 1–21.
Young N. (1989) *Am. J. Ind. Med.* **15**, 495–498.
Zeidberg L.D. *et al.* (1967) *Archives of Environmental Health,* **15**, 214–224.

CHAPTER 5

PARTICULATE AIR POLLUTION

R.L. Maynard

The views expressed in this chapter are those of the author and should not be taken as those of the UK Department of Health.

1. Introduction

Ambient air has always contained particles. Interest in particulate air pollution and its health effects has revived to an extraordinary extent in recent years and it is now the most intensively researched and hotly debated aspect of air pollution toxicology. Epidemiological studies have shown that current levels of particles have significant effects on health both in terms of their day-to-day variations and in terms of lifetime exposures. Control measures currently being adopted in many countries are expected to reduce ambient levels of particles but, of course, these cannot be reduced to zero and so we may expect effects to continue. As comparatively easy options for reducing levels are exhausted, costs of further reductions will rise and the examination of the benefits of these measures will intensify. It is already probably fair to say that no epidemiological research, with the exception of that related to the adverse effects of tobacco smoking, has been exposed to such rigorous examination. This examination has not been entirely disinterested as the costs to industry of accepting both the findings and the measures to reduce levels of particles are high. It is, as always, true to say that the final costs of reducing levels will be borne by the consumer; though not, necessarily, by the polluter.

This pressure for examination of the results of research into health effects, and the search for alternative explanations for the findings, has given rise to a large and rapidly expanding literature. Much of this is constructive and a number of colloquia have been held and their proceedings published (Phalen & Bates, 1995; Lee & Phalen, 1996). Authors of monographs have considered the problem (Wilson & Spengler, 1996; Dockery & Pope, 1997) and several excellent review papers have appeared (Pearce & Crowards, 1996; Seaton, 1996). Recently, the methods of meta-analysis have been applied to the findings of the epidemiological studies, again not without criticism (Spix *et al.*, 1998). Despite all re-examination, a consensus has emerged that current levels of particles *are* damaging to health. This chapter provides only an overview of the evidence upon which this is based and points to new lines of research. Perhaps the most important remaining question is: "What component or components of the ambient aerosol are responsible for the damage to health?" Answering this will not be easy. It is possible that in different areas, different components play the toxicological leading role. The prize for unravelling the problem and developing appropriate control strategies will however be very great. Investment in the area is increasing: the US Government has recently agreed to spend nearly 50 million dollars on the problem.

2. Historical Aspects

Pollution by particles has always occurred. Wind-blown dust, volcanoes, sea spray and the photochemical formation of particles in the atmosphere predate the appearance of life on the planet: it is likely that sunsets were red before man described and admired them as such. The development of fire as a source of warmth and a means of cooking, undoubtedly, added to the means of generating particles and deep caves with fires and burning torches must have been heavily polluted with smoke. Indoor air pollution of this sort has not disappeared and in some parts of the world, is the major source of personal exposure to particles; especially for women and children spending much time indoors (Smith, 1996). The development of towns and cities added to the problem. The reader will appreciate that Dickensian London was enveloped, at least in winter, in a pall of smoke. Whether this was damaging to health was disputed at the time, especially, by those operating industries that produced smoke. In many developed countries, the problem of smoke pollution has abated; however, in developing countries, such as China, it remains a major concern (Chen *et al.*, 1994).

In addition to smoke generated by burning of coal for heating and cooking, the production of large quantities of wood smoke by a slash and burn approach to forest clearance has recently become a problem. During the past few years, fires have raged in the Far East and many people have been affected. The lack of rainfall, perhaps caused by the El Niño phenomenon, has added to the problems of uncontrolled fires. Other sources of particulate air pollution continue to cause concern, for example, the high levels of ash and finer particles generated by volcanic activity on the island of Montserrat. Recent measurements on the island have recorded levels of PM_{10} of more than 8000 $\mu g/m^3$ during ash falls (see Fig. 1).

Concern about the effects of combination of smoke and sulphur dioxide produced by the domestic use of coal was brought into sharp focus by the London smog of 1952 (Ministry of Health, 1954). The smog occurred in early December of 1952 during a period of unusually cold weather when coal consumption was probably high. A stable synoptic weather pattern in south-eastern England produced a temperature inversion over London and smoke and sulphur dioxide accumulated at ground level. Concentrations of smoke, measured as Black Smoke (see below), rose until the filters used for collecting particles overloaded. Peak concentrations were of the order of 7000 $\mu g/$

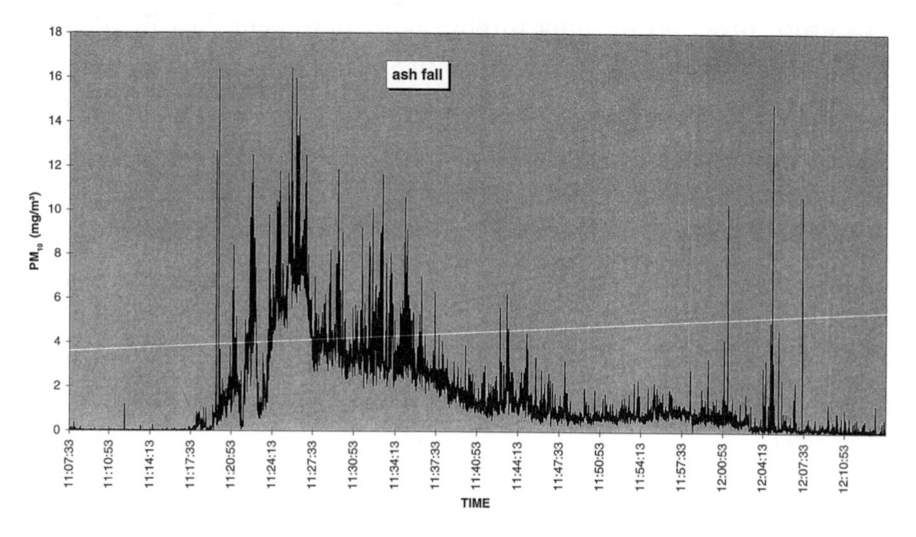

Fig. 1. Concentration of particles (PM_{10}) recorded with DUST TRAK monitor during ash fall on Montserrat 1997.

m^3 and concentrations of sulphur dioxide exceeded 4000 $\mu g/m^3$. These figures are the more remarkable when one recalls that these were not merely transient peaks: concentrations remained high for 4–5 days. The effects of the episode were striking: more than 4000 extra deaths, i.e. deaths over and above those expected for the time of year, occurred. The elderly were most severely affected though infants also suffered. Remarkably, the mechanism of effect is still unknown in any detail. It is clear that the combination of particles and sulphur dioxide was deadly for those already suffering from cardio-respiratory diseases but post-mortem examination failed to reveal any very dramatic and specific signs. It was simply observed that those with chronic diseases affecting the heart and lungs suffered an exacerbation of their symptoms and some died. This is remarkable in that sulphur dioxide is a known bronchoconstrictor agent and those suffering from asthma are known to be more sensitive to this compound than other individuals (Fry, 1953). It was suggested that the acidity of the aerosol was a factor in producing the effects though this has not been proven. Recent reanalyses of data collected later in London have provided support for this theory (Quality of Urban Air Review Group, 1993).

3. Concentrations of the Ambient Aerosol

Particles occur in the air in a range of sizes. Particles of more than 10 μm diameter are rapidly deposited under the influence of gravity and do not travel far from their sources. Wind-blown dust and sea spray contain many such particles. Particles generated by the condensation of hot gases emitted, for example, by motor vehicles are small: < 100 nm in diameter and these are also short-lived. This is not because they deposit under the influence of gravity but because they agglomerate and form particles of between 0.2 and 2 μm diameter. Particles of this size are remarkably stable in the atmosphere and can travel for thousands of kilometres before they are eventually deposited (Quality of Urban Air Review Group, 1993). The distribution of sizes of particles in the air is shown in Fig. 2.

The same factors that control deposition of particles in the air control deposition of particles in the respiratory tract. Particles of greater than 10 μm diameter hardly pass the upper airways: the nose, mouth and larynx. Particles with diameters of about 4 to 10 μm deposit in the conducting airways of the lung under the influence of impaction and sedimentation. Impaction accounts for the deposition of the larger particles at branch points in the airway. The deposition of particles carrying carcinogenic chemicals at these sites may explain the frequent occurrence of bronchogenic carcinoma at these sites. Particles of about 0.5 μm diameter are the least well-deposited in the lung. Smaller particles are well-deposited by diffusion. As particle size decreases, the importance of diffusion as a mechanism of deposition increases. Particles of diameter of about 20 nm are very effectively deposited in the gas exchange part of the lung: the alveoli and alveolar ducts; particles of less than 10 nm diameter are not. This is because these extremely small particles diffuse so rapidly in air that they are deposited in the

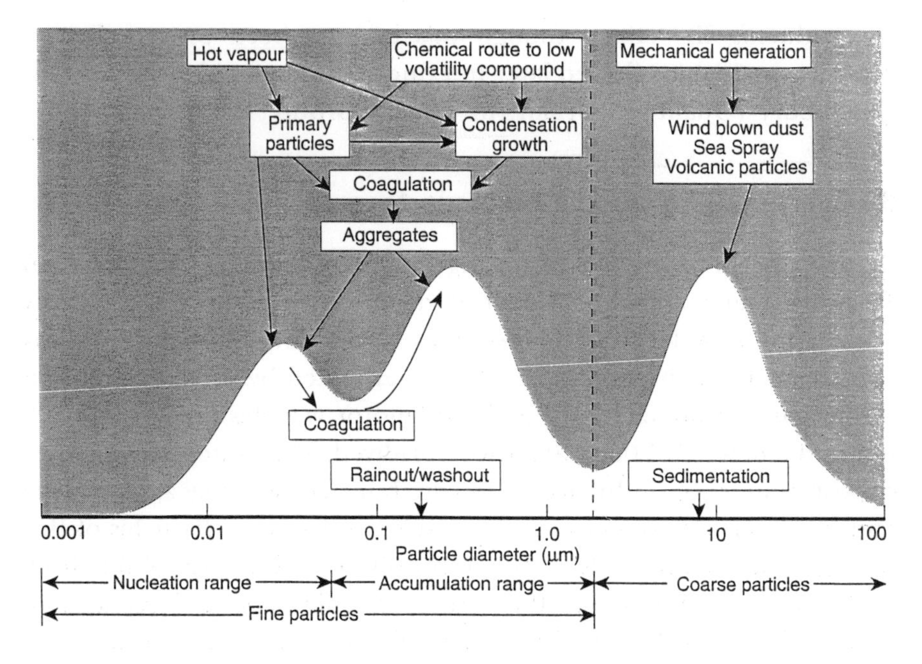

Fig. 2. Schematic representation of a typical size distribution for atmospheric particles, including some formation pathways (reproduced with permission from the Quality of Urban Air Review Group, 1993).

upper airways: for example, in the nose. Particles have been classified according to their deposition pattern in the respiratory system. For our purposes, the following are the more important terms:

> **Thoracic particles:** those capable of penetrating beyond the larynx and generally taken as less than 10 μm in diameter.
>
> **Respirable particles:** those capable of penetrating beyond the ciliated part of the airways, i.e. into the gas exchange zone. Such particles are taken as being of less than 4 μm in diameter. Occasionally, a High Risk Respirable Fraction is mentioned. This relates to particles liable to reach the gas exchange zone of the sick and infirm or children and can be taken as those particles of diameters less than 2.5 μm.

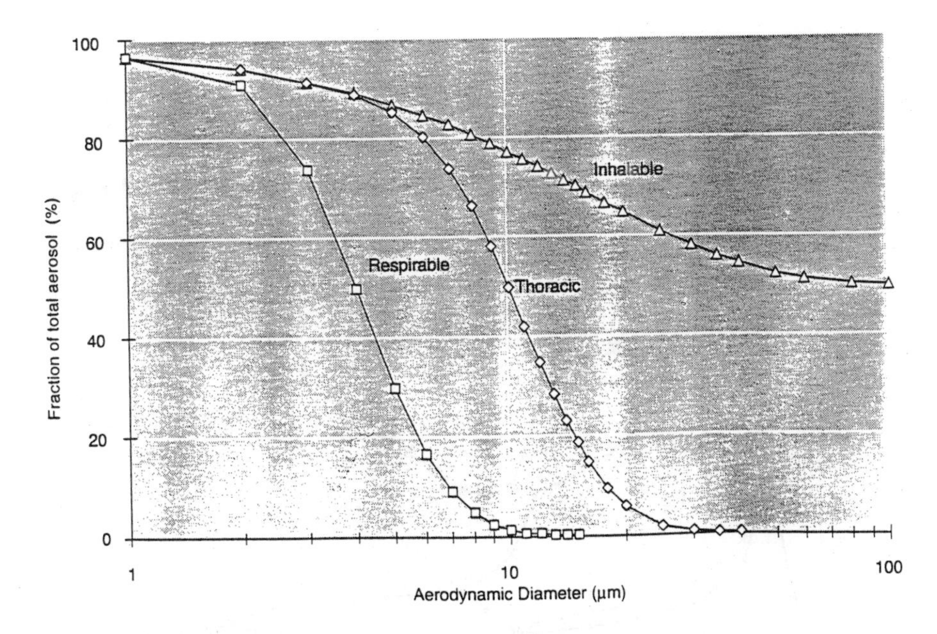

Fig. 3. Typical curves defining respirable, thoracic and inhalable particle fractions (from the Quality of Urban Air Review Group Report, 1993).

These conventions are shown in Fig. 3. The above definitions are a little less precise than aerosol physicists might prefer. When we speak of thoracic particles we really mean the thoracic fraction and should define it as the mass fraction capable of penetrating beyond the larynx and it is further defined by a cumulative log-normal distribution curve with a median diameter of 10 μm and a geometric standard deviation of 1.5 (Quality of Urban Air Review Group, 1996).

The pattern of deposition of particles in the lung is shown in Fig. 4. It is generally accepted that the particles most likely to cause damage to the lung are those that are deposited in the gas exchange zone. This part of the lung lacks the efficient particle clearance mechanisms of the conducting airway: there are no cilia to sweep the surface of the alveoli and no blanket of mucus to trap the particles. Alveolar clearance is dependent upon macrophages and these clear particles more slowly

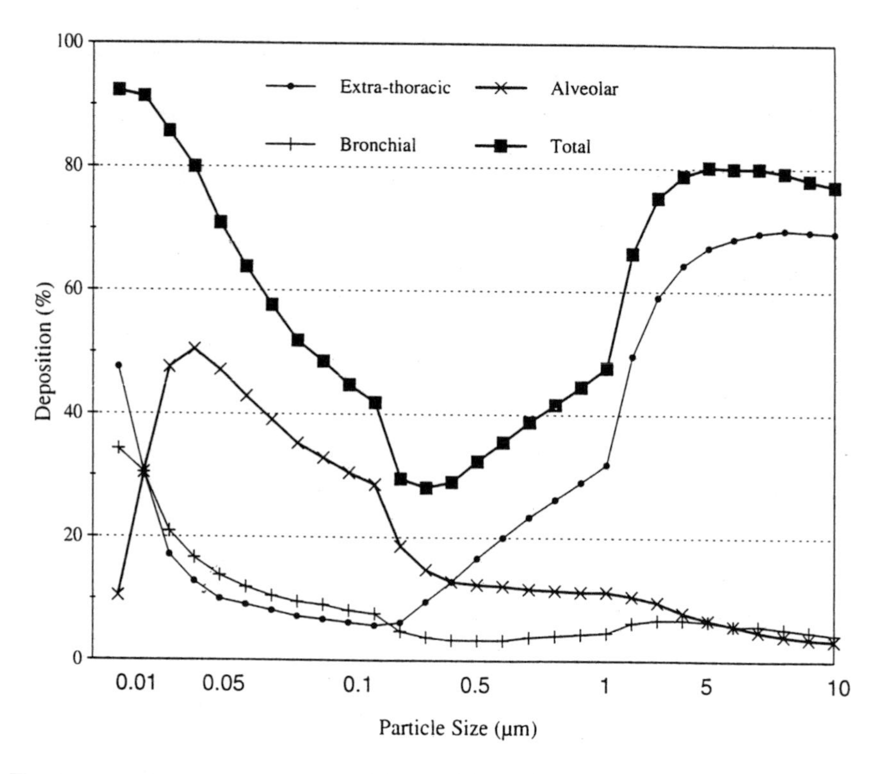

Fig. 4. Lung deposition versus particle size (from the Dept of Health Report on Non-Biological Particles and Health, 1995).

than do the cilia. Ciliary clearance removes about 90% of particles deposited in the conducting airways within 6 hours; but removal of 90% of particles deposited at the alveolar level may take up to one year (Dockery & Pope, 1997). Macrophages take up particles by phagocytosis and travel to the junction between the gas exchange zone and the conducting airways. Here, they "step" onto the muco-ciliary escalator and are transported out of the lung to be swallowed or coughed out. Recent studies have suggested that very small particles are not caught by the macrophages quickly enough to prevent them from travelling across the lining fluid of the alveoli and the alveolar epithelium to enter the interstitium. Here, they may set up an inflammatory reaction.

This is the basis of the "ultrafine hypothesis" that has been put forward to explain the findings of current epidemiological studies.

It should be recalled that the ambient aerosol is chemically very complex. A wide range of inorganic and organic compounds have been identified in airborne particles; indeed just about every chemical element that has been looked for has been found (Quality of Urban Air Review Group, 1993). The complexity presents the toxicologist with a challenge: which are the important active components? Some recent ideas on this will be discussed later. Some of the components of the ambient aerosol are soluble and might be expected to dissolve in the lining fluid of the airways. The alveoli, however, are comparatively dry as a result of the presence of a surfactant, a component of the fluid-repelling lipid-rich lining layer. The behaviour of water-soluble particles at the alveolar level is not well understood. That some dissolve, at least in the water filled corners of the alveoli, seems likely. Moreover, it is hardly possible that water-soluble particles could pass through the cytoplasm of alveolar epithelial cells without dissolving. Typically, soluble particles are represented by the inorganic salts formed in the atmosphere by the neutralisation of acid; ammonium bisulphate and sulphate, and a range of nitrates are common components. Insoluble components such as metallic oxides, on the contrary, might pass through without dissolving. The accumulation of particles in the pulmonary lymph nodes proves that some make the journey intact.

Occupational medicine has shed much light on the effects of particles on the lung. Studies on silicosis, for example, have shown that some substances can damage the lung, set up a brisk inflammatory reaction and stimulate a fibrotic response. The problem in the air pollution area is that doses of particles are so much smaller than seen in the occupational setting and that specific lesions attributable to air pollutants have not been identified. Some work has suggested that exposure to urban air pollution may stimulate a low-grade centri-acinar inflammatory reaction (Sherwin, 1991), but this hardly seems enough to account for the increase in deaths from cardiovascular diseases which have been associated with variations in concentrations of particles. It is likely that some more subtle mechanism is responsible but as already stated, finding this mechanism is the challenge.

4. The Description and Measurement of Particles

Particles can be weighed or counted. A wide range of techniques has been developed and the reader should seek details of individual methods in the relevant literature (Hinds, 1982; Quality of Urban Air Review Group, 1996). For our purposes, we shall only consider two techniques as they are the ones which have been most widely used in air pollution epidemiology.

4.1. Black Smoke

The first is the Black Smoke, BS, (sometimes referred to as the British Smoke) method. Air is drawn at a low flow rate through a filter paper. Particles are deposited on the paper and, if they themselves are dark in colour, will blacken the paper. The extent of blackening can be measured by examining the light reflecting properties of the stained filter paper. Degrees of reduction in reflectance may be related to the mass of particles in the air. Calibration against a gravimetric (weighing) method is, of course, needed. This method, which proved so useful in the early days of air pollution, has been widely criticised. A major problem is the calibration which will clearly be unstable as the composition of the aerosol changes. The method will underread in an area where most of the particles are white and overread in an area where the aerosol is blacker than that used for calibration. Comparison of readings from one country to another may be unreliable. Despite this, much Black Smoke monitoring is still undertaken and recent epidemiological studies have shown clear associations between levels of particles monitored as Black Smoke and effects on health (Ponce de Leon *et al.*, 1996). Interestingly, the Black Smoke method can measure particles of diameter less than 4.5 μm.

4.2. Gravimetric Measurements

Continuously-reading gravimetric methods are replacing the Black Smoke method. In essence, such methods depend on air being passed through a size-selective sampling head and the collected particles being weighed. This can be done by allowing the particles to deposit on an oscillating beam and monitoring the change in the period of oscillation. Calibration is again necessary but once done, the method should produce reliable results despite changes in particle composition. Of course, weighing particles tells one nothing of their composition nor anything of their source. The Black Smoke method, on the other hand, provides information on source. In urban areas in the UK, the Black Smoke method reflects, today, diesel emissions; in the 1950s it reflected domestic use of coal.

The selectivity of the sampling head used on gravimetric devices such as the Tapered Element Oscillating Microbalance (TEOM), widely used in the UK, is defined in terms of its percentage acceptance of particles of a given size. For example, a PM_{10} (PM = particulate matter) head allows 50% of particles of exactly 10 μm aerodynamic diameter to pass through, i.e., nearly all particles of about 5 μm diameter and almost no particles of about 20 μm diameter. The "cut" at 10 μm is thus sharp. As an approximation we can say that a PM_{10} monitor reports the mass of particles of less than 10 μm diameter. Similar definitions apply to $PM_{2.5}$ and $PM_{1.0}$ monitors.

Particles can be collected for weighing or chemical examination by using the sort of size-selective heads described above coupled with a higher airflow and deposition of the collected material on filters. The Graseby Anderson Hi-Vol, which samples with a flow rate of 1130 l/min, is an example of this sort of instrument. In the United States, the Environmental Protection Agency defines a reference method for monitoring particles which involves collection and weighing of total suspended particulates (TSPs). The upper size range of particles collected by this method is of the order of 25–45 μm diameter (Dockery & Pope, 1997).

4.3. Number Concentration

Particles can also be counted. A range of devices which rely on the dust cloud interfering with light impinging on a sensor has been developed. Such devices count particles and the counts may be converted into mass concentrations after appropriate calibration.

4.4. Collection of Size-Fractions

If detailed examination of particles is required, then, collection of different size-fractions of the ambient aerosol may be undertaken. This is often done by using a cascade impactor (Hinds, 1982). This device consists of a series of filters arranged in layers, i.e., the cascade. As air passes through the system, it is deflected and particles are deposited by inertial impaction. The largest particles are deposited early in the sequence and thus, appear on the upper filter plates; small particles make their way to the lower plates.

5. The Development of Particle Epidemiology

The recent surge of interest and spate of papers dealing with the effects of particles on health has obscured to some extent the contribution made by earlier workers. It is worth recalling that in London in the 1960s, a series of seminal papers were published which addressed the problem of effects of day-to-day variations in concentrations of particles and sulphur dioxide (Waller, 1971; Lawther *et al.*, 1979).

These studies were panel studies in which a group of patients suffering from chronic bronchitis were provided with diaries and asked to note their clinical conditions each day — noting the comparison with the previous day. Thus, fluctuations in symptoms were identified. A simple scoring system allowed the results to be plotted against concentrations of smoke and sulphur dioxide, each measured on a daily basis. The results showed that in the early 1960s, there was a clear relationship between levels of air pollution and symptom scores. However, as levels of pollution in London fell in the late 1960s and early 1970s, the previously clear relationship disappeared. The authors looked at the daily average concentrations and noted that as long as the Black Smoke concentrations were below 250 $\mu g/m^3$ and the sulphur dioxide concentrations were below 500 $\mu g/m^3$, fluctuations in concentration did not seem to be associated with changes in scoring of health status. These figures played an important part in establishing the World Health Organisation Guidelines for smoke and sulphur dioxide published in 1987 (World Health Organisation, 1987).

The studies just referred to were panel studies. Other epidemiological techniques have also been used to look at the effects of particles on health. These include studies of episodes of air pollution, important examples being the study of the 1952 London episode (Ministry of Health, 1954) and that of the 1991 London episode (Anderson *et al.*, 1995).

5.1. Time-Series Studies

In recent years, time-series studies of the effects of air pollutants on health have become popular, in part, due to the greater availability of the computing power needed to deal with the data collected by these methods. Developments in the statistical methodology, or at least in its availability, have also taken place. An excellent overview of these studies has recently been published by two of the principal authors in this area, Dockery and Pope (1997). Further comprehensive accounts have been provided by Schwartz *et al.* (1996) and the UK Committee on the Medical Effects of Air Pollutants (Dept of Health, 1995). It should perhaps be noted again, before discussing the time-series studies, that the statistical methodology used in these studies has been subjected to the most rigorous examination and original data have been taken and examined by independent authorities. The reports published by the US Health Effects Institute are a particularly important contribution in this respect (Health Effects Institute, 1995).

The essential characteristic of the time-series studies which have been recently applied to the problems of air pollution epidemiology is that they model the relationship between day-to-day levels of pollutants and health endpoints such as daily deaths or hospital admissions. One might readily imagine that if levels of pollutants varied substantially from day-to-day, so would these endpoints, always assuming that the two are causally related. Of course, it might be that even though levels of pollutants and the health endpoints move together, they might not be causally related: for example, levels of pollutants rise in cold weather and so does the number of deaths occurring each day. It might be that the apparent association between pollutants and deaths is entirely due to changes in the weather.

Allowing for the effects of such confounding factors, co-variates of both the levels of pollutants and the endpoints studied has been one of the challenges of the time-series approach. The studies, however, do not require allowance to be made for confounding factors such as smoking, social class or occupational exposures. These are unlikely to change day-to-day and the *day*, it should be recalled, is the unit of analysis in the time-series approach. One caveat should be added to the last statement: if the results of studies from different cities are to be compared, then these factors may come back into play. One could imagine that in a city with a high background level of chronic lung disease, daily variations in levels of air pollutants would have a greater effect than in a city without such a burden. In recent studies undertaken in Europe, significant differences have been found between Eastern and Western cities in this respect (Spix *et al.*, 1998). These have not yet been fully explained.

Building statistical models of associations involves the technique of regression analysis. A more detailed account is provided in Chapter 3. In the simplest case, two variates are studied and their association is described by a best-fitting equation. The equation describes the statistical model that best fits the data. Events such as daily deaths are best treated not as normally distributed variates but as being more likely to conform to a Poisson distribution. This distribution was in fact deve loped to model deaths due to horse-kicks in the Prussian army (Moroney, 1951). Schwartz (1994) and Kinney and colleagues (1995) have, however, shown that similar estimates of effects are produced if normal

Gaussian methods are applied. In practice, either approach is acceptable.

Co-variates such as temperature can be controlled for by either understanding the effects of temperature on health and allowing for them in a deterministic way or by stratifying the daily data by temperature and using dummy variables. This has been a common means of allowing for month of the year and thus for seasonal effects. Recent studies have used non-parametric approaches to adjust for the effects of factors affecting long-term fluctuations in endpoints.

A great deal of effort has been put into devising better ways of controlling for the effects of weather. Some critics of the time-series results have suggested that had weather been better controlled for, then no associations with pollutants would have been found. Dockery and Pope (1997) have pointed out that Mackenbach and colleagues (1993) demonstrated the feasibility to "adjust away" the relationship between a pollutant, in this case sulphur dioxide, and a health effect by adjusting for "14 weather variables (nine measures of temperature lagged up to 15 days, four measures of precipitation and humidity lagged up to five days, and wind speed lagged up to five days)". It can be argued that the assumptions involved in this complex process of allowing for temperature were more worrying than those required in the acceptance of the effects of the pollutant. Interestingly, recent work has shown that adjusting for synoptic weather conditions does not change the results of the analysis significantly when compared with those obtained by adjusting only for temperature. Indeed, Dockery and Pope (1997) point out that once season is allowed for, further allowance for weather has little effect on the discovered associations between some pollutants and health endpoints.

5.2. Pollution Mixtures

Nobody is exposed to one pollutant alone and this complicates the analyses. Multiple regression techniques are required to take into account different pollutants and methods have been developed to test the importance of each pollutant as an explanatory variable of the effect measured. Often, particles are the strongest explanatory variable and if removed from the model, the association with other indices of pollution disappears.

5.3. Lags

One further important factor to consider is the lag-structure of the analysis. It might seem obvious to seek associations between sameday variations in levels of pollutants and endpoints, but equally, one might argue that the pollutants might not affect health instantly. For example, there may be an association between today's death count and yesterday's pollution level; this is called lagging by one day . In practice, a variety of lags are tried. Not all studies have agreed as to which lag structure produces the strongest association. A good discussion has been provided by Lipfert (1994).

5.4. Autocorrelation

One last factor that needs to be taken into account is autocorrelation, i.e., the unexplained associations that occur between the number of events such as deaths and hospital admissions occurring day-to-day over a short period. It is not the case deaths on day one that cause deaths on day two but that the determinants of these deaths are correlated from day one to day two. Modern statistical methods allow for autocorrelation.

5.5. Expression of Results

How then are the results of the time-series studies expressed? After all the statistical manipulation, a regression coefficient which defines the equation linking the daily concentration of the studied pollutant and the study's endpoint is produced. This is expressed as a relative risk (RR) and indicates the percentage, above 100, increase in effects expected for a given increase in concentration of pollutant. For example, a 5% increase is expressed as an RR of 1.05.

The results of a number of studies are shown in Table 1.

It will be noticed that a range of measures of particulate pollution has been studied. The results are expressed in terms of the percentage increase in deaths expected for a 10 $\mu g/m^3$ increase in PM_{10}. Now, a

Table 1. Studies of effects of PM_{10} on daily mortality (from Dockery & Pope, 1997).

Location and period	Reference[a]	Particulate measure	Mean PM_{10}	Percentage change in daily mortality for each 10 $\mu g/m^3$ increase in PM_{10} (CI)
St Louis, MO 1985–86	Dockery *et al.*, 1992	PM_{10} (previous day)	28	1.5% (0.1%, 2.9%)
Kingston, TN 1985–86	Dockery *et al.*, 1992	PM_{10} (previous day)	30	1.6% (−1.3%, 4.6%)
Birmingham, AL 1985–88	Schwartz, 1995	PM_{10} (3 day mean)	48	1.0% (0.2%, 1.9%)
Utah Valley, UT 1985–89	Pope *et al.*, 1992	PM_{10} (5 day mean)	47	1.5% (0.9%, 2.1%)
Los Angeles, CA	Kinney *et al.*, 1995	PM_{10}	–	0.5% (0.0%, 1.0%)
Chicago, IL	Ito *et al.*, 1995	PM_{10}	–	0.5% (0.1%, 1.0%)
São Paulo, Brazil 1990–91	Saldiva *et al.*, 1995	PM_{10}	82.4	1.3% (0.7%, 1.9%)
Santiago, Chile 1989–91	Ostro *et al.*, 1996	PM_{10}	115	0.7% (0.4%, 0.9%)
Amsterdam, The Netherlands, 1986–92	Verhoef *et al.*, 1996	PM_{10} (same day)	38	0.6% (−0.1%, 1.4%)
Combined		All PM_{10}		0.75% (0.59%, 0.91%)

[a]Original references are found in the account by Dockery and Pope (1997).

1% increase in risk, or a relative risk of 1.01, is a small effect and this has led critics of the approach to doubt the results. Some epidemiologists have argued that relative risks of less than 1.2, i.e. a 20% increase, in effects should be ignored as it is not possible to establish their reliability (Gamble & Lewis, 1996). One could easily agree that larger effects are easier to be sure about than smaller ones, but many people are exposed to air pollution and even a small increase in risks, if true, would imply a considerable public health effect and a need for action.

In addition to studies of the association of concentrations of particles with mortality, associations with a wide range of other health endpoints have been reported. These include admissions to hospital, asthma

Table 2. Effects of PM_{10} on indicators of ill health (from Dockery & Pope, 1994).

	Percentage change in health indicators per 10 $\mu g/m^3$ increase in PM_{10}
Increase in daily mortality	
Total deaths	1.0
Respiratory deaths	3.4
Cardiovascular deaths	1.4
Increase in hospital usage	
All respiratory admissions	1.8
Emergency department visits	1.0
Exacerbation of asthma	
Asthmatic attacks	3.0
Bronchodilator use	2.9
Hospital admissions	1.9
Increase in respiratory symptoms reports	
Lower respiratory	3.0
Upper respiratory	0.7
Cough	1.2
Decrease in lung function	
Forced expired volume	0.15
Peak expiratory flow	0.08

attacks, use of medication for asthma, respiratory symptoms and lung function. A small selection of the results is shown in Table 2.

Criticism of the findings, or rather the interpretation of the findings of the time-series studies has focused on two points:

(a) Are the associations causal?
(b) Do the associations matter?

The first, the question of causality, has been debated at length. Bradford Hill defined in 1965 a series of tests which might be applied to an association in order to discover whether it was likely to be causal. It should be noted that Bradford Hill advised caution in the application of his tests or criteria. The criteria include one critical test: the temporal sequence of effect and suggested cause must make sense. If this is contravened, then one may conclude that the association cannot be causal. The other criteria: strength of association, consistency from study to study, coherence, specificity, biological gradient, biological plausibility, the results of natural experiments, and the possibility of reasoning of natural experiments and the possibility of reasoning by analogy, are not absolute tests in the sense of the temporal association test. Each of these criteria has been hotly debated with respect to the time-series studies. Some criteria are met with ease as the results are very consistent while others, such as strength of association, are not.

Some authors have focused on the question of coherence. This means that if the daily death rate from respiratory diseases increases, so should the number of people admitted to hospital for the treatment of such diseases and the number of patients consulting their doctors. Bates has discussed this in detail and has argued that the observed coherence of the data strongly supports a causal association (Bates, 1992).

One of the criticisms often levelled at epidemiological techniques is that they are not experimental. Hardly a fair criticism as they are not intended to be so but a common one from uninformed experimentalists. However, one fortuitous natural experiment that sheds light on the particle problem has been recorded. Many studies of the effects of particles on health have been undertaken in the Utah Valley of the United States. Particle concentrations in the valley can be high and local steel-works make a large contribution to the ambient aerosol. In 1986, the steel mill was closed for 13 months due to a strike and the daily average particle concentration (PM_{10}) in the valley fell from 50 to 35 $\mu g/m^3$. On the basis of a 1% increase in deaths for a 10 $\mu g/m^3$ increase in PM_{10} one might have expected the mortality rate to fall by 2.3%. It fell by 3.2% (Pope *et al.*, 1992). This is a most important observation and encourages those who believe that reducing levels of particles will be of benefit to health. This is discussed further below.

The likelihood of the association reported from the time-series studies being causal was discussed by the UK Department of Health Committee on the Medical Effects of Air Pollutants in 1995 (Dept of Health, 1995). The Committee concluded that:

> "... in terms of protecting public health it would be imprudent not to regard the associations as causal".

This forceful statement was carefully drafted and includes a proper precautionary approach to risks to public health.

The second of the criticisms: does it matter? is in some ways less easy to resolve than that regarding causality. The problem is best treated in two parts:

(i) how many people are affected?
(ii) how much are they affected?

The results of the time-series studies provide regression coefficients linking levels of pollutants, in this case, particles, with number of health endpoints occurring per day. The coefficient, of course, defines the slope of the line which has been fitted to the data. On examination of the regression line, it is apparent that linear models fit the data well and that there is no obvious threshold of effect. This is a most important point. Given that one does not know in advance the shape of the true relationship between variables, one should be careful in interpreting fitted regression lines. It is certainly true to say that a range of different models could be fitted. Anderson *et al.* (1995) have pointed out that in looking at the effect of ozone on hospital admissions, the data appear to show a threshold of effect at about 50 ppb, eight-hour average concentration. However, a straight line could certainly have been fitted to the data and an acceptably high correlation coefficient would have been obtained (Stedman *et al.*, 1997).

5.6. Thresholds of Effect

A threshold is the highest level of exposure, defined for our purposes as a concentration with an appropriate averaging time, which produces no health effect. The question of thresholds of effect has been discussed in a number of ways. One interesting observation points out that time-series studies are by their very nature unlikely to demonstrate thres holds of effect. One might guess that there would be inter-person variation in the threshold of effect of air pollutants, but still believe that there is some level of pollutant, albeit low, to which nobody would respond. Can this level be identified by time-series studies?

The answer is probably that it cannot. It should be recalled that time-series studies do not involve monitoring of individual exposure to pollutants but, generally, the concentration recorded at a single monitoring site in a town or city. Across that town or city, we may expect there to be a range of exposure to pollutants. Consider the cycling messenger: he will be exposed to large amounts of vehicle-generated particles, much more than the old person sitting in a suburban garden. There are thus two distributions in play; a distribution of exposure and a distribution of sensitivity. In combination, these make the demonstration of a threshold at a population level unlikely. This is important when considering standard setting for pollutants. Both the following statements may be true:

(i) exposure to $x\,\mu g/m^3$ PM_{10} is safe;
(ii) exposure to particles on a day when the monitored PM_{10} level is $x\,\mu g/m^3$ is not.

If we accept that the time-series studies do not indicate a threshold of effect and, on examination of the regression line, find that it passes towards the origin, we can calculate the total effect of particles on health by simply dealing with the annual average concentration. In the UK, for example, the annual average concentration is about 25 $\mu g/m^3$, thus, there are 2.5% more deaths per annum than there would be without the effects of particles, i.e. total deaths p.a. = about 400 000 with deaths due to particles to be about 10 000.

The above calculation has been deliberately simplified. However, adopting a more detailed approach and considering concentrations of pollutants on a grid square approach and local death rates do not produce a significantly different result (Dept of Health, 1998). More important than arguing about whether the answer should be 10 000 or 8000 or 12 000 is the need to discuss what the answer means.

In the simple calculation given above, it was suggested that the 10 000 deaths were due to particles. It would have been more accurate to say that the 10 000 deaths were associated with particles. Who actually died? We have no way of knowing. Studies have shown that the risk to the elderly and those suffering from cardio-respiratory disease is greater than that for the general population. We should also consider what we mean when we say that a death is associated with particles as opposed to being caused by particles. It seems likely that a proportion of the elderly and sick may have their deaths brought forward as a result of exposure to particles. By how much are their deaths brought forward? We do not know and cannot tell from the standard time-series analyses. It could be as little as a day or as long as many years: all that the time-series studies tell us is that more people die on days when levels of particles are raised.

It seems reasonable to think that the greater the extent by which deaths are brought forward, the more important in public health terms is the effect. This advancing of deaths, culling of the susceptibles, or harvesting, has been the subject of intense debate. If only those expected to die within a few days of an increase in levels of pollutants were affected one might expect to see a dip in the daily death figures following an air pollution episode. Such dips have been difficult to identify, even after the London smog of 1952. At the time of writing, at least two groups in the United States are working on statistical methods designed to estimate the extent by which deaths are brought forward. At the moment, all we have are best guesses, i.e. deaths, in general, are likely to be advanced by days to months. A few may be more markedly affected.

Advancing of deaths by months is not a trivial effect and given that many people are affected, this cannot be ignored in public health terms. In discussing the importance of the effects of particles, we have focused on deaths. We should not forget, however, the effects on hospital admissions. Here, the problem is, in fact, more complex in that we do not know how many of the extra admissions to hospital that can be calculated for a given level of pollution would have occurred anyway, perhaps a little later. The report published by the UK Department of Health (1998) summed up the estimates of effects for the UK. The committee that produced the report were notably cautious in including only data of which they were reasonably confident. For example, they excluded from their final statements calculations based on levels of nitrogen dioxide and carbon monoxide on the grounds that the data were not sufficiently convincing. With regard to ozone, both threshold

Table 3. Numbers of deaths and hospital admissions for respiratory diseases affected per year by PM_{10}* and sulphur dioxide in urban areas of Great Britain (from Dept of Health, 1998).

Pollutant	Health outcomes	GB urban
PM_{10}	Deaths brought forward (all causes)	8100
	Hospital admissions (respiratory) brought forward and additional	10 500
SO_2	Deaths brought forward (all causes)	3500
	Hospital admissions (respiratory) brought forward and additional	3500

*PM_{10}: Particulate matter generally less than 10 μm in diameter.
Estimated total deaths occurring in urban areas of GB per year = c430 000.
Estimated total admissions to hospital for respiratory diseases occurring in urban areas of GB per year = c530 000.

Table 4. Numbers of deaths and hospital admissions for respiratory diseases affected per year by ozone in both urban and rural areas of Great Britain during summer only (from Dept of Health, 1998).

Pollutant	Health outcomes	GB, threshold = 50 ppb	GB, threshold = 0 ppb
Ozone	Deaths brought forward: all causes	700	12 500
	Hospital admissions (respiratory) brought forward and additional	500	9900

and no-threshold approaches were adopted. The results are shown in Tables 3 and 4.

Many commentators have concluded that the effects do matter: many people may have been affected and the effects are not likely to be trivial.

6. Other Evidence Suggesting that Exposure to Ambient Levels of Particles Damage Health

6.1. The Cohort Studies

Important evidence has been published in two papers which report the results of following cohorts of individuals living in cities across the United States (Dockery *et al.*, 1993; Pope *et al.*, 1995). Such studies are quite different from the time-series type in that they seek to compare cities rather than days. The first of the studies, the Six Cities Study (Dockery *et al.*, 1993), followed a cohort of over 8000 adult individuals living in six cities, which differed significantly in their annual average concentrations of pollutants, for 14–16 years between 1974 and 1991. In this sort of study, concern about the types of confounding factors which we found could be ignored in the time-series studies, is paramount. Details of occupational history, educational attainments, social position, smoking history and the like play a large, indeed probably the main, part in controlling the life expectancy of people living under different conditions.

A study which seeks to compare cities and which only studies six cities, suffers from having only six data points for analysis. In a second study, based on data from a large cohort maintained by the American Cancer Society (Pope *et al.*, 1995), 552 000 people were followed from 1982 to 1989. Adjustment for confounding factors was undertaken with special attention being given to smoking history.

The results of these studies are expressed in terms of adjusted mortality rates:

> *Six Cities Study*: The ratio of the adjusted mortality rate in the most polluted city (Steubenville, Ohio, mean fine particle concentration of 29.6 $\mu g/m^3$) to that in the least polluted city (Portage, Wisconsin, mean fine particle concentration of 11.0 $\mu g/m^3$) was 1.26 with the other four cities occupying intermediate positions in line with their pollution levels.
>
> *American Cancer Society*: If the American Cancer Society cohort is examined for the ratio of adjusted mortality rates with regard to fine particles concentrations, then data are a little limited in that monitoring was found to have been carried out in only 50 metropolitan areas. However, the ratio of most polluted (33.5 $\mu g/m^3$) to least polluted (9.0 $\mu g/m^3$) was 1.17.

Interestingly, in both studies, closer correlations were found between adjusted mortality rates and fine particles or sulphates than with more all-embracing measures of particles or measures of concentrations of gaseous pollutants. These results suggest that long-term exposure to particles, especially fine particles, has a significant effect on health. These studies have not been ignored by the critics. One of the most worrying criticisms has been that current death rates may be telling us more about the effects of exposure to pollutants long ago than about the effects of current levels. Let us assume that the concentrations of pollutants in Steubenville, Ohio, were very much higher during the early years of life of those included in the cohort. One could hypothesise that exposure to such pollution for the first few years of life had a lasting effect on lung function. The higher mortality in Steubenville than in Portage might then be due to this long-lasting effect and not to current exposure. If this were so, then calculating a mortality ratio based on current comparisons of pollutant concentrations would be misleading. Unfortunately, data on concentrations of pollutants from long ago were not available and so this could not be explored.

The cohort studies have been recently examined in an attempt to discover by how much of the shortening of life implied by the higher mortality rates can be attributed to particulate air pollutants. This is a rapidly developing research area and not much has yet been published. One paper which used a life-table-based analysis suggests that, in The Netherlands, current levels of particulate air pollution may be producing an average loss of life expectancy of about a year (Brunekreef, 1997). This has not been confirmed but expert opinion seems to tend towards estimates ranging from six months to a year or two. A further paper from Finland suggests a similar effect (Nevalainen & Pekkanen, 1998).

6.2. Gathering the Threads and Asking Questions

All the evidences discussed above suggest that particles damage health. It will be recalled that one of the criteria that Bradford Hill included in his list for examining associations for likelihood of causality was biological plausibility. Hill stressed that this question should not be pressed too hard in that it was dependent on the state of biological knowledge at the time of asking. Despite this warning, a number of critics of current views of the effects of particles on health have pressed the point and it is worth some discussion.

Let us state the problem in the baldest terms.

The time-series epidemiological data require us to believe that on a day when the concentration of particles increases by 10 $\mu g/m^3$, the number of people dying rises by 1%. Let us assume that all the inhaled particles are deposited in the lung and that we breathe 20 m^3 of air per day, then the extra deposited dose would be 200 μg of particles.

The question to ask is: "Can such a small extra dose of particles not known to be made up of any exceedingly toxic material, bring forward deaths in even very sick people?"

Until a few years ago, it is likely that most inhalation toxicologists would have answered no; today, many would be less certain. The reason for this is that several new strands of evidence have appeared.

Firstly, studies by Oberdörster *et al.* (1990, 1992, 1994), Ferin *et al.* (1991), and more recently, Donaldson's group in Edinburgh (Gilmour *et al.*, 1996; Li *et al.*, 1996) have shown that very small particles have unexpected toxicological properties. Titanium dioxide, aluminium oxide and carbon black are all very much more toxicologically active in animal models when presented as particles of less than 100 nm diameter than when presented as particles of several hundred nanometres diameter. The reasons for this increase in toxicity are unknown but two suggestions have been made. The first is that such particles find their way rapidly into the interstitium of the lung and set up an inflammatory response there. This response may spill over into the alveoli. Also, such particles appear to be generators of hydroxyl free radicals. Iron may play a role in this process. These findings regarding ultrafine (< 100 nm diameter) stimulated Seaton *et al.* (1995) to propose the "Ultrafine Hypothesis" to explain the effects of ambient particles on health. The authors argued that an interstitial inflammatory reaction in the lung could affect the pulmonary capillary endothelium, perhaps, increase sequestration of inflammatory cells in the lung, and less clearly, affect the production of clotting factors such as fibrinogen and Factor VII by the liver. The great value of this hypothesis is that it suggests a link between inhaled particles and effects on the cardiovascular system. The hypothesis has received some support from the German observation that blood viscosity increased during a period when numbers of particles in the air were raised (Peters *et al.*, 1997).

If ultrafine particles are the key to the effects of particulate air pollution, then the consequences will be large. Such particles weigh very little but they occur in the air in large numbers. It may be that measures such as PM_{10}, or even $PM_{2.5}$ or $PM_{1.0}$, are merely acting as surrogates for particle number counts and it is upon these that regulatory toxicologists should be focusing. Some limited data that suggest that numbers of particles (always likely to be dominated by very small particles) are related to health endpoints and symptoms of respiratory disease, have been produced by Peters *et al.* (1997).

The second strand of evidence that increases the biological plausibility of the epidemiological findings is that which has been produced by studies of the effects of concentrated ambient particles on sick animals. Studies involving animal models of both chronic bronchitis and pulmonary hypertension have shown that these animals are very much more sensitive to particle challenge than are normal animals (Godleski *et al.*, 1996; Killingsworth *et al.*, 1996). Extrapolation to sick people is tempting.

Despite these toxicological advances, we are still uncertain as to the toxic component or components of the ambient aerosol. Some have argued cogently that the epidemiological evidence linking effects to sulphate concentrations means that inhaled acid is a key factor (Lippmann & Thurston, 1996). The present author finds this difficult to accept on the grounds that the dose of acid is likely to be small and the buffering power of the lining fluid of the lung should deal with ambient acid concentrations. Others argue that transition metals hold the key; iron, cobalt, vanadium, manganese and nickel are all found in the ambient aerosol and could catalyse free radical formation (Zhang *et al.*, 1998). Yet others believe that the chemistry of the particles may be less important than their physical characteristics, i.e. size and surface area. The solution to these problems still seems some way off.

7. Conclusion

Particles were among the first air pollutants recognised as being damaging to health. In many Western countries, the reduction of concentrations of smoke was felt to have solved this problem. This is now no longer thought to be the case and even the low particle concentrations of today may well be damaging to health. The mechanism of the effect of particles on the body is elusive, the cost of controlling the number concentration of small particles is likely to be large, but the prize in terms of reduction of damage to health may be great. That particulate air pollution presents a range of exciting challenges is beyond dispute.

References

Anderson H.R. *et al.* (1995) *Thorax*, **50**, 1188–1193.

Bates D.V. (1992) *Environ. Res.* **59**, 336–349.

Bradford Hill A. (1965) *Proc. R. Soc. Med.* **58**, 295–300.

Brunekreef B. (1997) *Occup. Environ. Med.* **54**, 781–784.

Chen X.L. *et al.* (1994) *Chinese J. Chronic Dis. Prev. Control*, **2**, 259.

Department of Health. (1995) *Non-Biological Particles and Health. Committee on the Medical Effects of Air Pollutants.* London: HMSO.

Department of Health. (1998) *Quantification of the Effects of Air Pollution on Health in the United Kingdom. Committee on the Medical Effects of Air Pollutants.* London: The Stationery Office.

Dockery D.W. *et al.* (1993) *New Engl. J. Med.* **329**, 1753–1759.

Dockery D.W. & Pope C.A. (1994) *Annu. Rev. Publ. Health*, **15**, 107–132.

Dockery D.W. & Pope C.A. (1997) In *Topics in Environmental Epidemiology*, eds. Steenland K. & Savitz D.A. pp. 119–166, Oxford: Oxford University Press.

Ferin J. *et al.* (1991) *J. Aerosol. Med.* **4**, 57–68.

Fry J. (1953) *Lancet*, **i**, 235–236.

Gamble J.F. & Lewis R.J. (1996) *Environ. Health Perspect.* **104**, 838–850.

Gilmour P.S. *et al.* (1996) *Occup. Environ. Med.* **53**, 817–822.

Godleski J.J. *et al.* (1996) In *Proceedings of the Second Colloquium on Particulate Air Pollution and Human Health*, eds. Lee J. & Phalen R.F. pp. 4(136)–4(143), May 1–3, Park City, Utah.

Health Effects Institute. (1995) *Particulate Air Pollution and Daily Mortality. The Phase I Report of the Particle Epidemiology Evaluation Project.* Cambridge, MA: Health Effects Institute.

Hinds W.C. (1982) *Aerosol Technology. Properties, Behaviour and Measurement of Airborne Particles.* London: J. Wiley & Sons.

Killingsworth C.R. *et al.* (1996) In *Proceedings of the Second Colloquium on Particulate Air Pollution and Human Health*, eds. Lee J. & Phalen R.F. pp. 4(238)–4(244), May 1–3, Park City, Utah.

Kinney P.L. *et al.* (1995) *Inhalation Toxicol.* **7**, 59–69.
Lawther P.J. *et al.* (1970) *Thorax*, **25**, 525–539.
Lee J. & Phalen R.F. (eds.) (1996) *Proceedings of the Second Colloquium on Particulate Air Pollution and Human Health*, May 1–3, Park City, Utah.
Li X.Y. *et al.* (1996) *Thorax*, **53**, 1216–1222.
Lipfert F.W. (1994) *Air Pollution and Community Health. A Critical Review and Data Sourcebook.* New York: Van Nostrand Reinhold.
Lippmann M. & Thurston G.D. (1996) *J. Expo. Anal. Environ. Epidemiol.* **6**, 123–146.
Mackenbach J.P. *et al.* (1993) *J. Epidemiol. Community Health*, **47**, 121–126.
Ministry of Health. (1954) *Mortality and Morbidity During the London Fog of December 1952. Reports on Public Health and Medical Subjects No 95.* London: HMSO.
Moroney M.J. (1951) *Facts from Figures.* pp. 98, Harmondsworth, Middlesex: Penguin Books Ltd.
Nevalainen J. & Pekkanen J. (1998) *Sci. Total Environ.* **217**, 137–141.
Oberdörster G. *et al.* (1990) *J. Aerosol. Sci.* **21**, 384–387.
Oberdörster G. *et al.* (1992) *Environ. Health Perspect.* **97**, 193–199.
Oberdörster G. *et al.* (1994) *Environ. Health Perspect.* **102**, 173–179.
Pearce D. & Crowards T. (1996) *Energy Policy*, **24**, 609–619.
Peters A. *et al.* (1997) *Am. J. Respir. Crit. Care Med.* **155**, 1376–1383.
Peters A. *et al.* (1997) *Lancet*, **349**, 1582–1587.
Phalen R.F. & Bates D.V. (eds.) (1995) *Proceedings of the Colloquium on Particulate Air Pollution and Human Mortality and Morbidity*, January 24–25 1994, Irvine, California. *Inhalation Toxicol.* **7**, 1–163 and 577–588.
Ponce de Leon A. *et al.* (1996) *J. Epidemiol. Community Health*, **50**(Suppl 1), S63–S70.
Pope C.A. *et al.* (1992) *Arch. Environ. Health*, **47**, 211–217.
Pope C.A. *et al.* (1995) *Am. J. Respir. Crit. Care Med.* **151**, 669–674.
Quality of Urban Air Review Group. (1993) *Urban Air Quality in the United Kingdom. First Report of the Quality of Urban Air Review Group.* London: DoE.
Quality of Urban Air Review Group. (1996) *Airborne Particulate Matter in the United Kingdom. Third Report of the Quality of Urban Review Group.* Birmingham: University of Birmingham.
Schwartz J. (1994) *Am. J. Epidemiol.* **139**, 589–598.
Schwartz J. *et al.* (1996) *J. Epidemiol. Community Health*, **50**(Suppl 1), S3–S11.
Seaton A. *et al.* (1995) *Lancet*, **345**, 176–178.
Seaton A. (1996) *J. R. Soc. Med.* **89**, 604–608.
Sherwin R.P. (1991) *J. Toxicol. Clin. Toxicol.* **29**, 385–400.
Smith K.R. (1996) *Natl. Med. J. India*, **9**, 103–104.

Spix C. *et al.* (1998) *Arch. Environ. Health,* **53**, 54–64.

Stedman J.R. *et al.* (1997) *Thorax,* **52**, 958–963.

Waller R.E. (1971) *J. R. Coll. Phys. Lond.* **5**, 362–368.

Wilson R. & Spengler J. (eds.) (1996) *Particles in Our Air.* Cambridge, MA: Harvard University Press.

World Health Organisation. (1987) *Air Quality Guidelines for Europe 1987.* Copenhagen: WHO Regional Office for Europe.

Zhang Q,. *et al.* (1998) *J. Toxicol. Environ. Health,* **53**, 423–438.

CHAPTER 6

ALTERNATIVE FUELS

Jeffrey S. Gaffney & Nancy A. Marley

1. Introduction

Increasing attention has been given recently to the use of alternative automotive fuels as a method of improving urban air quality and reducing greenhouse gas emissions by reducing combustion-related pollution while simultaneously minimising the dependence on foreign oil imports. Alternative fuels are defined as those fuels that are not petroleum-based liquid fuels. (Examples of petroleum-based liquid fuels would be gasoline and diesel). These unconventional fuels are currently being evaluated as environmental or economic improvements over the conventional fossil-based fuels. For example, a perfect alternative fuel would be a potentially "clean" fuel that when combusted, would result in the reduction of emissions of primary pollutants and would therefore, result in reduced formation of secondary urban and regional air pollutants. The perfect fuel would provide a more efficient and economical means of powering automobiles and trucks than conventional gasoline or diesel and could also be produced domestically resulting in a reduced dependence on foreign oil imports. The alternative fuels that have been given the most attention in recent years are natural gas (methane, ethane and propane), liquefied petroleum gas (propane and butanes) and oxygenated fuels. The principal oxygenated fuels that have received attention are methanol and ethanol (and their gasoline blends), with some attention being given recently to methyl *tert*-butyl ether (MTBE)/gasoline blends. In addition, biofuels, and in particular, biodiesel fuels, have also been given attention as a means of reducing the use of fossil fuels and lowering carbon dioxide emissions. Many of the alternative fuels proposed today were considered in the past for automotive usage (Kreucher, 1995) but for economic reasons were not able to compete successfully with conventional petroleum-based gasoline and diesel fuels. As foreign oil dependence, environmental regulations, and increasing costs for petroleum have arisen as important issues, the current technoeconomic aspects of biomass conversion to alternative liquid fuels have been assessed in detail (e.g. Elliot *et al.*, 1990; Lönner & Törnqvist, 1990; Turhollow & Kanhouwa, 1993; Mitchell *et al.*, 1995).

The government of the United States has promoted research and development of technology for alternative fuel usage through the administration of the Alternative Fuels Utilization Program of the U.S. Department of Energy. This programme includes research, development and demonstration of technologies for alternative fuel vehicles (AFV). Two important programme elements — engine optimisation and atmospheric reactions — are managed at the National Renewable Energy Laboratory in Golden, Colorado. These two programme elements are aligned with the two primary reasons for this federally funded research: the need to enhance our energy security position and the need to control emissions for improved air quality. The goal of this research is to develop light-duty and heavy-duty vehicle utilisation technologies that will be commercially superior to technologies that rely on conventional petroleum-derived transportation fuels (National Renewable Energy Laboratory, 1995).

There is considerable concern that, even with mandated vehicle emission regulations (Tables 1 and 2), several urban areas in the United

Table 1. Gasoline-fueled automobile exhaust emission standards for the United States overall and the state of California. Standards are for light-duty vehicles[a] in grams per mile at five years and 50 000 miles* or ten years and 100 000** miles (Chang *et al.*, 1991; Gushee, 1992; Calvert *et al.*, 1993)

United States	1994*		2003**	
Hydrocarbons	0.41		—	
Non-methane hydrocarbons	0.25		0.125	
Carbon monoxide	3.4		1.7	
Oxides of nitrogen	0.4		0.2	
California	1993* (CVs)[b]	1994* (TLEVs)[b]	1997* (LEVs)[b]	1997* (ULEVs)[b]
Non-methane hydrocarbons	0.25	0.125	0.075	0.04
Carbon monoxide	3.4	3.4	3.4	1.7
Oxides of nitrogen	0.4	0.4	0.2	0.2
Formaldehyde	—	0.015	0.015	0.008

[a]For light-duty vehicles of < 3750 pounds loaded vehicle weight.
[b]CV = conventional vehicle; TLEV = transitional low-emission vehicle; LEV = low-emission vehicle; ULEV = ultra-low-emission vehicle.

States will remain out of compliance with the National Ambient Air Quality Standards (NAAQS) well into the next century (Chang *et al.*, 1991). As part of the Clean Air Act Amendments of 1990 (CAAA), the 101st Congress of the United States mandated oxygenated and reformulated gasoline programmes in the worst non-attainment areas, in an attempt to speed up progress towards compliance with the NAAQS. Specifically, the oxygenated fuels programme requires the use of fuels containing 2.7% oxygen by weight (e.g. methanol, ethanol and MTBE) in winter time to reduce carbon monoxide (CO) emissions. The reformulated gasoline programme requires the use of fuels containing 2% oxygen, and reduced concentrations of benzene and other aromatics in the summer to reduce the atmospheric formation of ozone (Calvert *et al.*, 1993). In addition, there is a clean-fuel fleet programme that requires new motor vehicle fleets to include some percentage of

Table 2. Diesel engine exhaust emission standards for the United States overall, the state of California, Europe, Korea, and Japan. Standards are for new light-duty vehicles in grams per kilometre (Zelenka *et al.*, 1996).

	Particulates	CO	Hydrocarbons	NO_x
United States				
1996	0.1	4.2	0.31	1.25
California				
1996 (CVs)[a]	0.08	4.2	0.31	1.0
TLEVs[a]	0.08	4.2	0.156	—
LEVs[a]	0.08	4.2	0.090	0.1
ULEVs[a]	0.04	2.1	0.055	0.05
Europe				
1996	0.1	1.0	0.9[b]	—
1999	0.04	0.5	0.5[b]	—
Japan				
1994	0.2	2.1	0.4	0.5
2000	0.08	2.1	0.4	0.4
Korea				
1996	0.08	2.11	0.25	0.62
2000	0.05	2.11	0.25	0.62

[a]CV = conventional vehicle; TLEV = transitional low-emission vehicle; LEV = low-emission vehicle; ULEV = ultra-low-emission vehicle.
[b]Values are for hydrocarbons + NO_x.

low-emission vehicles (LEVs) beginning in 1998. These vehicles include those that can operate on reformulated gasoline, methanol, compressed natural gas (CNG), liquefied petroleum gas (LPG) or other "clean" fuels (Gushee, 1992). However, a quantitative assessment of the resulting air quality benefits of these mandated programmes has not been completed and it is not completely certain that the reformulated or blended fuels will help to abate urban air pollution (Tanner *et al.*, 1988;

Gaffney & Marley, 1990; Chang *et al.*, 1991; Calvert *et al.*, 1993; Anderson *et al.*, 1995; Popp *et al.*, 1995, National Research Council, 1996).

In addition to these federally mandated programs, some states also have issued their own alternative fuel programmes. Emission standards for gasoline-fueled new cars and light-duty trucks (< 3750 pounds) are tighter in the state of California than those required by the 1990 CAAA (see Table 1). The low-emission vehicle/clean-fuel requirements in California are designed around the use of conventional vehicles (CV), transitional low-emission vehicles (TLEVs), low-emission vehicles (LEVs), ultra low-emission vehicles (ULEVs) and zero-emission vehicles (ZEVs). Vehicle manufacturers are required to meet a fleet-averaged non-methane organic gas (NMOG) standard with a combination of the above vehicle types (Chang *et al.*, 1991).

The mandate for cleaner air and reduced emissions of carbon dioxide and other greenhouse gases produced by the combustion of fossil fuels makes alternative automotive fuels particularly attractive because of their biomass source and their potential to reduce air pollutant emissions. However, it is not clear to what extent these fuels are cleaner than gasoline (Chang *et al.*, 1991; Gushee, 1992; Calvert *et al.*, 1993), or whether the use of alternative fuels may lead to other environmental problems (Gaffney & Marley, 1990). In response to increasing concerns, a comprehensive assessment of the use of alternative fuels was drafted by the United States government under the direction of the White House Office of Science and Technology Policy (OSTP). This assessment addresses public health, air quality, fuel economy and engine performance from the use of alternative fuels, and has been critically reviewed by the National Research Council in terms of scientific credibility, comprehensiveness and internal consistency (National Research Council, 1996). In general, the assessment concludes that there is evidence that the levels of some major pollutants (e.g. CO) have been reduced through the use of oxygenated fuels, while others, e.g. oxides of nitrogen [NO_x], remain unchanged, and still others, e.g. aldehydes, are increased and that further studies to include other effects, such as local meteorology, are needed to assess adequately the overall impacts of alternate fuel usage on air quality.

This chapter will focus on issues of urban and regional air quality as they relate to the use of these alternative transportation fuels in automotive engines. Particular attention will be given to the use of oxygenated fuels and oxygenated-fuel/gasoline blends that have been mandated by the CAAA. The use of natural gas, LPG and hydrogen will also be considered. Some focus will also be given to other alternative fuels and their approaches, such as the use of biodiesel fuels, which may aid in the improvement of urban air quality. Lastly, the use of fuel cells will be discussed as a future substitute for internal combustion and its associated problems.

2. Air Quality Issues and Alternative Fuels

Considerable attention has been given to alternative transportation fuels as a means of reducing primary pollutant emissions and the subsequent formation of secondary pollutants in the atmosphere, specifically as this relates to the criteria pollutants. Criteria pollutants are those that have been identified as representing a direct environmental health risk (e.g. CO and ozone) or being strongly connected indirectly to an environmental health risk (e.g. NO_x). Although other important pollutants are associated with the combustion of both petroleum and alternative fuels, the criteria pollutants are the ones that are subject to regulation by the CAAA. Very recently, the Clean Air Act regulations have been changed to lower the exposure standards for ozone and to add air quality standards for fine particulate matter below 2.5 μm in diameter (PM-2.5). These changes in the current regulations will add extra impetus to the use of alternative fuels, particularly, in urban areas that are not currently in compliance with the NAAQS. In this section, we will outline some of the chief chemical species of concern that are currently being monitored as criteria pollutants, as well as overview some of the non-criteria chemical species that are likely to become important concerns if alternative fuels become more widely used.

2.1. Primary Pollutants

Primary pollutants are those that are produced within the combustion engine and emitted directly into the atmosphere at the tailpipe. The primary criteria pollutants that are currently regulated as direct health-related hazards are CO and lead. Lead is added to gasoline as an octane enhancer and is emitted from the tailpipe as particulates. These particulates are in the form of inorganic lead halides (> 90%) and to a lesser extent, organo-lead compounds (Finlayson-Pitts & Pitts, 1986). Approximately, one-half of the mass of these particles are greater than 5.0 μm in diameter and fall out relatively close to the source while the smaller particles may remain suspended for long periods of time (Ter Haar *et al.*, 1972). Airborne lead is of concern because of its extreme toxicity, particularly in children. With the mandated use of unleaded fuels, the levels of airborne lead particulate matter have been lowered drastically in the United States. However, leaded fuels are still in use in other areas of the world and can be a substantial health problem. The use of unleaded fuels can and is being initiated in many regions to alleviate this primary pollutant problem. Unlike the problems associated with other primary pollutants of internal combustion, this one has a relatively simple solution. However, with the removal of lead from gasoline, other additives have been used to enhance octane, such as aromatics or branched alkylated compounds, and these may lead to other atmospheric problems especially in vehicles without catalytic converters.

Carbon monoxide is of concern because it is a direct toxin. It acts efficiently to bind with haemoglobin in the blood, forming a very stable carboxyhaemoglobin complex. In this manner, CO prevents the haemoglobin from functioning in its normal oxygen-binding capacity. At levels of low parts per million, CO exposure can lead to stress for many individuals who suffer from respiratory or coronary dysfunctions. The addition of catalytic converters to automobiles has reduced substantially the emissions of CO from the combustion of conventional gasoline and diesel fuels. However, CO emissions during the "cold start" of engines, when the catalyst is cold and has not reached operating temperature, can still be substantial. Alternative fuels, especially oxygenated fuels, have been found to reduce the emissions of CO during the cold start. For this reason, oxygenated fuels have been mandated in the United States in those areas where atmospheric CO concentrations have continued to exceed health standards.

The indirect primary pollutants that are regulated include NO_x, hydrocarbons and sulphur dioxide (SO_2). The NO_x is generally emitted from spark-emission and diesel engines as nitric oxide (NO), with smaller amounts of nitrogen dioxide (NO_2) and other nitrogen oxide species. The NO_x, $hydrocarbons_x$ and SO_2 are emitted as primary pollutants but are converted in the atmosphere to the secondary pollutants, ozone and sulphuric acid aerosols, which are also criteria pollutants. Because both ozone and sulfuric acid aerosols are associated directly with an environmental health risk, their precursors are considered to be indirect pollutants.

In addition to the tailpipe emissions of hydrocarbons from incomplete combustion, they are also emitted to the atmosphere from loss of unburned fuel. This loss of unburned fuel can occur through (1) spillage during fueling; (2) diurnal evaporative losses (occurring as the fuel tank is heated during the day, followed by cooling at night, resulting in "breathing" of air and trapped hydrocarbons); (3) hot-soak evaporative losses (occurring when the engine is shut off and residual heat is transferred to the fuel system); and (4) running evaporative losses (Calvert *et al.*, 1993). The evaporative losses can be high at high ambient temperatures or for vehicles that run abnormally hot. It is possible to control these fuel losses by trapping them into carbon-containing canisters, followed by purging them into the engine for combustion. However, this method of control has not been perfected for field use as yet and these emission sources can become important, especially as the emission standards for hydrocarbons are lowered.

A number of non-criteria pollutants that are primary emissions deserve mention. These non-criteria pollutants include aldehydes, ketones and alcohols. These oxygenated hydrocarbons are likely to become more important as oxygenated fuels become more widely used.

In the case of aldehydes, particularly formaldehyde, their toxicity is likely to demand future control strategies. The state of California has already added an emission standard for formaldehyde beginning in 1994 (see Table 1). In addition, aldehydes are reactive species, both photochemically and with hydroxyl radical (OH), and can lead to the formation of secondary atmospheric oxidants (ozone, hydrogen peroxide, peroxyacyl nitrates, and organic peroxides and peracids).

2.2. Secondary Pollutants

As noted in the previous paragraphs, secondary pollutants are those that are formed in the atmosphere from the reactions of primary pollutants. The secondary pollutant that has received the most attention is ozone (National Academy of Sciences, 1991). Ozone is formed in the lower troposphere from the solar photolysis of nitrogen dioxide via reactions [1] and [2]:

$$NO_2 + hv \rightarrow NO + O(^3P) \quad [1]$$

$$O(^3P) + O_2 + M \rightarrow O_3 + M \quad [2]$$

where hv is a photon in the solar wavelength region of 280–400 nm, and M is a third body (e.g. nitrogen, oxygen or argon). Note that the decomposition of NO_2 is reversible, and the back reaction of ozone with nitric oxide is very fast (reaction [3]):

$$O_3 + NO \rightarrow NO_2 + O_2 \quad [3]$$

Therefore, without conversion of NO to NO_2 via some other route, atmospheric ozone formation would be limited by the NO back reaction.

The photolysis of ozone produces excited oxygen atoms [$O(^1D)$], and their subsequent reaction with water vapour produces the hydroxyl radical (OH) (reactions [4] and [5]):

$$O_3 + hv \rightarrow O_2 + O(^1D) \quad [4]$$

$$O(^1D) + H_2O \rightarrow 2OH \quad [5]$$

The hydroxyl radical is very reactive with most volatile hydrocarbons (RH), leading to the production of peroxy radicals (RO_2). These peroxy radicals can then react with NO to form NO_2 without loss of ozone (reactions [6] through [8]).

$$OH + RH \rightarrow R + H_2O \quad [6]$$

$$R + O_2 \rightarrow RO_2 \quad [7]$$

$$RO_2 + NO \rightarrow RO + NO_2 \quad [8]$$

Subsequent reactions of the alkoxy radicals (RO) lead to further formation of aldehydes (or ketones) and HO_2 radicals, thus allowing for the catalytic conversion of NO to NO_2 and the subsequent production of ozone (Finlayson-Pitts & Pitts, 1986):

$$RO + O_2 \rightarrow R'CHO(R'COR'') + HO_2 \quad [9]$$

$$HO_2 + NO \rightarrow NO_2 + OH \quad [10]$$

It can be seen from these reactions that the existence of oxides of nitrogen and hydrocarbons in the presence of sunlight can lead to the formation of the secondary pollutant, ozone. Therefore, in order to control the formation of ozone in the atmosphere, the emissions of hydrocarbons and NO_x are currently regulated in vehicle exhaust. However, other secondary pollutants of concern are also formed from these reactions. These secondary pollutants include other oxidants (e.g. hydrogen peroxide, which can react with SO_2 to form sulphuric acid aerosol), aldehydes, organic acids, nitric acid (formed by reaction of OH with NO_2) and peroxyacyl nitrates (PANs). These secondary pollutants are of concern because of their potential roles in acidic deposition, as plant toxins, and as health hazards in their own right.

Another important aspect of the hydrocarbon/NO_x system, which leads to the formation of ozone and other products of photochemical smog, is that the atmospheric reactivities of organic compounds vary dramatically. Table 3 gives some relative reaction rates for a number of organics with OH radical and ozone. The organics listed in Table 3

Table 3. Hydrocarbon relative reactivities for OH reactions (relative to propane) and ozone reactions (relative to propene) at room temperature (Gaffney & Levine, 1979; Finlayson-Pitts & Pitts, 1986; Atkinson, 1994; Teton *et al.*, 1995).

Compound	OH relative rate[a]	Ozone relative rate[a]
Methane	7×10^{-3}	—
Ethane	0.2	—
Propane	1	—
n-Butane	2.2	—
Ethene	7.2	0.16
Propene	21.7	1
1-Butene	25.8	1
trans-2-Butene	53.3	17.7
n-Hexane	4.7	—
Cyclohexane	6.2	—
n-Octane	7.4	—
Benzene	1.1	—
Toluene	5.2	—
m-Xylene	20.8	—
Isoprene	83.3	1.3
α-Pinene	44.2	7.6
Methanol	0.8	—
Ethanol	2.7	—
Methyl *tert*-butyl ether	2.6	—
Ethyl *tert*-butyl ether	7.3	—

[a]Propane = 1.

are those that are typically emitted from gasoline (cyclohexane, octane and aromatics), natural gas (methane) and LPG (propane, butane and butenes), as well as oxygenated fuels (methanol, ethanol and MTBE), along with some naturally occurring hydrocarbons (isoprene and alpha-pinene). Note that the only organics that are reactive with ozone at room temperature are the very reactive olefins and natural hydrocarbons. Liquefied petroleum gas can contain considerable concentrations of alkenes (propene and the butenes as shown in

Table 3), thus increasing the atmospheric reactivity of the hydrocarbons that are released from this alternative fuel (Blake & Rowland, 1995).

The formation of ozone in the atmosphere from the various organics is dependent upon these initial reaction rates, as well as the rates of subsequent reactions with NO and oxygen. These further lead to the formation of peroxy and alkoxy radicals. Estimations are made of the ozone formation potential for these hydrocarbon species in an urban environment by studying the organic compounds in smog-chamber systems (under varying concentrations of hydrocarbons and NO_x), followed by using chemical modelling to predict the amount of ozone formed. The ozone-forming potentials of different hydrocarbons have been calculated as reactivity adjustment factors by using the maximum incremental reactivity (MIR) scale developed by Carter (1989, 1990). These so-called ozone production factors are usually referred to in units of grams of ozone per gram of non-methane volatile organic compound (VOC) emitted into the atmosphere. Table 4 shows some proposed reactivity factors developed by the California Air Resources Board (CARB) for a number of low-reactivity fuels (California Air Resources Board, 1991). The European equivalent to the MIR is the photochemical ozone-creation potential (PCOP) which is given in moles of ozone

Table 4. Alternative fuel reactivities in grams of ozone per gram of non-methane volatile organic emitted estimated for organic-limited or nitrogen-oxide-limited conditions (California Air Resources Board, 1991; Gushee, 1992).

	Organic-limited	NO_x-limited
Fuel	Reactivity	Reactivity
Gasoline	3.4	1.3
Methanol	0.6	0.3
Ethanol	1.3	0.7
Methane	0.01	0.01
Ethane	0.3	0.2
Propane	0.5	0.3

produced per mole of VOC emitted scaled to ethylene (moles of ozone produced per mole of ethylene) (Simpson, 1995).

Note that these estimated reactivity factors for ozone formation are strongly dependent upon the uncertainties in the peroxy radical formation rates, as well as the uncertainties in the exhaust emission composition and the variability in secondary product yields (Yang *et al.*, 1996), particularly for species such as the PANs and other oxidants associated with ozone formation. Thus, as we improve our information regarding these uncertainties in emission rates and in the chemical mechanisms, these reactivity indices will require revision and should be considered as our "best guess" estimates of atmospheric reactivity.

Of course, ozone is not the only oxidant formed from these reactions; other oxidants such as hydrogen peroxide and organic oxidants, including the PANs, are also produced from the atmospheric oxidation of VOC emissions. The PANs are particularly interesting because they are known to be potent phytotoxins and eye irritants. They are produced directly from the reaction of peroxyacyl radicals with NO_2, both of which are in equilibrium with the PANs (Gaffney *et al.*, 1989).

$$RC=O-OO+NO_2 \leftrightarrow RC=O-OO-NO_2 \quad [11]$$

The most prevalent of the peroxyacyl nitrates is the methyl derivative, peroxyacetyl nitrate (PAN). This secondary pollutant is produced along with ozone in the process of urban photochemical smog formation.

Peroxyacetyl nitrate is coupled to aldehyde formation because the peroxyacetyl radical is formed primarily by the abstraction reaction with OH radical. For example, for the case of acetaldehyde:

$$CH_3CHO+OH \rightarrow CH_3CO+H_2O \quad [12]$$

$$CH_3CO+O_2 \rightarrow CH_3CO-O_2 \quad [13]$$

Subsequently, the CH_3CO-O_2 radical can either react with NO_2 to form PAN or react with NO in an urban atmosphere to convert it to NO_2 and form ozone via a reaction similar to reaction [8]. Under low NO conditions on regional scales, the CH_3CO-O_2 radical can also react with HO_2 to form peracetic acid (similar to the HO_2 radical self-reaction to form hydrogen peroxide and oxygen):

$$CH_3CO - O_2 + HO_2 \rightarrow CH_3CO_3H + O_2 \quad [14]$$

Peracetic acid is reasonably soluble in water and is a strong oxidant and phytotoxin (Gaffney *et al.*, 1989).

With the exception of formaldehyde, almost all of the higher aldehydes have the potential to form PANs. In addition, the aldehydes and ketones can be photolysed directly in the urban air, leading to increased levels of organic peroxy radicals (RO_2) and subsequently ozone and other oxidants, such as the peracids. Emissions of oxygenated hydrocarbons, particularly aldehydes and ketones, are not currently regulated, but it is anticipated that they will be in the future because of their photochemical activity, potential toxicity, and ability to form the PANs and other organic oxidants and air toxics.

2.3. Air Toxics

Air toxics are those compounds that are of concern at low levels primarily because of their potential carcinogenic properties. The air toxics which have been targeted by the CAAA of 1990 are benzene, 1,3-butadiene, formaldehyde, acetaldehyde and polycyclic organic matter (POM) (Health Effects Institute, 1996). Also of concern, when considering the total changes in carcinogenicity of vehicle emissions with the use of alternative fuels, are the primary emissions of MTBE or other oxygenates. Other species certainly will be added to this list as further studies identify other carcinogens and toxic substances in the exhaust. For example, formic acid is a likely candidate, along with other aromatics such as toluene and the xylenes.

Benzene dominates the total emissions (65% to 80%) of air toxics in the current fleet of gasoline-fueled vehicles (Gorse *et al.*, 1991). Benzene is also the only mobile source air toxic, thus far, classified as a known human carcinogen by the United States Environmental Protection Agency (USEPA) because of its ability to cause leukaemia in humans at relatively high exposure levels (US Environmental

Protection Agency, 1994). The USEPA has added that benzene may produce developmental effects at exposure levels as low as 1 ppm. Three air toxics, formaldehyde, acetaldehyde and 1,3-butadiene, are classified as probable human carcinogens based on limited evidence from epidemiological and/or animal studies. Methyl *tert*-butyl ether has been tentatively classified as a possible human carcinogen on the basis of inhalation studies (Health Effects Institute, 1996).

Other air toxics of concern are POM compounds, which are defined by the CAAA as a class of organic compounds having more than one benzene ring and a boiling point of 100°C or higher. This includes polycyclic aromatic hydrocarbons (PAHs), substituted PAHs (e.g. nitro-PAHs and alkyl-PAHs) and heterocyclic compounds (e.g. azo-arenes, thio-arenes and lactones). Polycyclic organic matter compounds with five or more membered rings are usually associated with particles such as carbonaceous soot. Those POM compounds having four or fewer rings are semivolatile (Health Effects Institute, 1993). Nitroarenes, nitrolactones and oxy-polycyclic aromatic compounds can be produced as primary pollutants associated with carbonaceous soot, principally diesel soot, or as secondary pollutants produced by heterogeneous reactions of gas-phase POM on aerosol surfaces in VOC/NO_x-polluted atmospheres (Pitts *et al.*, 1980; Pitts, 1993). Many POM compounds are mutagenic and some are carcinogenic. The best known of these is benzo(α)pyrene, a powerful human carcinogen (Phillips, 1983). It has been known for some time that PAHs are potential carcinogens and the nitroarenes show a much higher mutagenic activity than other PAHs. The combination of PAHs in both particulate and vapour phases appears to be responsible for a substantial portion of the mutagenic activity of the atmosphere (Pitts *et al.*, 1978). Thus, POM compounds may add to the total carcinogenicity of vehicle emissions to an extent which is currently undetermined.

2.4. Carbonaceous Soot and Secondary Fine Aerosols

Another important pollutant is carbonaceous soot which is a primary emission product in diesel exhaust and is known to be a potential carcinogen (National Academy of Science, 1972). There has been a substantial increase in the proportion of new diesel cars sold throughout Europe in the last decade, primarily because of the better fuel economy of diesel-powered vehicles (Tancell *et al.*, 1995). This increased use of diesels is of concern because of the human health risks associated with diesel exhaust emissions. The particulates associated with diesel exhaust are very small (< 1 μm) and therefore, can lodge deep in the lung and increase the risk of lung cancer. Along with their small size, these particulates have a very large surface area onto which other organic contaminants present in the diesel exhaust can adsorb. Therefore, associated with the soot particles are a number of potential air toxics, including PAHs and nitroarenes.

Carbonaceous soot is produced from the incomplete combustion of gasoline and diesel fuels as well as lubricant engine oils. This material is a primary pollutant with properties that vary significantly depending upon the engine operating conditions, the engine type and configuration, and the fuel composition (Wolff & Klimisch, 1982). The carbonaceous aerosol or soot can be separated operationally into organic and elemental carbon components by thermal separation techniques (e.g. Tanner *et al.*, 1982). Carbonaceous soot has also been determined as black carbon by the use of optical absorbance methods (Rosen *et al.*, 1980).

Diesel engines are a dominant source of automotive soot emissions. This soot is typically less than 1 μm in diameter, with a typical mean aerodynamic diameter of 0.3–0.6 μm. In this size range, the particles have long atmospheric residence times because the diffusional and depositional losses are at a minimum (Finlayson-Pitts & Pitts, 1986). Carbonaceous soots are in a size range that optically scatters light quite efficiently and they are also strong absorbers of incoming solar radiation. They, therefore, also impact visibility and the atmospheric radiative balance. Diesel emission standards vary around the world, but in general, the particulates are limited to a maximum total permitted weight generated during a defined test cycle (Table 2). However, because of their small size range, they will also be subject to the new

PM-2.5 standard regulating overall urban air quality in the United States.

Another important contribution to the atmospheric fine particulates is due to the formation of secondary organic aerosols from the atmospheric oxidation of organic compounds. Although the oxidation of small hydrocarbons in the atmosphere contributes to ozone formation, the oxidation of hydrocarbons containing seven or more carbon atoms contributes to secondary aerosol formation because the vapour pressure of the products is sufficiently low to allow them to partition into the particulate phase (Grosjean & Seinfeld, 1989; Grosjean, 1992). The organic-aerosol-forming potential of a hydrocarbon is dependent on its atmospheric reactivity (similar to the ozone-forming potential), and the volatility of the product. It has been suggested that the aromatic content of fuel is predominantly responsible for its organic-aerosol-forming potential (Odum *et al.*, 1997). This implies that the use of oxygenated fuels with smaller molecular weights may reduce the secondary organic-aerosol-forming potential.

3. Impacts of Alternative Fuel Usage on Urban Air Quality

As indicated in the previous sections, which outline the principal primary and secondary air quality issues, alternative fuel usage will affect our air quality problems in a number of ways. Principally, it is hoped that alternative fuels will lead to reduced emissions of primary pollutants (i.e. hydrocarbons, NO_x, CO and SO_2) and consequently, to reduced formation of secondary pollutants in the atmosphere. In addition, the hydrocarbons released from a "clean" alternative fuel through evaporative loss would be less reactive than those from gasoline or diesel, thus reducing the atmospheric effects of evaporative emissions, especially on urban scales. However, it is still unclear whether the use of the currently popular alternative fuels will lead to an overall improvement of urban air quality. Although reductions have been noted in the concentrations of some atmospheric pollutants, current studies often show conflicting results. Here, we attempt to review and summarise the results to date for a number of alternative fuels, particularly the oxygenated fuels mandated by the CAAA.

3.1. Oxygenated Fuels

Oxygenated fuels are hydrocarbon fuels that contain oxygen in their molecules. The oxygenated fuels currently being used include the alcohols (ethanol and methanol) and ethers (e.g. MTBE). These oxygenated fuels (or "oxyfuels") can be derived from the reprocessing of fossil fuels or from fermentation or pyrolysis of biomass materials. The production of methanol from coal or natural gas is an example of the fossil fuel conversions that have been proposed and researched. However, biomass-based fuels have the added advantages of overall reduction of CO_2 emissions and sustainable supply. Some of the renewable non-food biomass feedstocks that can be used are hardwood, corn and municipal solid waste (Lynd *et al.*, 1991).

The use of oxygenated fuels as an automotive power source is certainly not new. Indeed, ethanol was considered an ideal fuel for the very first internal combustion engines and was used as such until it was replaced by petroleum-based fuels because of cost and availability considerations (Kreucher, 1995). During the Depression, the Farm Chemurgic Movement aimed to solve the farm crisis by producing a 10% alcohol/gasoline blended fuel on a nationwide basis, thus creating an agricultural/chemical-based revival of the rural economy (Wright, 1993). Similar economic driving forces have surfaced again in the mid-1970s and today, with the added motivation of our current environmental concerns, have produced a renewed incentive towards the use of oxygenated fuels. The possibility of generating renewable biofuels that would reduce foreign oil dependence and not contribute further to increases in atmospheric CO_2 and other greenhouse gas levels has created specific interest in ethanol and (to a lesser extent) methanol as gasoline alternatives.

Alcohols, principally methanol and ethanol, along with their gasoline blends, have received considerable attention as alternative fuels.

Indeed, with the current push for ULEVs, numerous papers have been published examining the vehicular emissions of these fuels, both neat and as blends. The principal drawback from an air quality standpoint is the production of aldehydes during combustion. During cold-start situations, alcohols crack to produce aldehydes, principally formaldehyde in the case of methanol and acetaldehyde in the case of ethanol. It is the aldehyde emissions from alcohols that make their potential impacts on air quality different from the non-oxygenated alternative fuels. On the positive side, the use of alcohols and alcohol/petroleum blends in diesel engines has been shown to reduce emissions of the potentially carcinogenic carbonaceous soot particles (Gaffney *et al.*, 1980; Sapienza *et al.*, 1985; Wang *et al.*, 1997). However, in comparison with diesel soot, the particulate emissions from alcohol fuel combustion have not been well characterised with regard to any changes in chemical, physical or health-related properties.

The major motivation for the mandated oxyfuel programme in the United States has been to reduce atmospheric concentrations of CO, notably in Denver, Colorado and Albuquerque, New Mexico, where CO has been a problem. The most common oxygen-containing additives used in these areas are MTBE and ethanol. The ethanol/gasoline blend known as "gasohol" is typically a 10% blend of ethanol in gasoline that is used as a means of reducing CO levels in these non-attainment areas. Exhaust emission studies of various oxygenates added to gasoline have reported a reduction of CO emissions by at least 10% and up to more than 20% (Health Effects Institute, 1996). A recent study in the Caldecott tunnel in the San Francisco Bay area, measured a decrease in CO emission rates of 21 ± 7% with the use of MTBE in gasoline (Kirchstetter, 1996). Other studies have reported up to a 9% reduction in CO emissions with the use of 15% MTBE and a 13% reduction in CO emissions-accompanied by a 5% increase in NO_x, with the use of gasohol (Reuter *et al.*, 1992). However, results vary with the age and condition of the vehicle. There is also evidence that the beneficial effects of oxygenates may disappear at very low temperatures (+ 20 to − 20°F), where CO emissions from gasoline are much higher (Health Effects Institute, 1996).

This issue has generated some controversy. Field studies in Denver have indicated a significant downward trend in ambient CO concentrations during the oxygenated fuel periods averaging a 13% decrease in recent years (Anderson *et al.*, 1995, 1997). However, the magnitude of this reduction (5–15%) is much lower than the 20–30% predicted by EPA's mobile source emission model, and it is further accompanied by an increase in NO_x of 14%. Studies in Albuquerque also reported a decrease in the number of CO exceedances (times that atmospheric CO levels were above that required by the NAAQS) since the use of gasohol. However, this trend began before the oxyfuel programme was implemented and it is therefore difficult to assign credit to the use of ethanol fuels (Popp *et al.*, 1995). Other field studies conducted in areas participating in the winter-time oxyfuel programme, have shown a slightly greater decrease in the mean daily CO concentrations over time (20.5%) as compared to non-participating areas (10.3%) (Mannino & Etzel, 1996). During this period of study, the frequency of CO exceedances decreased from 66% to 28% in the areas using oxygenated fuels while the exceedances in non-participating areas decreased from 52% to 11% (National Research Council, 1996). There was a larger decrease of CO exceedances in conventional-fuel areas than in areas using oxygenated gasolines.

One complication in the analysis of field data has been the simultaneous implementation of more than one mitigation factor. The gradual replacement of the older precatalyst vehicles with newer catalyst-equipped automobiles has resulted in a substantial fraction of the declining CO levels observed in many of the cities under study (Zhang, 1996). It has been estimated that the "dirtiest" 10–20% of the cars may be responsible for 50–80% of the ambient CO (Keislar *et al.*, 1995). Emission results of the mandatory emission test programme in Colorado (Fig. 1) have shown that older vehicles (1982 models) produce 12 to 14 times the amount of CO as newer vehicles (1995 models) using either oxygenated or conventional fuels. The largest decrease in CO emissions with the use of oxyfuels also occurs in these older vehicles (Anderson *et al.*, 1997).

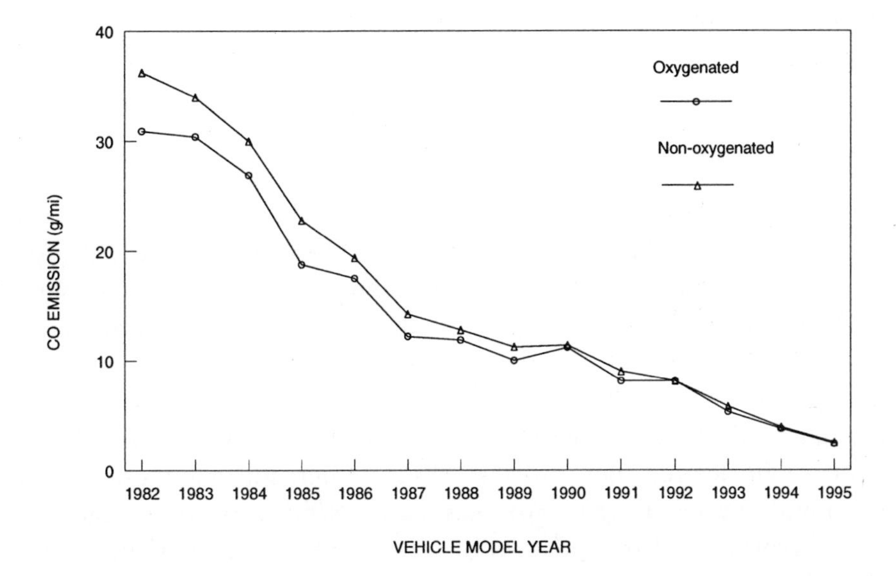

Fig. 1. Carbon monoxide emissions, as a function of vehicle model year, obtained from the required emission test programme in Colorado for 1995 (Anderson *et al.*, 1997). Data were obtained for the oxygenated period; January, February, November and December (Oxygenated) and the non-oxygenated period; April through September (Non-oxygenated) and were restricted to those vehicles which passed the test on the first attempt.

Changing meteorological conditions could also contribute to the observed CO reductions in some areas so that the analyses of long-term trends will be necessary to rule out short-term meteorological effects (Wolfe *et al.*, 1993; National Research Council, 1996). Studies conducted in Provo, Utah have attempted to remove the effects of varying local meteorology by using the ratio of excess (background corrected) CO to excess CO_2 to determine changes in CO emissions before and during the use of oxyfuels (Keislar *et al.*, 1997). In addition, nearby Salt Lake City, which does not use oxyfuels, was used as a control. A decrease in CO/CO_2 of $15 \pm 20\%$ was observed at traffic sites while a more statistically significant decrease of 11% to 33% was observed in parking garages. This indicates a CO benefit with the use of oxyfuels during cold-start conditions but not necessarily during normal driving conditions.

Methanol

Methanol has been strongly considered as an alternative fuel for some time. This is primarily because methanol can be produced fairly cheaply by catalytic reduction of CO with hydrogen. Methane or coal can serve as the feedstock for this inorganic synthetic procedure. Neat methanol as a fuel has some interesting properties that could lead to new engine and transmission designs (Gray & Alson, 1989). However, most of the methanol studies have focused on methanol blends for use in existing fleets or in flexible fuel vehicles (FFVs), which still make use of conventional engine designs. For reasons of safety and engine material (i.e. corrosion and degradation of plastic fuel lines), methanol/gasoline blends of up to 85% have been studied for use. The convention has been to refer to the methanol/gasoline blends as M85 for 85% methanol, M50 for 50% methanol, and so on up to neat gasoline (M0).

Comparison studies of gasoline-fueled conventional vehicles and M85-fueled variable fuel vehicles (VFVs) have been difficult to interpret because the exhaust emissions are not the same for both vehicle types when tested on gasoline alone. Studies conducted by the Auto/Oil Air Quality Improvement Research Program (1992a) showed total organic emissions were statistically equivalent for both fuels. However, because the exhaust emissions from an alcohol-fueled vehicle contain oxygenated hydrocarbons, the NMOG (expressed in grams per mile) are weighted by the presence of the heavier oxygen atoms present in the compounds when compared with NMOG in gasoline emissions. To compensate for this, regulations for methanol-fueled cars are expressed as organic mass in "hydrocarbon equivalents" (OMHCE) which is calculated by substituting $CH_{1.85}$ for the oxygenated carbon mass (Federal Register, 1989). When expressed in this manner, the total organic emissions were 37% lower with M85 than with M0. In the same study, CO was 31% lower with M85 than with M0, while NO_x was 23% higher.

As expected, exhaust methanol emissions were higher with both M0 and M85 and formaldehyde emissions were about 5 times higher with M85 when compared with M0. The emissions of benzene 1,3-butadiene and acetaldehyde were lower with M85 than M0 by 84%, 93% and 70%, respectively. The decreases corresponded roughly to the dilution effect of methanol in M85. The study concluded that it was difficult to determine whether the overall emissions data for M0 in gasoline-fueled vehicles or for M85 in FFV/VFV vehicles was better. In a related study (Auto/Air Quality Improvement Research Program, 1992b), it was determined that the transition to and use of M85 in FFV/VFV vehicles would be more costly relative to conventional gasoline, particularly in the short to medium term.

For a prototype FFV (Gabele, 1990), methanol blends were found to reduce the exhaust emissions of total hydrocarbons and OMHCE but have little effect on the toxic hydrocarbons (benzene and butadiene). However, the formaldehyde, methanol and methane content of the exhaust gas differed dramatically with fuel composition (Table 5). Little

Table 5. Exhaust emissions (in grams per mile) for a 1987 prototype Ford LTD Crown Victoria FFV (24 000 miles) using 85% methanol fuel (M85), as compared with gasoline alone (M0), for three standard driving cycles (Gabele & Knapp, 1993).

Compound	M0–HFET[a]	M0–NYC[b]	M0–UDDS[c]	M85–HFET[a]	M85–NYC[b]	M85–UDDS[c]
Hydrocarbons	0.27	0.89	0.62	0.04	0.19	0.16
OMHCE	0.27	0.89	0.62	0.07	0.44	0.34
Methanol	0	0	0	0.06	0.48	0.35
Formaldehyde	0.0049	0.0043	0.0090	0.0310	0.0807	0.0750
Acetaldehyde	0.0013	0.0023	0.0026	0.0013	0.0015	0.0010
NO_x	0.80	2.60	0.95	0.50	1.12	0.71
CO	0.61	1.90	1.62	0.27	1.87	2.10
Benzene	1.5	1.7	2.3	1.6	1.3	2.4
Butadiene	0	0	0.07	0	0	0
mpg[d]	26.9	8.9	18.1	16.6	5.5	11.0

[a]Highway Fuel Economy Test Cycle.
[b]New York City Cycle.
[c]Urban Dynamometer Driving Schedule (includes cold start).
[d]Miles per gallon.

effect was seen on the exhaust emission rates of CO and NO_x. When cold start was included in the driving cycle (UDDS of Table 5), CO emissions were greater for M85 than for M0. The measured exhaust emissions of this vehicle were most affected by catalyst replacement and cold starts. Replacement of the catalyst after 25 000 miles resulted in a reduction of formaldehyde emissions (including cold start) from 75.0 mg/mile to 23.2 mg/mile and a reduction of CO emissions from 2.1 g/mile to 0.79 g/mile. The emissions of formaldehyde from methanol-fueled vehicles, therefore, can be reduced by the use of properly functioning catalytic converters (Sodhi *et al.*, 1990); however, during cold starts, methanol blends will still emit formaldehyde.

Formaldehyde is a carcinogen and an eye irritant as well as being a photochemically active compound. As a primary pollutant, formaldehyde can photolyse directly (reactions [15] and [16]) or it can react with OH radicals (reaction [17]) (Calvert & Pitts, 1967).

$$H_2CO + h\nu \rightarrow H_2 + CO \qquad [15]$$

$$\rightarrow H + HCO \qquad [16]$$

$$H_2CO + OH \rightarrow H_2O + HCO \qquad [17]$$

Both pathways produce HCO, which reacts to form HO_2 and CO (reaction [18]).

$$HCO + O_2 \rightarrow HO_2 + CO \qquad [18]$$

In addition, photolysis (reactions [15] and [16]) generates a second HO_2 radical by reaction of H with O_2. The reaction of the HO_2 with NO in urban air can then lead to its conversion to NO_2 and the formation of ozone upon photolysis of NO_2.

Methanol itself has a lower atmospheric reactivity than gasoline (see Table 4). In comparison, methanol's combustion product, formaldehyde, reacts much more rapidly to form ozone (Atkutsu *et al.*, 1991). Indeed, the relative reactivity of formaldehyde with OH is 7.8 compared with propane (Atkinson, 1994), while methanol is 0.8 on that same scale. Thus, the ozone-forming potential of methanol-fueled emissions is very dependent upon the formaldehyde emissions, which increase with increasing methanol content (Gabele, 1990). Formaldehyde can also photolyse quite efficiently in the troposphere (Finlayson-Pitts & Pitts 1986), thus increasing its ozone-forming potential, particularly in an urban environment.

If the primary emissions of formaldehyde occur in the presence of aqueous aerosols, the highly soluble formaldehyde can be found to be present in relatively high concentrations in the form of the hydrated species, methylene glycol (Klippel & Warneck, 1980; Finlayson-Pitts & Pitts, 1986). In addition, further aqueous reactions with SO_2 emissions can lead to the formation of hydroxymethane sulphonate ions, while aqueous oxidation will form formic acid, both of which are highly toxic (Boyce & Hoffmann, 1984; Finlayson-Pitts & Pitts, 1986). These aqueous-phase reactions of formaldehyde, as well as the other aldehydes, and the resulting concentrations in inhalable aerosols are an aspect that must also be considered when evaluating the air quality impacts of oxygenated fuels, particularly methanol.

Ethanol

Of all the oxygenated fuels, ethanol has had the greatest worldwide application. In 1975, Brazil started a programme called Proalcool to convert their nation's fuel economy to one of sugar-based ethanol. This programme is by far the largest alternative fuel programme in the world and was initiated to eliminate Brazil's dependence on foreign oil imports. In 1993, about 4.3 million vehicles in Brazil (40% of its car population) were run on neat ethanol. Ethanol fuels have received considerable attention in studies of the economics of fuel production; some of which have found these fuels to be costly, particularly when corn is used as a feedstock (e.g. Pimentel, 1991; Grimsted & Moran, 1994). The Proalcool programme has had to be subsidised heavily by the Brazilian government to be cost competitive with gasoline (*Chemical and Engineering News*, 1993). Other economic analyses that include tax structures and local economic situations have led to a more positive perspective on ethanol (VanDyne *et al.*, 1996). The economics of ethanol fuel production may improve with the use of other sources, such as lignocellulose, as improvements in ethanol fermentation are developed (Lynd, 1990; Lynd *et al.*, 1991).

Ethanol, similar to methanol, can crack during the combustion process to form acetaldehyde in the exhaust emissions. Dynamometer studies on the use of gasohol in motor vehicles report an average decrease in total hydrocarbon emissions of 5%, a decrease in CO emissions of 13% with an increase in NO_x emissions of 5%. The same studies showed a decrease in the emissions of the air toxics, benzene and 1,3-butadiene, by 12% and 6%, respectively, while acetaldehyde emissions increased by 159% (Health Effects Institute, 1996). Although the atmospheric reactivity of ethanol is much lower than that of gasoline (Table 4), no significant change was reported in the overall atmospheric reactivity (MIR) of the exhaust emissions from gasohol when the higher reactivity of acetaldehyde is included. Acetaldehyde reacts with OH at a relative rate of 13.2 compared with propane (Atkinson, 1994). Ethanol is 2.7 on the same scale. The evaporative emissions of hydrocarbons also increased by the addition of ethanol to the fuel. Diurnal hydrocarbon emissions increased by 30% while hot soak hydrocarbon emissions increased by 50%. The atmospheric reactivity of these evaporative emissions represents an increase of 30% for diurnal and a decrease of 19% for hot soak when compared to gasoline alone.

Studies on the use of an 8.8% ethanol/gasoline blend in light-duty passenger vehicles found a general reduction in the emissions of total hydrocarbons, CO, benzene and 1,3-butadiene over unoxygenated gasoline fuel (Stump *et al.*, 1996). Three different vehicles were tested over varying operating temperatures, and overall results were variable. Table 6 gives the average differences in exhaust emissions observed for one vehicle in this study, a 1984 Buick Century, operated with and without a dual-bed three-way catalyst and fuel ratio controls. For CO, overall reductions in emissions are seen, which are larger at lower ambient temperatures. However, these reductions were obtained at the expense of increased aldehyde emissions in all cases, especially without

Table 6. Exhaust emission changes for an 8.8% ethanol/gasoline fuel blend, as compared with gasoline alone, for a 1984 Buick Century (52 015 miles) operated at three different ambient temperatures, with and without a dual-bed three-way catalyst and a "closed-loop" oxygen sensor for air-fuel ratio control (Stump *et al.*, 1996). Emissions are reported in grams per mile ethanol blend – grams per mile gasoline alone.

	With catalyst			Without catalyst		
Compound	90°F	75°F	40°F	90°F	75°F	40°F
Hydrocarbons	+ 0.1	+ 0.2	– 0.2	– 0.4	– 0.4	– 0.4
Methanol	+ 0.012	+ 0.010	+ 0.016	+ 0.070	+ 0.080	+ 0.070
Ethanol	+ 0.014	+ 0.016	+ 0.058	0.140	+ 0.240	+ 0.180
Formaldehyde	+ 0.006	+ 0.006	– 0.001	+ 0.003	+ 0.020	+ 0.007
Acetaldehyde	+ 0.008	+ 0.009	+ 0.009	+ 0.030	+ 0.030	+ 0.030
NO_x	+ 0.5	+ 0.6	0	+ 0.1	+ 0.3	+ 0.4
CO	– 1	0	– 12	– 1	– 5	– 9
Benzene	– 0.001	+ 0.001	– 0.017	+ 0.050	– 0.030	– 0.040
Butadiene	0	0	– 0.003	– 0.012	– 0.010	+ 0.002

the use of a catalyst. Similar results from a higher emitting vehicle showed larger decreases for total hydrocarbon (– 4.2 g/mile at 75°F) and CO (– 33 g/mile at 75°F) emissions with a correspondingly higher emission of NO_x (+ 2 g/mile at 75°F) and comparable acetaldehyde emissions (+ 25 mg/mile at 75°F). This study also determined that, similar to methanol, most of the emissions in catalyst-equipped cars occur during the cold start.

A number of field studies have examined the question of increased levels of aldehydes caused by ethanol/gasoline blend usage. In Brazil, when 20–30% ethanol/gasoline blends were used in the motor vehicle fleets with little or no use of catalytic converters, studies have found increased atmospheric levels of acetaldehyde and formaldehyde (Tanner *et al.*, 1988; Grosjean *et al.*, 1990). Indeed, these Brazilian field studies found that the acetaldehyde levels were higher than the formaldehyde levels, in contrast to the typical situation where formaldehyde is the predominant aldehyde found in urban air. Because even higher acetaldehyde levels were found in auto emissions and in a Rio de Janeiro tunnel, the explanation for this observation is that the ethanol/gasoline blend usage leads to elevated primary acetaldehyde emissions (Grosjean *et al.*, 1990). Field studies in Albuquerque, New Mexico have also looked at the question of whether atmospheric aldehyde levels are higher when 10% ethanol/gasoline blends are used (Popp *et al.*, 1995; Gaffney *et al.*, 1997). These studies also indicated higher atmospheric aldehyde levels, albeit at a lower level than was observed in Brazil, for 10% ethanol/gasoline blend usage in vehicles with catalytic converters. Elevated levels of acetaldehyde occurred more frequently in the winter months, when ethanol was used, than in the summer months. These elevated occurrences also corresponded more frequently to peak traffic hours when the formaldehyde:acetaldehyde ratio was 1.4, indicating an anthropogenic source of the acetaldehyde. Similar studies conducted in Denver, Colorado have concluded that oxygenated fuel usage has not had a major effect on ambient concentrations of acetaldehyde (Anderson *et al.*, 1995; 1997). However, the data showed a significant increase in reported acetaldehyde concentrations since 1994, when ethanol had obtained 80% of the oxyfuel market in the Denver area.

As indicated earlier, aldehydes are quite photochemically reactive and can lead to the formation of secondary atmospheric pollutants. Therefore, similar to the case for methanol, the emissions of acetaldehyde strongly affect the ozone-forming potential of the exhaust emissions. Although formaldehyde, as a primary pollutant of methanol combustion leads to increased production of ozone, hydrogen peroxide, formic acid and CO, depending upon the atmospheric levels of NO/NO_2 (Atkutsu *et al.*, 1991; McNair *et al.*, 1992), primary acetaldehyde from ethanol leads to the formation of these same products as well as PAN, acetic acid and peracetic acid. The formation of PAN has been used as an indicator compound for primary acetaldehyde emissions (Tanner *et al.*, 1988; Popp *et al.*, 1995; Gaffney *et al.*, 1997). Under typical urban conditions, the primary hydrocarbon emissions react in the atmosphere to form acetaldehyde as a secondary pollutant (Finlayson-Pitts & Pitts, 1986). Acetaldehyde can react subsequently by

OH abstraction to form the peroxyacetyl radical and then PAN by addition of NO_2 (see reactions [9]–[11] above). This process takes time to occur, so that the formation of PAN in an urban area is usually delayed when the reactivity of the primary hydrocarbon emissions is low to moderate. Because of this time delay and coupled meteorology, most downtown areas do not have high levels of PAN. If the acetaldehyde is emitted as a primary pollutant, reaction with OH radical and formation of PAN can proceed directly, leading to higher levels of the PANs in the main urban areas. This was observed in Rio de Janeiro (Tanner *et al.*, 1988) when 20–30% ethanol/gasoline blends were in use, consistent with the high atmospheric acetaldehyde levels that were observed. The elevation of urban PAN levels has also been seen in Albuquerque, New Mexico in winter time, when use of a 10% ethanol/gasoline blend is mandated, as compared with summer-time periods, with minimal ethanol fuel usage (Gaffney *et al.*, 1997). Canadian assessment of the 10% ethanol/gasoline blend estimates that ethanol usage will lead to 0.4–1.6% increase in ozone, 1–5% increase in formaldehyde, approximately 2.7% increase in acetaldehyde and 2.9–4.5% increase in PAN, with an approximate 15% reduction in CO (Singleton *et al.*, 1997). This estimate is in fairly good agreement with field data obtained in the United States (Gaffney *et al.*, 1997).

Thus, the use of ethanol/gasoline blends will likely lead to enhanced primary emission of aldehydes, particularly acetaldehyde, which in turn will lead to increases in urban concentrations of PAN. The increase in PAN may have impacts on regional scale ozone because PANs can act to transport NO_2 over long distances (Gaffney *et al.*, 1993). Because reaction [9] is reversible, the major pathway for the thermal decomposition of PAN leads to the formation of peroxyacetyl radical and NO_2. This decomposition depends upon the temperature and NO_2 concentrations and leads to the formation of ozone in the presence of NO. Under low-NO conditions, decomposition of PAN leads to the formation of hydrogen peroxides, and organic oxidants, such as peracetic acid and organic hydroperoxides over regional areas (Gaffney *et al.*, 1987; Gaffney & Marley, 1992).

Methyl tert-*butyl ether (MTBE) and other oxygenated fuels*

Methyl *tert*-butyl ether has been used as a fuel additive to increase oxygen content and reduce CO and reactive hydrocarbon emissions while maintaining fuel octane number. Methyl *tert*-butyl ether is manufactured from isobutene and methanol to make the ether (Maxwell & Naber, 1992). Higher analogues, such as ethyl *tert*-butyl ether (ETBE), made from ethanol, and *tert*-amyl methyl ether (TAME), made from isopentene, can also be used as additives to increase the oxygen content of fuel. However, ETBE and TAME have not received as much attention as MTBE because of the higher costs of their starting materials, ethanol and isopentene. The methanol and isobutene needed for MTBE production can be produced from natural gas via syngas catalytic techniques, which may lower production costs (Maxwell & Naber, 1992). Methanol and ethanol can be obtained from renewable biomass sources as well.

Methyl *tert*-butyl ether obtains its reduced reactivity with OH from the *tert*-butyl group (Teton *et al.*, 1995). Thus, MTBE reacts with OH at room temperature at a relative rate of 2.6 compared with propane (Table 3), and the higher analogue, ETBE, reacts at a relative rate of 7.3. The major oxidation products identified from MTBE and ETBE in air are formaldehyde, and the formates and acetates (Japar *et al.*, 1990; Smith *et al.*, 1991; Tuazon *et al.*, 1991; Wallington & Japar, 1991). Di-*tert*-butyl ether (DTBE) has also been examined as a low-reactivity compound (Langer *et al.*, 1995), with a reactivity similar to that of MTBE, and *tert*-butyl acetate has been identified as the major oxidation product in the presence of NO_x. The other anticipated reaction products are the corresponding formates. Both the acetates and the formates are expected to undergo hydrolysis in wet aerosols and precipitation to form *tert*-butyl alcohol and the corresponding acids. Reactions with nitrate radical (NO_3), important as a night-time oxidative pathway (Finlayson-Pitts & Pitts, 1986), have been examined for ETBE (Langer & Ljungstrom, 1995). This study found that the night-time loss of ETBE due to nitrate radical could be equivalent to the day-time loss by OH. Products from the reaction initiated by NO_3 included *tert*-butyl formate, *tert*-butyl acetate, formaldehyde and methyl nitrate. Atmospheric lifetimes for ETBE were estimated to be approximately 30 hours for a moderately polluted urban situation (Langer & Ljungstrom, 1995). Heterogeneous photooxidation reactions of MTBE and ethanol adsorbed onto fly ash particles have been demonstrated and proposed as a significant loss mechanism for these vapours in urban air, leading to the production of formaldehyde and acetone (Idriss & Seebauer, 1996; Idriss *et al.*, 1997).

Similar to other oxygenated fuels such as methanol and ethanol, MTBE and TAME can lead to increased formaldehyde emissions over non-oxygenated fuels (Kaiser *et al.*, 1991). This increase is expected because both additives contain a methoxy group. Similarly, the use of ETBE and diisopropyl ether (DIPE) as fuel additives can lead to increased emissions of acetaldehyde. In emission studies of MTBE using dynamometers, considerable variation of results has been noted with different fuels and different fleets (Health Effects Institute, 1996). Often, no statistically significant effects were detected on formaldehyde emissions from the addition of 15% MTBE to the fuel. Adding MTBE in the Auto/Oil AQUIRP fuel formulations increased formaldehyde emissions by 27% in the current fleet but did not affect formaldehyde emission levels of the older fleet (Gorse *et al.*, 1991). However, the addition of MTBE to California reformulated gasoline resulted in an increase of formaldehyde emissions of 21% in the current fleet, 16% in the older fleet and 11% in the Federal Tier I (existing technology) Fleet (Auto/Oil Air Quality Improvement Research Program, 1995). In addition, in studies on emissions in the California Caldecott tunnel, formaldehyde was increased by 13% ± 6% during the period when MTBE was in use as compared to periods when oxygenates were not added to the fuel (Kirchstetter *et al.*, 1996). In these same tunnel studies, an increase of isobutene of 87% was also reported during periods when MTBE was used. Isobutene is also a partial combustion product of MTBE which has not been considered by most studies. Similar tunnel studies in the California South Coast Air Basin have reported a formaldehyde:isobutene molar ratio of 1 in emissions where MTBE was in use (Gertler *et al.*, 1997; Zielinska *et al.*, 1997).

Under normal conditions, formaldehyde is produced in the secondary atmospheric reactions of gasoline emissions during the photochemical oxidation reactions driven by OH (Finlayson-Pitts & Pitts, 1986, 1993). Therefore, the levels of formaldehyde usually increase later in the day in an urban area, unless primary emissions are substantial. Field studies in Mexico City examined formaldehyde levels before and after the use of MTBE/gasoline blends (Humberto Bravo *et al.*, 1991). After the introduction of MTBE, the atmospheric levels of formaldehyde increased by approximately 3 ppb, and the time of peak concentrations changed from midday to the morning hours, indicating a primary source of formaldehyde. Ozone levels also increased in the morning and evening hours, with higher levels occurring more frequently after the introduction of MTBE. These field studies indicate that, similar to ethanol/gasoline blends, MTBE use will lead to primary emissions of the photochemically reactive formaldehyde, as well as isobutene, thus increasing the atmospheric reactivity of the emissions and diminishing the ability of the oxygenated fuel to reduce secondary ozone formation.

Results of emission studies of total hydrocarbons, CO and NO_x have also produced varying results with different fuels and fleets (Health Effects Institute, 1996). The addition of MTBE to reformulated gasoline resulted in a decrease in total hydrocarbon emissions of 20% and a decrease of CO emissions of 26% with an increase in NO_x emissions of 2% (Noorman *et al.*, 1993). Similar results were reported when DIPE or TAME was used as the oxygenate in the same fuel. Another study reported decreases of 7% for total hydrocarbon emissions, 9% for CO emissions, and no significant increase in NO_x emissions with 15% MTBE fuel (Reuter *et al.*, 1992). Studies on the use of a 9.5% MTBE/gasoline blend in light-duty passenger vehicles, similar to those reported for ethanol, found no general pattern of reduction of tailpipe emissions when compared with the unoxygenated fuel (Stump *et al.*, 1994). Three different vehicles were tested over varying operating temperatures. Table 7 gives the exhaust emission differences observed for the same light-duty vehicle as used in the ethanol emission studies

Table 7. Exhaust emission changes for a 9.5% MTBE/gasoline fuel blend, as compared with gasoline alone, for a 1984 Buick Century (52 015 miles) operated at three different ambient temperatures, with and without a dual-bed three-way catalyst and a "closed-loop" oxygen sensor for air-fuel ratio control (Stump *et al.*, 1996). Emissions are reported in grams per mile ethanol blend – grams per mile gasoline alone.

	With catalyst			Without catalyst		
Compound	90°F	75°F	40°F	90°F	75°F	40°F
Hydrocarbons	+0.1	+0.08	−0.09	−0.29	−0.34	−0.57
Methanol	+0.023	+0.014	+0.036	+0.104	+0.088	+0.047
Ethanol	+0.018	+0.003	+0.004	+0.006	+0.006	+0.005
MTBE	+0.014	+0.016	+0.038	+0.240	+0.223	+0.297
Formaldehyde	+0.008	+0.012	+0.001	+0.005	+0.011	+0.004
Acetaldehyde	+0.002	+0.003	0	−0.002	0	−0.003
NO_x	−0.09	+0.04	−0.39	−0.29	+0.06	+0.28
CO	−0.22	+0.58	−6.82	−1.64	−2.89	−2.53
Benzene	0	+0.001	−0.011	+0.018	−0.025	−0.034
Butadiene	−0.004	−0.001	−0.006	−0.010	−0.002	−0.005

presented above (Table 6). The CO emissions again appeared to be reduced more significantly at colder ambient temperatures; however, the reduction was not as large as when ethanol was the oxygenate. Formaldehyde emissions were greater with the fuel containing MTBE, for all vehicles tested.

3.2. Methane (Compressed Natural Gas) and Liquefied Petroleum Gas

Methane (natural gas), propane and butanes have all been proposed as clean alternatives to conventional liquid gasoline and diesel fuels. Natural gas is primarily composed of methane but can contain other light hydrocarbons (e.g. alkenes) in low concentrations. Natural gas is typically handled as a compressed gas and is usually referred to as compressed natural gas (CNG). Methane is an attractive fuel in many ways. It has a low atmospheric reactivity with OH radical (see Table 2), and therefore has a low atmospheric ozone-forming potential. Propane and butanes as liquefied petroleum gas (LPG) are also of reasonably low reactivity with regard to their atmospheric oxidation and low ozone-forming potential. Reductions in CO, reactive hydrocarbon emissions and nitrogen oxides are all feasible with the use of these fuels, particularly if engines are designed for the fuels and make use of three-way catalysts for emission control (Fowler *et al.*, 1991; Stodolsky & Santini, 1993; Tabata *et al.*, 1995; Chang & McCarty, 1996).

The atmospheric oxidation reactions of the light alkanes and methane are reasonably well understood (Finlayson-Pitts & Pitts, 1986, 1993). Abstraction reactions by OH radical are the sink for these hydrocarbons. Methane, propane and butanes have low reactivities with OH, leading to long atmospheric lifetimes and low ozone-forming potential on urban scales. However, the associated olefins, such as ethene, propene and the butenes (1- and 2-isomers) are quite reactive with OH, proceeding via an addition mechanism. Thus, the alkenes, present in LPG at low concentrations can be a considerable problem when leakage from LPG containers occurs. The usage of LPG in Mexico City and other foreign cities for heating and cooking uses has led to some concern regarding the potential for the formation of ozone, primarily because of the large concentration buildup and the olefin contents of these fuels (Blake & Rowland, 1995). In addition to emissions from tank leakage, the use of LPG as a transportation fuel would result in the release of uncombusted olefins in vehicles without catalysts or during cold start. This release would vary with vehicle condition and the amount of alkenes present in the fuel. Alkenes can also react with ozone as well as OH (see Table 3), leading to increases in secondary aldehyde production and the formation of PANs. Many of these problems can be resolved but may require reformulation of the LPG to reduce the alkene content if these fuels are to be used in vehicles worldwide.

Since CO emission rates are a function of the air/fuel ratio, vehicles which utilise gaseous fuels have a potential for lower CO emissions because they can operate on a stoichiometric air/fuel ratio during cold start, when CO emissions are at their highest. Studies comparing 11 vehicles tested on CNG, LPG, methanol, ethanol, reformulated

Table 8. Exhaust emissions (in grams per mile) for gasoline (a 1993 Lumina and a 1993 Taurus), CNG (two Dodge vans and a Chevrolet Sierra), and LPG (a Ford pickup and a Chevrolet pickup) (Gabele, 1995).

Compound	Gasoline		CNG			LPG	
	Lumina[a]	Taurus[b]	Dodge van[c]	Dodge van[d]	Sierra[e]	Ford pickup[f]	Chev. pickup[g]
Hydrocarbons	0.63	0.11	0.02	0.59	0.06	1.14	0.53
Formaldehyde	0.003	0.002	0.002	0.008	0.004	0.004	0.008
Acetaldehyde	0.002	0.001	0.001	0.001	0.001	0.001	0.002
NO_x	0.63	0.39	0.19	1.48	0.39	1.62	0.48
CO	17.02	2.29	0.55	1.46	5.95	1.52	
Benzene	0.047	0.008	0	0.002	0	0	8.57
Butadiene	0.002	0.001	0	0	0	0	0.001
							0

[a]23 870 miles; [b]9196 miles; [c]742 miles; [d]9833 miles; [e]4982 miles; [f]60 000 miles; [g]10 280 miles.

gasoline (using MTBE) and CAAA baseline gasoline fuels showed the LPG and CNG vehicles to have, in general, the lowest emission values (Gabele, 1995). Overall exhaust emission rates were a function of vehicle type and condition for conventional, CNG and LPG vehicles. However, in general, both CO and toxic emissions (benzene, 1,3-butadiene and aldehydes) were less for vehicles using CNG and LPG (Table 8).

Total reactive hydrocarbon emissions were also, in general, reduced with CNG and LPG than with gasoline or oxyfuel blends. The atmospheric reactivity of the organic emissions listed in Table 8 was decreased by 26–83% for the CNG vehicles and by 51–78% for the LPG vehicles when compared to the low-emitting conventional vehicle (Taurus). Although all CNG and LPG fuels were > 99% alkane, the olefin content of the hydrocarbon emissions from these fuels varied from 1.6–22% for the CNG vehicles and from 2–14% for the LPG trucks. This variation in olefin content of the exhaust corresponded roughly to their overall atmospheric reactivities. The major reactive component of the emissions is therefore the alkenes. This study concluded that the overall exhaust emission rates were still more a function of vehicle condition than of fuel type. Vehicles with properly operating engine and emission control systems consistently produced lower exhaust emission rates than vehicles with poorly tuned systems, regardless of the type of fuel used.

Other issues arise when considering the change from liquid fuels to CNG or LPG (Eckhoff, 1994). Methane is relatively non-toxic and is odourless and tasteless. Thus, odorants are typically added to alert bystanders to fuel leakage. Currently, the odorants used are mercaptans, which are reasonably reactive and lead to the formation of SO_2 and sulphate aerosols. Gas explosions from CNG and LPG, as well as pool fires in the case of LPG spillage, are also of concern. However, the single biggest drawback for methane lies in its very low reactivity. Its atmospheric lifetime has been estimated at approximately 10 years (Senum & Gaffney, 1985). Methane is a very strong greenhouse gas with many natural sources but fossil releases of methane are of concern when climate change is considered. The use of fossil-based CNG and LPG will also increase CO_2 emissions, albeit by lesser amounts than with the use of gasoline or diesel fuels. In addition, methane is the major source of stratospheric water vapour, and its increase in the troposphere would likely lead to an increase in stratospheric clouds and the associated catalytic destruction of ozone.

3.4. Fuel Cells

Fuel cells convert gaseous fuels (hydrogen, natural gas, gasified coal) directly into electricity by an electrochemical process. The key components are an anode, to which fuel is supplied; a cathode, to which the oxidant is supplied; and an electrolyte, which permits the flow of ions between the anode and cathode. Fuel cells operate like batteries but since the fuel and oxidant are not integral parts of the fuel cell, they do not need to be recharged and will continually produce power as long as fuel and oxidant is supplied. The chemical reaction within the fuel cell is exactly the same as when the fuel is burned, but because the fuel and oxidant are spacially separated, the flow of electrons can be intercepted and used in an external circuit. The chemical conversion process occurs at a much lower temperature than in an internal combustion engine, so there is no NO_x formation, and since there are no lubricating oils, there are also no hydrocarbon or CO emissions. In short, fuel cells can be used to operate zero-emitting vehicles (ZEV).

The most efficient fuel for fuel cells is hydrogen. It can be supplied directly from refueling stations, or indirectly from on-board generation systems. Due to infrastructure compatibility, safety considerations, and optimum driving range of the vehicles, on-board generation from gasoline or methanol is more widely favoured. Recent advances in this area have been quite promising and may lead to long-term changes in gasoline and methanol usage (Kartha & Grimes, 1994; Kordesch & Simader, 1995; Hohlein *et al.*, 1996; Klaiber, 1996). Increases in fuel economy of approximately 50% with very low CO, NO_x, and VOC emissions appear to be feasible with this approach. If gasoline is used as the fuel source, the system will still emit carbon dioxide. However, if methanol from biomass is used, minimal CO_2 increases are anticipated. The methanol reaction does produce CO, which has to be removed catalytically and will require CO-tolerant anodes (Schmidt *et al.*, 1994). Other environmental issues, such as water quality and fugitive emissions of the fuels, will of course, still be issues for gasoline- or methanol-powered fuel cell systems.

Fuel-cell-powered electric vehicles are particularly attractive as a "clean" alternative to gasoline and diesel engines because of their inherent efficiency, zero or near-zero emissions, fuel flexibility and quiet operations (Lloyd *et al.*, 1994). They can compete with conventional vehicles in all key aspects including vehicle driving range and refueling and have been recognised as part of the solution to the major air pollution problem in Southern California. However, the key to development of these technologies will be the development of the infrastructure and the commercialisation and replacement of the current fleets (Serfass *et al.*, 1994; Appleby, 1996; Chalk *et al.*, 1996). Even with the current environmental pressures, the establishment of alternative vehicles in the market will not occur overnight and will require government support considering the magnitude of investment required.

After all, internal combustion engines have the benefit of 100 years of evolution and infrastructure development. However, this approach appears to be the most promising long-term solution to the air quality problems currently facing us.

3.5. Biodiesel Fuels

Vegetable oils have been considered as fuel for diesel engines since the earliest days of the compression-ignition engine. However, long-term use of neat vegetable oils leads to severe engine problems because of their high viscosity, low volatility and tendency for polymerisation. The fatty acids produced from vegetable oils and animal fats have shown more promise as alternative diesel fuels as they reduce the problems associated with the use of the neat oils (e.g. Muniyappa *et al.*, 1996; Peterson, 1986). These so-called biodiesel fuels are typically the monoesters of rapeseed, soybean, safflower, peanut, sunflower, coconut, cottonseed or other vegetable oils (Peterson, 1986; Peterson *et al.*, 1991). These esters contain approximately 10% oxygen by weight and therefore may encourage more complete combustion in diesel engines than petroleum-based fuels. In addition, the monoesters have high cetane numbers and do not contain sulphur or aromatics. One limitation to the use of vegetable oil esters is their tendency to crystallise at low temperatures, which causes the diesel-blended fuels to separate under winter-time temperatures. The use of branched-chain esters, such as isopropyl esters, has been proposed as one solution to this problem (Chang *et al.*, 1996).

A number of studies have examined the emissions of these biodiesel fuels, as well as esterified oil/diesel fuel blends. The emission studies performed up to 1994 have been reviewed previously by Peterson and Reece (1996). More recent results have been reported and are reviewed briefly here (Isigigur *et al.*, 1994; Ali *et al.*, 1995a; 1995b; Chang *et al.*, 1996; Schumacher *et al.*, 1996). Chang and coworkers (1996) reported observations of substantial reductions in CO, hydrocarbon, soot and particulate emissions, while oxides of nitrogen were observed to increase when using a 50/50 blend of soybean oil methyl and isopropyl esters with No. 2 diesel fuel. The use of 50% methyl ester lowered CO emissions by 25.3%, lowered total hydrocarbon emissions by 7.4%, and increased NO_x emissions by 4.0%. The isopropyl ester/diesel blend resulted in a larger decrease in hydrocarbon emissions (29%) and a larger increase in NO_x emissions (12.1%). Ali, Eskridge and Hanna (1995a) reported little, if any, effects on emissions for biodiesel fuels when using a Cummins N14-410 engine, and they found exhaust emissions effects similar to those observed by Chang *et al.* (1996) when other types of diesel engines were studied (Ali *et al.*, 1995b). Results from Schumacher and colleagues (1996) also found reductions in particulate matter, total hydrocarbons and CO with the use of soybean/ diesel blend. In addition, the reductions in CO and total hydrocarbons were linear with the addition of the biodiesel, indicating more complete combustion of the biodiesel fuel. The increase in NO_x emissions was also linear with the amount of biodiesel used and a 20/80 biodiesel/ diesel blend was determined to be optimum, representing a compromise between the increased NO_x and the reduction of other emissions. An inverse correlation has been reported between the emissions of NO_x and particulates with the use of both diesel and biodiesel (Barenescu,1994; Holmberg & Peeples, 1994; Peterson & Reece, 1996). If the engine is optimised for the reduction of NO_x emissions, the total particulates increase correspondingly. Similarly, for every unit reduction in particulates achieved, a unit increase in NO_x will result. However, the severity of this trade-off is greater for petrodiesel than for biodiesel.

Most studies have focused on the reduction of emissions of the criteria pollutants (i.e. CO, NO_x, hydrocarbons and particulate matter); while the non-criteria emissions such as aldehydes, aromatic hydrocarbons and PAHs, were not considered. However, elevated levels of aldehydes have been reported from the use of cottonseed oils, sunflower seed oils, and their methyl esters as compared with No. 2 diesel oil (Geyer *et al.*, 1984). Reduction in the formation of aldehydes from the methyl-esterified rapeseed oil, as compared with the raw oil, has been reported (Mittelbach *et al.*, 1985). It is clear from these studies that, similar to the oxygenated alternative fuels, the potential emissions of aldehydes and PAHs, along with other non-criteria pollutants, need to be investigated in order to assess the overall air quality impacts of the use of biofuels in diesel engines.

4. Regional and Global Issues of Alternative Fuel Usage

In the push for clean alternative fuels to reduce primary emissions and the formation of secondary pollutants, the focus has been on urban air quality. However, it is possible that the use of some alternative fuels may trade a decrease in urban levels of the criteria pollutants for their overall increase on a regional scale. For example, the increase in primary emissions of aldehydes and the subsequent increase in urban levels of PAN may lead to higher ozone levels down-wind of the urban areas as the NO_2 is transported over a larger area (Gaffney *et al.*, 1997). In addition, the primary emission of less reactive hydrocarbons from oxygenated fuels will react on regional scales, leading to secondary aldehyde formation and an increase in regional-scale oxidant and aerosol production, particularly with increased tropospheric ultraviolet-B radiation arising from stratospheric ozone depletion (Gaffney & Marley, 1994).

Besides the noted effects on air quality, many of the alternative fuels may lead to improvements or increased problems with regard to global climate change. Carbon dioxide and methane, both components of vehicle exhaust, are the two predominant greenhouse gasses. Vehicles fueled by CNG will emit methane. However, methane emissions from all other types of vehicles are usually very low and can be neglected. The potential global climate impact of the combustion of alternative fuels (with the exception of CNG) is proportional to their relative CO_2 emissions, as presented in Table 9. The total global warming effects of CNG use must also include methane emissions. This can be accomplished by converting the methane emissions into CO_2 equivalents using their relative global warming potentials. Methane has approximately a 10 times higher global warming potential than CO_2 (Lashof & Ahuja, 1990). However, its atmospheric lifetime is 25 times shorter. If the relative atmospheric lifetimes are not considered, CNG vehicles

Table 9. Energy content and global warming impacts of fuels (Chang *et al.*, 1991).

Fuel	Energy content	Fuel economy	Energy economy	CO_2 production
	(BTU/gallon)	(miles/gallon)	(BTU/mile)	(grams/mile)
Gasoline	115 000	34	3353	315
Diesel	102 000–156 000	39–48	3257–2614	315–252
RFG[a]	114 000	34	3353	315
Methanol[b]	64 000	22	2923	272
Ethanol	82 000	28	2923	243
CNG (methane)	26 000	8	3353	301
LPG (propane)	88 000	31	2829	229

[a]Reformulated gasoline containing 15% MTBE.
[b]Natural gas used as the feedstock.

would have approximately the same global warming impact as gasoline vehicles (Chang *et al.*, 1991).

The total CO_2 released from the use of any transportation fuel is the sum of the CO_2 produced in direct combustion and that released in the process used to manufacture and transport the fuel. Those fuels that use less fossil carbon than the petroleum fuel they are replacing will lead to improvements in the carbon dioxide budget. The use of methanol derived from natural gas has a small overall CO_2 benefit while methanol derived from coal produces higher CO_2 emissions. The use of biomass-derived fuels from renewable sources are likely to result in a reduction of CO_2 emissions, provided that significant fossil fuel is not used in the production process. In addition, the energy content of the fuels must also be considered, along with the fuel economy. The oxygenated fuels have a lower energy content than gasoline (Table 9) and therefore require more fuel to travel the same distance. The important values to consider when determining the overall greenhouse impact of any fuel is the amount of CO_2 produced per mile, which will also vary with the fuel economy of the vehicle. All of these factors need to be evaluated, along with the air quality and economic factors, in selecting the alternative fuels of choice.

Other climate impacts can arise from the primary emissions or secondary pollutants formed from alternative fuels. Primary emissions of the oxygenates can act as greenhouse species in both the vapour and the aerosol phases (Marley *et al.*, 1993). Once dissolved in aerosols, these species can act as strong infrared absorbers, leading to local heating (Gaffney & Marley, 1992). The aerosols formed from these primary emissions can also contribute to cooling due to light scattering. The chemical interactions of peroxy radicals formed from the aldehyde emissions of oxygenated fuel usage will likely lead to increased acidic aerosol production on urban scales.

5. Other Environmental Concerns Regarding Alternative Fuels

As can be seen from the previous discussions, the principal considerations of the use of alternative fuels with regard to air quality is the atmospheric reactivities of the fuels and emission products as they contribute to urban ozone formation and atmospheric CO levels. However, other concerns arise from the toxicity of the fuel itself and the exhaust emissions, particularly in regard to potential water pollution problems and atmospheric exposures via vapours and aerosols. The potential health effects of oxygenated fuels have been reviewed in detail recently and will only be summarised briefly here (Costantini, 1993; Health Effects Institute, 1996; National Research Council, 1996). Essentially, introduction of oxygenated fuels will likely reduce CO and benzene emissions while increasing exposures to the primary aldehydes and the oxygenated fuels themselves, as well as the secondary atmospheric products such as peroxyacetyl nitrate. A comprehensive risk assessment of the use of alternative fuels would require a comparison of the risks of exposure to environmental contaminants of conventional gasoline and those of the alternative fuel.

The major pathway of exposure to the oxygenates is assumed to be inhalation. The low reactivities of MTBE and methanol in the atmosphere lead to long atmospheric lifetimes and an increased probability of inhalation exposure. However, limited data are available for the atmospheric concentrations of the oxygenates in urban areas, so that general exposures are difficult to determine. Data gathered for service station attendants and consumers exposed to MTBE during refueling indicate that median 1- to 2-minute exposures ranged from 0.3 to 6 ppm with a maximum of 10 ppm (Health Effects Institute, 1996). Limited measurements of ethanol during refueling indicated levels less than 1 ppm. Atmospheric levels of MTBE in Milwaukee, Wisconsin were determined in one study to be 0.13 ppb and ETBE levels were 0.04 ppb while service station perimeters generally had MTBE levels of 100 ppb (Allen & Grande, 1995).

Methanol and other oxygenates, used as fuel additives, are quite water-soluble and are known to be toxic if ingested at moderate to high concentrations, leading to blindness and ultimately, death. Because of their low atmospheric reactivity and relatively long lifetimes, and because they can partition into water, fuel oxygenates are expected to occur in precipitation in direct proportion to their atmospheric concentrations. This leads to dry and wet depositions as major atmospheric sinks, resulting in surface and groundwater contamination possibilities (Pankow *et al.*, 1997). Therefore, fuel oxygenates in the atmosphere provide a non-point, low-concentration source to surface and groundwaters. In addition, point-source releases can occur from leaking underground storage tanks, pipelines and refueling facilities. There is also some evidence that the presence of alcohols, such as ethanol and methanol, in gasoline fuels will lead to an enhanced water solubility of other less soluble components such as benzene (Piel, 1989). Thus, the use of the water-soluble oxyfuels may lead to potentially much more serious groundwater problems than the less water-soluble petroleum-based fuels (Gaffney & Marley, 1990).

Available data suggest that MTBE is sometimes present in precipitation, stormwater runoff, groundwater and drinking water (National Research Council, 1996). Methyl *tert*-butyl ether has been detected in 27% of urban wells sampled by the US Geological Survey (1995) in areas using oxyfuels. The highest concentration (23 mg/l) was determined in Denver, Colorado. Because of the widespread presence of MTBE in the samples collected, it is concluded that a major method of groundwater contamination was from wet deposition. Methyl *tert*-butyl ether concentrations of 200 000 μg/l have been observed in shallow groundwaters down-gradient from underground storage tanks (Davidson, 1995). Ethanol and methanol are more soluble in water than MTBE and may occur in even higher concentrations in similar situations. However, MTBE is more difficult to biodegrade and is more likely to persist in the groundwater and travel further from the source. This suggests that oxygenate contamination of drinking water represents a possible route of exposure. Yet the majority of states do not have any current requirements for monitoring fuel oxygenates in water.

Methyl *tert*-butyl ether has received particular attention recently because of its increased usage both as an oxyfuel and as an octane enhancer. Primary focus on MTBE has been on health effects of the MTBE/gasoline blends, particularly after residents of Fairbanks, Alaska, reported the occurrence of acute health effects, including headache, nausea, dizziness and breathing difficulties. Blood levels of MTBE were determined to be higher in workers during the oxygenated fuel programme (1.8 μg/l) than after the programme was suspended (0.24 μg/l) (Moolenaar *et al.*, 1994). Potential health effects of MTBE/gasoline blends include headache, nausea, sensory irritation and acute reversible neurotoxic effects. Experimental evidence indicates that MTBE is carcinogenic in rats (Health Effects Institute, 1996). In addition, the primary metabolites of MTBE — formaldehyde and *tert*-butyl alcohol — are also carcinogenic in animals.

Some toxicological studies have examined the potential effects on rats of methanol-fueled engine exhaust emissions with and without catalytic converters (Maejima *et al.*, 1992; 1993; 1994). These studies indicated that the predominant effects (e.g. nasal lesions and blood chemistry effects) were mainly due to formaldehyde exposure. Since the oxygenated fuels and their associated primary aldehyde emissions are quite soluble in water, the concentration of these species in wet aerosols, fogs and clouds will be increased by the use of oxyfuels (Gaffney *et al.*, 1984; Gaffney & Marley, 1992). When examining the toxicological effects of inhalation of these pollutants, the aqueous properties need to be considered too. Thus, highly water-soluble species, such as formaldehyde, will likely occur in both vapour and wet aerosol phases, and their solubilities will tend to lead to nasal and upper respiratory effects in humans, as has been observed (Maejima *et al.*, 1992).

Although PAN is known to be a potent plant toxin and was identified as a potential criteria pollutant along with other oxidants in the 1960s and 1970s (Gaffney *et al.*, 1989), little is known regarding its human health effects besides it being a strong lachrymator (eye irritant). It has been proposed as a possible carcinogen and has shown positive mutagenicity in Ames tests (Gaffney *et al.*, 1989). Further toxicological studies are needed for these compounds, as well as aldehydes, peracids and other organic oxidants that are likely to be increased with the use of oxygenated fuel/gasoline blends.

6. Conclusions and Suggestions for Future Work on Alternative Fuels

It has been clear for some time that if we were to achieve ultra-low levels of emissions from automotive sources, conventional petroleum fuels used in spark-ignition or diesel engines will have difficulties meeting the proposed standards. The search for alternative fuel strategies will continue into the future, and the most reliable, economical and environmentally safe fuels will win out in the marketplace. With regard to the alternative fuels and their effects on urban air quality and regional-scale tropospheric chemistry, the following summary can be made.

In general, oxygenated fuels (methanol, ethanol, MTBE, etc.) and their gasoline blends will lead to increases in cold-start aldehyde emissions. Increases in primary aldehyde emissions will increase the photochemical reactivity of the exhaust gas emissions and will lead to the formation of ozone and PANs (except for methanol) in all cases. The use of oxygenated fuels and their gasoline blends may lead to some reductions in CO and hydrocarbon emissions, with some increases in NO_x emissions. The use of alcohols and biofuels in diesel engines may have some advantages in reducing carbonaceous soot emissions. However, a more systematic and complete characterisation will be required for the particulates and their toxicity, particularly for biodiesel fuels (esterified plant oils), before an adequate assessment of the use of oxygenated fuels for soot reduction can be made.

Most of the emissions of CO, aldehydes and exhaust hydrocarbons occur during the cold start of the engine when the catalyst is not sufficiently heated to operating temperatures. Research aimed at improvement of catalyst performance and preheated catalytic system designs will be important in reducing emissions from both gasoline and alternative fuels in these types of engines.

While combustion of fossil-derived fuels leads to increased levels of carbon dioxide, the combustion of biomass-derived fuels will likely lead to net carbon dioxide emission reductions. From a global environmental perspective, the biomass-derived fuels that act to recycle carbon dioxide are attractive as a means of reducing combustion-related CO_2 emissions. However, a complete assessment of the bio-derived fuels needs to include the fossil fuels used in their production to determine net carbon dioxide reductions.

Compressed natural gas is certainly a cleaner fuel than gasoline. Its biggest drawbacks are in its handling and distributing as well as associated safety issues. Methane will be of concern as a greenhouse gas with a fairly long lifetime, and so the main use of natural gas is likely to be in stationary power plants where the fugitive emissions can be better controlled. Liquefied petroleum gas is not as clean as methane but is easier to handle, although one of the major issues with both of these is the NO_x emissions. Indeed, all of the internal combustion systems that make use of a spark-ignition or diesel configuration will lead to emissions of NO_x.

With all considered, the use of a fuel-cell-powered electric vehicle with an on-board hydrogen generation system would seem to be the longer-term "best" answer to the air quality problems facing most major urban centres in the world. The approach would make use of liquid fuels such as methanol or gasoline, which would use the current distribution systems. The fuel cell systems would yield zero or near-zero levels of hydrocarbon, aldehyde, and, most importantly, NO_x emissions, and reduced levels of CO emissions when the proper catalytic controls are in place. In addition, the increased projected fuel economy would lead to lower carbon dioxide emissions, particularly if biomass-derived fuels are used to generate hydrogen. Little or no particulate emissions are anticipated from this approach. As pointed out, the development of the market supply will take some time. As the demand and need for cleaner and more economical motor vehicle power trains increases worldwide into the 21st century, it is anticipated that other innovative engineering solutions will also arise. These will likely incorporate similar approaches or alternatives to combustion. With improvements in superconductive materials and power transmission, it is likely that electric vehicles that are battery-powered may also fill a substantial niche in conjunction with fuel cell vehicles. The former will make use of the cleaner electric power from hydroelectric and nuclear power plants, resulting in zero atmospheric emissions.

Acknowledgements

The authors wish to acknowledge the current support of the US Department of Energy's Atmospheric Chemistry Program under contract W-31-109-ENG-38 and the past support of the National Renewable Energy Laboratory as part of the Alternative Fuels Utilization Program of the U.S. Department of Energy. This submitted manuscript has been authored by a contractor of the US Government under contract No. W-31-109-ENG-38. Accordingly, the US Government retains a non-exclusive, royalty-free license to publish or reproduce the published form of this contribution, or allow others to do so, for US Government purposes.

References

Ali Y. *et al.* (1995a) *Bioresource Technol.* **53**, 243–254.
Ali Y. *et al.* (1995b) *Bioresource Technol.* **52**, 185–195.
Allen M. & Grande D. (1995) *Reformulated Gasoline Air Monitoring Study*, Publication Number AM-175-95, State of Wisconsin, Madison WI.

Anderson L.G. *et al.* (1995) In *Alternative Fuels and the Environment,* Chapter 5, ed. Sterrett F.S. pp. 75–102, Lewis Publishers, Boca Raton, FL.

Anderson L.G. *et al.* (1997) *Paper 97-RP139.05, Air & Waste Management Association 90th Annual Meeting.* p. 11, Toronto, Canada.

Appleby A.J. (1996) *J. Power Sources,* **69**, 153–176.

Atkinson R. (1994) *J. Phys. Chem. Ref. Data,* Monograph No. 2, 131–134.

Atkutsu Y. *et al.* (1991) *Atmos. Environ.* **25A**, 1383–1389.

Auto/Oil Air Quality Improvement Research Program (1992) *Gasoline Reformulation and Vehicle Technology Effects on Exhaust Emissions.* pp. 1–18, Technical Bulletin Number 17, Coordinating Research Council, Atlanta, Georgia.

Auto/Oil Air Quality Improvement Research Program (1992) *Emissions and Air Quality Modeling Results from Methanol/Gasoline Blends in Prototype Flexible/ Variable Fuel Vehicles.* p. 24, Technical Bulletin Number 7, Coordinating Research Council, Atlanta, Georgia.

Barenescu R. (1994) In *Commercialization of Biodiesel: Establishment of Engine Warranties,* National Center for Advanced Transportation, University of Idaho, Moscow, Idaho.

Blake D.R. & Rowland F.S. (1995) *Science,* **269**, 953–956.

Boyce S.D. & Hoffmann M.R. (1984) *J. Phys. Chem.* **88**, 4740–4764.

California Air Resources Board (1991) *Proposed Reactivity Adjustment Factors for Transitional Low-Emission Vehicles,* Sept. 27, Technical Support Document California Air Resources Board, Sacramento, CA.

Calvert J.G. & Pitts J.N., Jr. (1967) *Photochemistry,* J. Wiley & Sons, New York.

Calvert J.G. *et al.* (1993) *Science,* **261**, 37–45.

Carter W.P.L. & Atkinson R. (1989) *Environ. Sci. Technol.* **23**, 864–880.

Carter W.P.L. (1990) *Development of Ozone Reactivity Scales for Volatile Organic Compounds,* April, Statewide Air Pollution Research Center, Riverside, CA.

Chalk S.G. *et al.* (1996) *J. Power Sources,* **61**, 7–13.

Chang D.Y.Z. *et al.* (1996) *JAOCS,* **73**, 1549–1555.

Chang T.Y. *et al.* (1991) *Environ. Sci. Technol.* **25**, 1190–1197.

Chang, Y.-F. & McCarty J.G. (1996) *Catalysis Today,* **30**, 163–170.

Chemical and Engineering News, (1993) Oct. 18, 13–15.

Costantini M.G. (1993) *Environ. Health Perspectives Suppl.* **101**(Suppl. 6), 151–160.

Davidson J.M. (1995) *Proceedings of the Petroleum Hydrocarbons and Organic Chemicals in Groundwater: Prevention, Detection, and Remediation Conference.* pp. 285–301, November 29–December 1, Houston, Texas.

Eckhoff R.F. (1994) *Marine Pollution Bulletin,* **29**, 304–306.

Elliot D.C. *et al.* (1990) *Biomass,* **22**, 251–269.

Federal Register, *Standards for Emissions from Methanol-Fueled Motor Vehicles and Motor Vehicle Engines*, Federal Register 40CFR Part 86, Final Rule, 54 (68), April 11, 1089.
Finlayson-Pitts B.J. & Pitts J.N., Jr. (1986) *Atmospheric Chemistry*, J. Wiley & Sons, New York.
Finlayson-Pitts B.J. & Pitts J.N., Jr. (1993) *Chemistry & Industry*, **October**, 796–800.
Fowler T. *et al.* (1991) *Fuel*, **70**, 499–502.
Gabele P.A. (1990) *J. Air Waste Manage. Assoc.* **40**, 296–304.
Gabele P.A. (1995) *J. Air Waste Manage. Assoc.* **45**, 770–777.
Gabele P.A. & Knapp K.T. (1993) *J. Air Waste Manage. Assoc.* **43**, 736–744.
Gaffney J.S. & Levine S.Z. (1979) *Int. J. Chem. Kinetics*, **IX**, 1197–1209.
Gaffney J.S. & Marley N.A. (1990) *Atmos. Environ.* **24A**, 3105–3107.
Gaffney J.S. & Marley N.A (1992) In *Applications and Appraisals* Volume 3: The Summer Volume, (coordinators) Schwartz S.E. & Slinn W.G.N. pp. 1735–1743, Hemisphere Publishing Corporation, Washington, D.C.
Gaffney J.S. & Marley N.A. (1994) *29th Intersociety Energy Conversion Engineering Conference: Technical Papers.* pp. 1134–1138, Part 3, Technical Paper AIAA-94-4151-CP.
Gaffney J.S. & Senum G.I. (1984) In *Gas-Liquid Chemistry of Natural Waters*, Vol. 1, Brookhaven National Laboratory Report BNL 51757, UC-11 (Environ. Contr. Technol. Earth Sci. — TIC-4500), ed. Newman L. pp. 5–1 to 5–7.
Gaffney J.S. *et al.* (1980) *Combustion Sci. and Technol.* **24**, 89–92.
Gaffney J.S. *et al.* (1987) *Environ. Sci. Technol.* **21**, 519–524.
Gaffney J.S. *et al.* (1989) In *Handbook of Environmental Chemistry*, 4/Part B, ed. Hutzinger O. pp. 1–38, Springer-Verlag, Berlin, Germany.
Gaffney J.S. *et al.* (1997) *Environ. Sci. Technol.* **31**, 3053–3061.
Gaffney J.S. *et al.* (1993) *Environ. Sci. Technol.* **27**, 1905–1910.
Gertler A.W. *et al.* (1997) *Paper 2090, The Fifth Chemical Congress of North America*, November, Cancun, Quintana Roo, Mexico.
Geyer S.M. *et al.* (1984) *Trans. ASAE*, **27**, 375–381.
Gorse R.A. *et al.* (1991) *SAE Technical Paper Number 912324*, Society of Automotive Engineers, Warrendale, PA.
Gray C.L., Jr. & Alson J.A. (1989) *Sci. Am.* **261**, 108–114.
Grimsted B. & Moran M. (1994) *Illahee (J. Northwest Environ.)*, **10**, 205–215.
Grosjean D. & Seinfeld J.H. (1989) *Atmos. Environ.* **23**, 1733–1747.
Grosjean D. (1992) *Atmos. Environ.* **26A**, 953–963.
Grosjean D. *et al.* (1990) *Atmos. Environ.* **24B**, 101–106.
Gushee D.E. (1992) *Chemtech*, **22**, **July**, 406–411, and **August**, 470–476.
Health Effects Institute (1993) *HEI Communications* Number 2, April, Health Effects Institute, Cambridge, MA.

Health Effects Institute (1996a) *The Potential Health Effects of Oxygenates Added to Gasoline: A Review of the Current Literature,* April, Health Effects Institute, Cambridge, MA.

Health Effects Institute (1996b) *The Economics of Methanol,* April, Health Effects Institute, Cambridge, MA.

Hohlein B. *et al.* (1996) *J. Power Sources,* **61**, 143–147.

Holmberg W.C. & Peeples J.E. (1994) *Biodiesel, A Technology, Performance and Regulatory Overview,* National Soydiesel Development Board, Jefferson City, MO.

Humberto Bravo A. *et al.* (1991) *Atmos. Environ.* **25B**, 285–288.

Idriss H. & Seebauer E.G. (1996) *J. Vac. Sci. Technol.* **14A**, 1627–1632.

Idriss H. *et al.* (1997) *Catalysis Today,* **33**, 215–225.

Isigigur A. *et al.* (1994) *Applied Biochem. Biotechnol.* **45/46**, 95–101.

Japar S.M. *et al.* (1990) *Int. J. Chem. Kinetics,* **22**, 1257–1269.

Kaiser E.W. *et al.* (1991) *J. Air Waste Manage. Assoc.* **41**, 195–197.

Kartha S. & Grimes P. (1994) *Physics Today,* **November**, 54–61.

Keislar R.E. *et al.* (1995) *Effect of Oxygenated Fuels on Ambient Carbon Monoxide Concentrations in Provo, Utah,* Final Report, Desert Research Institute, Reno, Nevada.

Keislar R.E. *et al.* (1997) *Int. J. of Vehicle Design,* **18**, in press.

Kirchstetter T.W. *et al.* (1996) *Environ. Sci. Technol.* **30**, 661–670.

Klaiber T. (1996) *J. Power Sources,* **61**, 61–69.

Klippel W. & Warneck P. (1980) *Atmos. Environ.* **14**, 809–818.

Kordesch K.V. & Simader G.R. (1995) *Chem. Rev.* **95**, 191–207.

Kreucher W.M. (1995) *Chemistry & Industry,* **7 August**, 601–604.

Langer S. *et al.* (1995), *Int. J. Chem. Kinetics,* **28**, 299–306.

Lashof D.A. & Ahuja D.R. (1990) *Nature,* **344**, 529–531.

Lloyd A.C. *et al.* (1994) *J. Power Sci.* **49**, 209–223.

Lönner G. & Törnqvist A. (1990) *Biomass,* **22**, 187–194.

Lynd L.R. (1990) *Appl. Biochem. Biotechnol.,* **24/25**, 695–719.

Lynd L.R. *et al.* (1991) *Science,* **251**, 1318–1323.

Maejima K. *et al.* (1992) *J. Toxicol. Environ. Health,* **37**, 293–312.

Maejima K. *et al.* (1993) *J. Toxicol. Environ. Health,* **39**, 323–340.

Maejima K. *et al.* (1994) *J. Toxicol. Environ. Health,* **41**, 315–327.

Mannino D.M. & Etzel R.A. (1996) *Air & Waste Manage. Assoc.* **46**, 20–24.

Marley N.A. *et al.* (1993) *Environ. Sci. Technol.* **27**, 2864–2869.

Maxwell I.E. & Naber J.E. (1992) *Catal. Lett.* **12**, 105–116.

McNair L. *et al.* (1992) *J. Air Waste Management Assoc.* **42**, 174–178.

Mitchell C.P. *et al.* (1995) *Biomass and Bioenergy,* **9**, 205–226.

Mittlebach M. *et al.* (1985) *Energy in Agriculture,* **4**, 207–215.

Moolenaar R.L. *et al.* (1994) *Arch. of Environ. Health,* **49**, 402–409.
Muniyappa P.R. *et al.* (1996) *Bioresource Technol.* **56**, 19–24.
National Academy of Science (1991) *Rethinking the Ozone Problem in Urban and Regional Air Pollution,* National Academy Press, Washington, D.C.
National Academy of Sciences (1972) *Particulate Polycyclic Organic Matter. Committee on Biological Effects of Atmospheric Pollutants,* National Academy Press, Washington, D.C.
National Renewable Energy Laboratory (1995) *DOE/NREL Research and Development 1995 Annual Report.* p. 106, NREL Report MP-425-8301, Golden, Colorado.
National Research Council (1996) *Committee on Toxicological and Performance Aspects of Oxygenated Motor Vehicle Fuels.* p. 106, National Academy Press, Washington, D.C.
Noorman M.T. (1993) *SAE Technical Series Paper Number 932668,* Society for Automotive Engineers, Warrendale, Pennsylvania.
Odum J.R. *et al.* (1997) *Science,* **276**, 96–99.
Pankow J.F. *et al.* (1997) *Environ. Sci. Technol.* **31**, 2821–2828.
Peterson C.L. (1986) *Trans. ASAE,* **29**, 1413–1422.
Peterson C.L. & Reece D. (1996) *Trans. ASAE,* **39**, 805–816.
Peterson C.L. *et al.* (1991) *Appl. Engineer. Agriculture,* **7**, 711–716.
Phillips D.H. (1983) *Nature,* **303**, 468–472.
Piel W.J. (1989) *Fuel Oxygenate Effects on Aromatic Solubility in Water, American Chemical Society National Meeting,* September 10–15, Miami Beach, FL.
Pimentel D. (1991) *J. Agricultural Environ. Ethics,* 1–13.
Pitts J.N., Jr. (1993) *Research Chem. Intermediates,* **19**, 251–298.
Pitts J.N., Jr. *et al.* (1978) *Science,* **202**, 515–519.
Pitts J.N., Jr., *et al.* (1980) *Science,* **210**, 1347–1349.
Popp C.J. *et al.* (1995) In *Alternative Fuels and the Environment,* ed. Sterrett F.S. pp. 61–74, Chapter 4, Lewis Publishers, Boca Raton, FL.
Reuter R.M. *et al. SAE Technical Paper Number 9203326,* Society of Automotive Engineers, Warrendale, PA, 1192.
Rosen H. *et al.* (1980) *Science,* **208**, 741–744.
Sapienza R.S. *et al.* (1985) In *Chemistry of Engine Combustion Deposits,* ed. Ebert L.B. pp. 263–272, Plenum Press, New York.
Schmidt V.M. *et al.* (1994) *J. Power Sources,* **49**, 299–313.
Schumacher L.G. *et al.* (1996) *Bioresource Technol.* **57**, 31–36.
Senum G.I. & Gaffney J.S. (1985) In *Geophysical Monograph 32,* eds. Sundquist E.T. & Broecker W.S. pp. 61–69, American Geophysical Union, Washington, D.C.
Serfasss J.A. *et al.* (1994) *J. Power Sources,* **49**, 193–208.

Simpson D. (1995) *J. Atmos. Chem.* **20**, 163–177.

Singleton D.L. *et al.* (1997) *National Research Council Report PET-1384-96S.* p. 65, July, Ottawa, Canada.

Smith D.F. *et al.* (1991) *Int. J. Chem. Kinetics,* **23**, 907–924.

Smith D.F. *et al.* (1992) *Int. J. Chem. Kinetics,* **24**, 199–215.

Sodhi D. *et al.* (1990) *J. Air Waste Manage. Assoc.* **40**, 352–356.

Stodolsky F. & Santini D.J. (1993) *Chemtech,* **23 October**, 54–59.

Stump F.D. *et al.* (1994) *J. Air Waste Manage. Assoc.* **44**, 781–786.

Stump F.D. *et al.* (1996) *J. Air Waste Manage. Assoc.* **46**, 1149–1161.

Tabata T. *et al.* (1995) *Applied Catalysis B: Environmental,* **7**, 19–32.

Tanner R.L. *et al.* (1988) *Environ. Sci. Technol.* **22**, 1026–1034.

Tanner R.L. *et al.* (1982) *Anal. Chem.* **54**, 1627–1630.

Tancell P.J. *et al.* (1995) *Environ. Sci. Technol.* **29**, 2871–2876.

Ter Haar G.L. *et al.* (1972) *J. Air Pollut. Control Assoc.* **22**, 39.

Teton S. *et al.* (1995) *Int. J. Chem. Kinetics,* **28**, 291–297.

Tuazon E.C. *et al.* (1991) *Int. J. Chem. Kinetics,* **23**, 1003–1015.

Turhollow A. & Kanhouwa S. (1993) *Appl. Biochem. Biotechnol.* **39/40**, 61–70.

U.S. Environmental Protection Agency (1994) *EPA 600/R-94/217,* Office of Research and Development, Washington, D.C.

U.S. Geological Survey (1995) *Fact Sheet Number 114-95, U.S. Geological Survey,* Denver, CO.

VanDyne D.L. *et al.* (1996) In *Proceed. Third Liquid Fuel Conference, Nashville, TN,* pp. 311–318, Sept, Amer. Soc. Agric. Engineer., St. Joseph, MI.

Wallington T.J. & Japar S.M. (1991) *Environ. Sci. Technol.* **25**, 410–415.

Wang W.G. *et al.* (1997) *Environ. Sci. Technol.* **31**, 3132–3137.

Wolff G.T. & Klimisch (eds) (1982) *Particulate Carbon: Atmospheric Life Cycle,* Plenum Press, New York.

Wolfe P. *et al.* (1993) *Paper 93-TP-41B.03, p. 14, Air Waste Management Association 86th Annual Meeting,* June, Denver CO.

Wright D.E. (1993) *Agricultural History,* **67**, 36–66.

Yang Y-J. *et al.* (1996) *Environ. Sci. Technol.* **30**, 1392–1397.

Zelenka P. *et al.* (1996) *Appl. Catalysis B: Environmental,* **10**, 3–28.

Zhang Y. *et al.* (1996) *Environ. Sci. Technol.* **30**, 1445–1450.

Zielinska, B. *et al.* (1997) *Paper 1434, The Fifth Chemical Congress of North America,* November, Cancun, Quintana Roo, Mexico.

CHAPTER 7

MECHANISM OF TOXICITY OF GASEOUS AIR POLLUTANTS

D.G. Housley & R.J. Richards

1. Introduction

There are many gaseous pollutants present in the atmosphere but the biological mechanisms of reaction of ozone (O_3), nitrogen dioxide (NO_2) and sulphur dioxide (SO_2) have received the greatest amount of attention. Studies on the effects of O_3 are far more extensive than for NO_2 and SO_2. Sulphur dioxide and NO_2 are both primary pollutants whereas O_3 is a secondary pollutant that is formed as a result of photolytic reactions.

Studies on air pollutant toxicity have mainly concentrated on the lower respiratory tract, although some studies of the upper respiratory tract (URT) and conducting airways have also been performed. Since the URT is exposed to the highest lumenal concentration, more studies are still required, especially in humans where nasal cavity structure differs widely from that of other animals. Different individuals show great variation in their responses to pollutants but the reasons for this variability and sensitivity are still unknown. However, since the toxicity of O_3, NO_2 and possibly SO_2 is believed to be mediated through their reactions with the epithelial lining fluids (ELF) (Postlethwait *et al.*, 1995), the composition of the ELF may dictate differences in individual response.

2. Chemical Properties and Sites of Reaction of Pollutant Gases

Sulphur dioxide is a colourless, soluble acidic gas that rapidly hydrates and subsequently dissociates to (bi)sulphate and (bi)sulphite. The hydration process is very rapid, with the final equilibria between species depending on pH, ionic strength of the solution and temperature (Manson, 1980).

Both O_3 and NO_2 are reactive oxidant gases (Redox potential = 2.07 V and 1.07 V, respectively) with limited solubility. Ozone can react with a wide range of biomolecules including proteins, lipids and nucleic acids because of its strong oxidising nature. Nitrogen dioxide being a weaker oxidant than ozone, reacts in a similar, but more limited way. Nitrogen dioxide is readily ionised to form the nitrite (NO_2^-) ion. It also has the capacity to react with water to form acidic solutions (Greenwood & Earnshaw, 1995). The reactions that occur within the respiratory tract are influenced strongly by these differing chemical properties.

Ozone and NO_2 primarily induce lesions at the junction of the terminal bronchioles and alveolar ducts (the central acinus) (Mustafa, 1990). The cellular responses at this site involve cellular necrosis, exfoliation, degranulation of secretory cells, influx of inflammatory cells followed by epithelial hyperplasia and hypertrophy (Wright *et al.*, 1990). Sulphur dioxide, being a water-soluble gas has a pattern of deposition that is mainly confined to the URT and conducting airways where it causes epithelial degeneration, hyperplastic and hypertrophic changes. It can also cause severe bronchoconstriction via reflex activity (Dept of Health, 1992).

The precise mechanisms underlying the focal nature of lesion development are not fully understood but may include (i) localised overriding of the lining fluid protective mechanisms; (ii) differential sensitivity of separate cell types; and/or (iii) the dosimetry of each gas. A large number of studies have concentrated on dosimetry of inhaled gases and aerosols. A great volume of dosimetry experiments (Hatch *et al.*, 1994; Ben-Jebria *et al.*, 1995) and mathematical modelling data (Miller & Overton, 1989) have been accumulated for O_3, with far less available for NO_2 and SO_2.

3. Dosimetry of Pollutant Gases

Due to its high water solubility and subsequent formation of bisulphite (which prevents gaseous lumen SO_2 and dissolved SO_2 from reaching equilibrium), SO_2 is absorbed predominantly (> 90%) in the nose and upper airways (Dahl, 1990). However, it has been speculated that the degree of uptake of SO_2 may be substantially reduced at low concentrations (Strandberg, 1964). The removal (scrubbing) of SO_2, as with other gases within the nasal cavity, will reduce the concentration of gas that reaches the lower respiratory tract (Bedi & Horvath, 1989). This absorption of gases within the nasal cavity may not mean that responses are confined to the URT, as other mechanisms, including neural reflexes may give rise to changes in lung function (Raphael *et al.*, 1988). The potential for the URT to protect the lung is substantially reduced by any factor (e.g. exercise) that causes a change from nasal to oral breathing (Ben-Jebria *et al.*, 1995).

Several models have been used to estimate local dosimetry of O_3; they indicate that predicting tissue "dose" is not simple. It has been suggested that gases which react quickly with mucus will cause lesions mainly in the nasal cavity, whereas those with low to moderate reactivity will show effects in the lower respiratory tract (Dahl, 1990). Although fewer models exist for NO_2 dosimetry, its oxidative capacity which is similar to that of O_3 means that several of the factors affecting O_3 uptake should also influence NO_2 dosimetry. Although the dosimetry of the two gases is similar, the extent to which they can compromise the ELF or cellular components may differ for each gas (Miller *et al.*, 1985).

The amount of O_3 and/or NO_2 acting at a given level of the respiratory tract is dependent upon the lumen concentration of each gas (Miller *et al.*, 1978). Thus, the nasal cavity receives the highest concentration due to it being the portal of entry. The proportion of pollutant gas that is subsequently absorbed may depend on several factors (Aharonson *et al.*, 1974; Miller & Overton, 1989; Miller *et al.*, 1993), which include:

(i) *Respiratory tract morphology.* This is implicated because it is thought that cells such as the type I pneumocytes which have a large surface area and low levels of antioxidants may be more sensitive to O_3 (Dept of Health, 1991). In a study by Hotchkiss *et al.* (1989), rats were exposed to 0, 0.12, 0.8 or 1.5 ppm O_3 for six hours. Injury was noted on the maxilloturbinates which are covered by a non-secretory transitional epithelium, whilst the lateral wall which is directly opposite and covered by mucus-secreting ciliated respiratory epithelium was unaffected. Given the very close proximity of these two sites, it is likely that they received the same dose of O_3, so differences in response are likely to be due to differences in sensitivity of the individual cell types.

(ii) *Physical and chemical properties of the tissue.* The uptake of both O_3 (Pryor, 1992) and NO_2 (Postlethwait *et al.*, 1995) is thought to be mediated by reactions occurring within the ELF. During mathematical modelling procedures, increases in tissue reactivity do not have as great an effect on O_3 uptake as increases in ELF reactivity (Miller *et al.*, 1985). The extent to which the composition of parenchymal tissue influences ELF composition and hence O_3 uptake in ELF is currently unknown.

(iii) *Route of breathing.* Ben-Jebria *et al.* (1995) showed that nasal breathing is approximately 60% more efficient at removing O_3 from the inhaled air than oral breathing. Approximately 40% of NO_2 is also deposited within the nasal cavity of dogs and rabbits

(WHO, 1987). In addition, the concentration of SO_2 reaching the lung may also be influenced by changes in the route of breathing (Bedi & Horvath, 1989). Oral breathing leads to greater penetration of SO_2 to the lung.

(iv) *Depth and rate of breathing.* The efficiency of O_3 uptake in the nasal cavity (Gerrity & Weister, 1987; Kimbell *et al.*, 1993) and in other regions (Arahonson *et al.*, 1974; Overton *et al.*, 1987; Hu *et al.*, 1994) decreases at higher flow rates, possibly due to their effects on boundary layer diffusion (Ultman, 1995). Increasing tidal volume decreases the concentration of O_3 absorbed in the conducting airway and mucus but increases alveolar absorption (Miller *et al.*, 1978). Ozone uptake is also inversely related to breathing frequency (Gerrity *et al.*, 1988), although this is not likely to be as important as tidal volume (Miller *et al.*, 1993). This means exercise may be more, or equally important, as species in affecting dose (Hatch *et al.*, 1994). For example, O_3 deposition in the URT is approximately 50% under normal breathing conditions but this falls to 10% during moderate exercise (Hu *et al.*, 1994). Differences between species have also been noted. Total respiratory tract uptake of O_3 in humans is 96%, whereas in rats it is 44% (Gerrity & Weister, 1987). Similar observations have also been made for NO_2 (Maynard & Waller, 1993), with uptake in isolated lung being proportional to tidal volume (Postlethwait *et al.* 1992). A greater concentration of NO_2 is noted in lung parenchyma than that found in the conducting airways following exercise (WHO, 1987).

(v) *Physical and chemical properties of pollutant gas.* As mentioned above, physical and chemical properties such as solubility influence the dosimetry and subsequent reactions.

(vi) *Airflow patterns in each region, including convective and diffusional processes.* Computer simulation of airflow-driven uptake of soluble gases in the rat nasal cavity has shown that airflow patterns may play an important role in the dosimetry of soluble gases (Kimbell *et al.*, 1993).

(vii) *Airway geometry.* Differences in airway geometry, including diffusion path length, affect uptake (Overton *et al.*, 1987; Kimbell *et al.*, 1993), as does airway volume (Hu *et al.*, 1994). Upper respiratory tract shape also affects airflow (Miller *et al.*, 1993), although the effects of airway geometry changes that occur during disease of the URT have yet to be investigated (Bascom *et al.*, 1990).

(viii) *Mucus flow patterns.* Mucus flow patterns are probably too slow to affect O_3 uptake (Miller *et al.*, 1993).

(ix) *Blood flow.* Blood flow is often assumed not to influence O_3 absorption (Miller *et al.*, 1993) but this may not be the case if changes in perfusion rates cause alterations in ELF biochemistry. However, in isolated lungs, perfusion rate was found not to influence uptake of NO_2 (Postlethwait & Bidani, 1990).

(x) *Physical and chemical properties of the ELF.* Differences between tissue dose and net dose, especially in the URT, suggest that extracellular reactions are important (Miller & Overton, 1989). Also, when mathematical constants predicting O_3 uptake are calculated using regional dose alone, large errors occur because ELF reactions need to be considered (Hu *et al.*, 1994). It has also been found that when ELF reaction rates are increased during mathematical modelling procedures, the net dose increases but the tissue dose decreases (Miller *et al.*, 1985). Other investigators have also predicted the importance of airway fluids (Kimbell, 1995). Factors that may influence absorption into the lining fluids include molecular diffusion and convection due to ciliary beating, solubility, chemical reactions and biochemical composition of the ELF. The amount of information available about the composition, turnover and replacement of the ELF is limited and requires further investigation. Both ELF composition and thickness have the potential to determine how readily O_3 could reach the underlying epithelium (Miller *et al.*, 1993). Hence, the efficiency of O_3 removal during chronic exposures may depend upon how readily the ELF can be replaced (Hu *et al.*, 1992). The importance of the ELF has also been stressed since calculations based on reactivity versus diffusivity indicate that O_3 is unlikely to reach the epithelium without reacting with ELF components (Pryor, 1992). Both O_3 and NO_2 are absorbed by a process of reaction-mediated absorption (reactive absorption), i.e. the higher the concentration of reactive substrates present, the more the concentration of pollutant gas will be absorbed. Ozone and NO_2 are eliminated during these reactions and it is the longer lasting secondary (2°) and tertiary (3°) reaction products that are believed to mediate the toxic effects and give rise to epithelial cell changes. This has been termed the "cascade" mechanism (Pryor *et al.*, 1995).

4. Mechanisms of Toxicity

The mechanisms of O_3 toxicity are due to its strong oxidising nature and its capacity to generate free radicals (Mustafa, 1990; Pryor & Church, 1991). These two properties mean that a wide range of biomolecules will react with O_3 and thus become susceptible to damage. Such susceptible molecules include unsaturated fatty acids which can react by direct oxidation or by free radical mechanisms. The free radical reactions can cause oxidation of biomolecules and can also give rise to the generation of non-radical toxins such as aldehydes (Mustafa, 1990). Ozone is not a radical itself but it has at least two mechanisms of reaction, one for unsaturated compounds and one for electron donors (Pryor, 1994). Also, the reactant compound may vary somewhat because O_3 can participate in multiple pathways of reaction under different conditions (Pryor, 1993). Although all biomolecules are potential targets of oxidative damage, the processes of protein oxidation and lipid peroxidation are most commonly assumed to mediate O_3 toxicity (Kelly *et al.*, 1995).

With respect to lipid peroxidation, the initial reaction is probably a non-radical addition reaction that gives rise to a 1,2,3-trioxalane. Trioxalane decomposes via a transition state to yield an aldehyde and carbonyl oxide. Recombination of these two products gives rise to the 1,2,4-trioxalane or Criegee ozonide (Fig. 1).

However, as mentioned previously, inhaled O_3 is believed to react within the ELF. Unsaturated fatty acids that react in this more aqueous environment are proposed not to lead to the Criegee ozonide but to give rise to hydrogen peroxide and aldehydes (Pryor & Church, 1991) (Fig. 1). Pryor, 1994 estimated that only 10% of O_3 reacts in the lung to form the Criegee ozonide. Pryor, therefore, proposed that up to 50% of the pulmonary damage caused by O_3 inhalation is caused by free radicals. Thus, antioxidants and particularly those in the ELF provide considerable protection (see below).

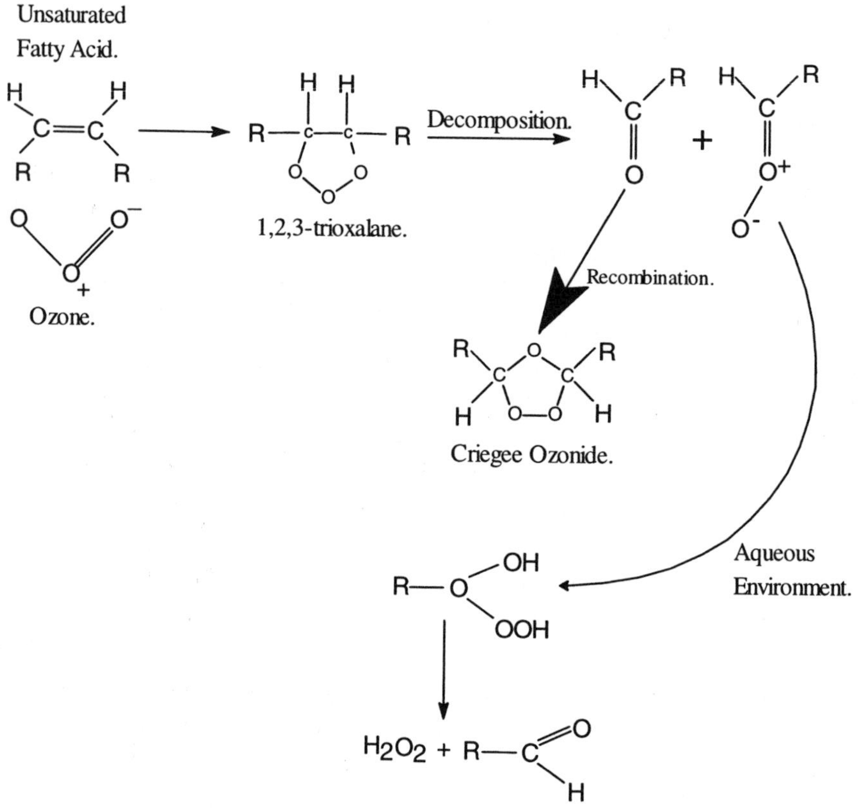

Fig. 1. Summary of the reaction between ozone and unsaturated fatty acids. Following an addition reaction, an aldehyde and carbonyl oxide are produced by a decomposition process. These two components can then recombine to form the Criegee ozonide. However, it is now thought that in an aqueous environment such as would occur in the ELF, the products are hydrogen peroxide and aldehydes.

Direct oxidative or free-radical-mediated reactions can also give rise to protein oxidation. Functional groups including sulphydryls, amines, alcohol and aldehydes can be oxidised, with the amino acids cysteine, methionine, tryptophan and tyrosine showing the greatest reactivity. Ozone has been shown to inactivate several proteins, with the detection of protein carbonyls (Cross *et al.*, 1992). It is these aldehydes, peroxides and protein carbonyls that are thought to constitute the secondary and tertiary products that mediate ozone toxicity. However, the question of whether O_3 reacts predominantly with the lipids (Uppu *et al.*, 1995) or the proteins (Cross *et al.*, 1992) is currently an area of debate.

The uptake of NO_2 from the airways is also dependent upon a reaction that is first order with respect to NO_2 concentration (Postlethwait *et al.*, 1995). Components of the ELF that can react with NO_2 include lipid, protein and low molecular weight antioxidants. With respect to lipid groups, NO_2 can react with unsaturated fatty acids, either by a process of addition or hydrogen abstraction. Although small amounts of insoluble and soluble NO_2 derived adducts can be detected, none of these is associated with the lipid species (Postlethwait & Bidani 1989). This indicates that NO_2 reacts with lipids by a process of hydrogen abstraction as opposed to addition (Pryor, 1981). The major products formed as a result of this abstraction are lipid radicals and nitrite anion (via HNO_2 dissociation, pK 3.4) (Postlethwait *et al.*, 1990). Although nitrite anion is the major product, it only accounts for 80% of inhaled NO_2 with the remainder being addition products. Addition products that can form include nitro, nitroester and nitroso (a minor component) products. Nitrogen-based acids could also be formed. Also, once formed, the nitrite anion can diffuse to other sites including the plasma, where it is thought to be converted to nitrate (Postlethwait & Bidani, 1989). Alternatively, nitrite can undergo further oxidation to other oxides of nitrogen (Postlethwait & Bidani, 1989). Both nitric and nitrous acid salts have been identified in blood and urine following exposure (WHO, 1987). Although nitrite anion has been shown to react with several amino acids at pH 2–5, the importance of these reactions at the higher pH of ELF is unknown (Kikugawa *et al.*, 1994). Nitrogen dioxide gas, however, can react with amino acids, occurring either in free solution or as part of protein molecules. Nitrogen dioxide has thus been reported to lead to the formation of dityrosine and 3-nitrotyrosine from tyrosine (Van der Vliet *et al.*, 1995); 5-nitroindole derivatives from tryptophan and cystine from cysteine. These reactions have also been observed to occur in bovine serum albumin, γ-globulin and α-crystallin, with non-sulphydryl cross links also occurring in these proteins (Kikugawa *et al.*, 1994). These reactions are summarised in Fig. 2.

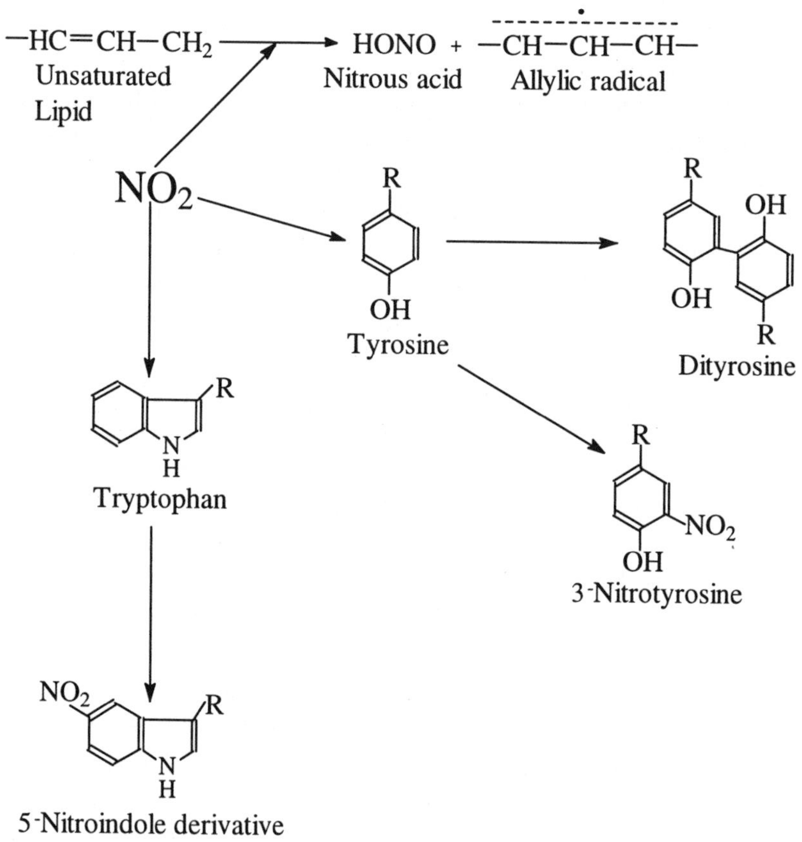

$-HC{=}CH{-}CH_2$ Unsaturated Lipid → HONO Nitrous acid + $-CH{-}CH{-}CH-$ Allylic radical

NO_2

Tyrosine → Dityrosine

Tyrosine → 3-Nitrotyrosine

Tryptophan → 5-Nitroindole derivative

Fig. 2. Some of the reactions that nitrogen dioxide could undergo with lipid and protein components of ELF.

Several key experiments on the uptake and fate of inhaled NO_2 have been carried out by Postlethwait *et al.* between 1989–95. Uptake of NO_2 into the lungs was shown to be an oxygen independent, free-radical-mediated process which was pH, substrate (both gas phase NO_2 and liquid phase substrate) and temperature dependant. Although absolute uptake continues to increase as NO_2 concentration increases, the fractional (percentage) uptake can decrease at high concentrations. Other physical forces such as convection may also influence uptake (Postlethwait *et al.*, 1990). A number of artificially prepared or animal lavage fluids have been used to examine the extracellular reactions of O_3 and NO_2. It was found that the kinetics of NO_2 uptake was similar to those of intact lung, again establishing the importance of the ELF in the uptake of pollutant gases. When lavage fluids were dialysed through membranes with a range of molecular weight cut-offs, it was found that components below 12 kilodaltons (kDa) molecular weight accounted for 86% of uptake. The most effective substrates for NO_2 have molecular weights between 100 and 1000 daltons (Postlethwait *et al.*, 1995). Following depletion of individual lavage components, it was found that reduced glutathione (GSH) and ascorbic acid (AA) were the principal reactive substrates. Bovine serum albumin had some function as an absorption substrate for NO_2, whilst tyrosine, tryptophan and taurine had minimal activity. Dipalmitoyl-phosphatidylcholine, the major component of pulmonary surfactant, was also shown to be an ineffective substrate, suggesting that the lipoprotein complex may not be important in clearing NO_2 from air. However, unsaturated fatty acids may have a high enough reactivity to play an important part (Postlethwait *et al.*, 1995).

Postlethwait *et al.* (1991), have utilised a model chamber system to investigate the passage time (or "breakthrough") of gases through porous "kim wipe" filters which had been treated with a single antioxidant or lavage components, thus mimicking the condition experienced by ELF (Fig. 3). Breakthrough times for NO_2 (10.9 ppm) were very fast (< 30 secs) when the "kim wipe" was soaked in buffer but slower when soaked in 1 mM GSH (1.25 mins). The NO_2 was similarly impeded

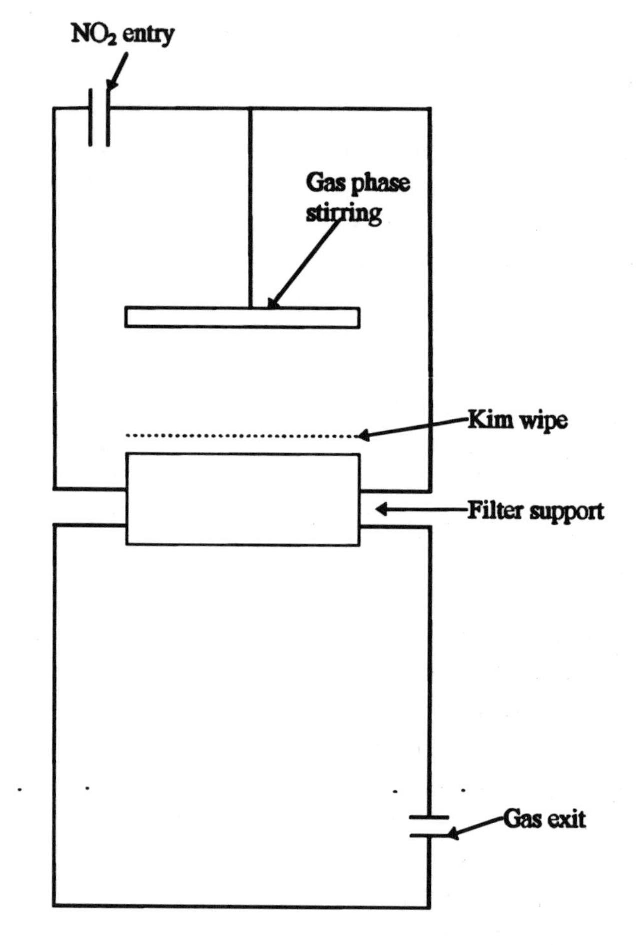

Fig. 3. Diagrammatic representation of model system used to investigate "breakthrough" of gases (Postlethwait *et al.*, 1991).

(2.0 mins) by diluted lavage fluid. Importantly, the passage of oxygen was not inhibited when the filters were coated with GSH or lavage components. Nitrogen dioxide was shown to react with antioxidants such as GSH, 40% of which needed to be depleted before gas breakthrough occurred. Thus, it was concluded that ELF is likely to impede the transfer of reactant pollutant gases and the substrates present in ELF are of high enough concentration to provide very efficient protection of the underlying epithelium (Postlethwait *et al.*, 1991).

It is probable that the toxicity of SO_2 is also mediated by the products that are formed following dissolution of SO_2 in the ELF, or products formed subsequently in the circulating plasma. Sulphur dioxide dissolves in water to form sulphurous acid which then dissociates to bisulphite and sulphite. Sulphite can react with all major classes of cellular molecules and the formation of thiols and S-sulphonates from disulphide bonds has been well documented. In proteins, this gives rise to cysteine sulphonate, while reaction with oxidised glutathione gives rise to glutathione S-sulphonate (Menzel *et al.*, 1986). Solutions of SO_2 can also slowly oxidise to form sulphuric acid which can then react with metals or ammonia to form sulphate salts (Dept of Health, 1992) (Fig. 4). Exhaled air contains high but variable levels of ammonia which could react with components such as sulphates and hence provide protection against inhaled SO_2. The presence of disulphides, both as oxidised glutathione and protein S-S bonds (as found in proteins such as albumin), could also give rise to protection by formation of S-sulphonates (Menzel *et al.*, 1986).

Alternatively, SO_2 may act as a strong reductant, although recent studies have indicated that vitamins C and E can reduce SO_2-induced toxicity in rat brain (DalÇik *et al.*, 1995) and liver (Karaöz *et al.*, 1995). This implicates the role of oxidant species (possibly derived from lipids) in mediating toxicity.

Thus, following inhalation of SO_2, the respiratory tract may be exposed to a wide variety of reaction products (Fig. 4). It is unclear which of these products may initiate biological effects in the respiratory tract although sulphite from foodstuffs or medications has been implicated in worsening asthma symptoms (Dept of Health, 1992). Details of the many compounds that can be produced following SO_2 inhalation and the mechanisms underlying the initiation of neural reflex activity have been reported previously (Dept of Health, 1992). A summary of the interactions of the individual pollutants with the lung is shown in Fig. 5.

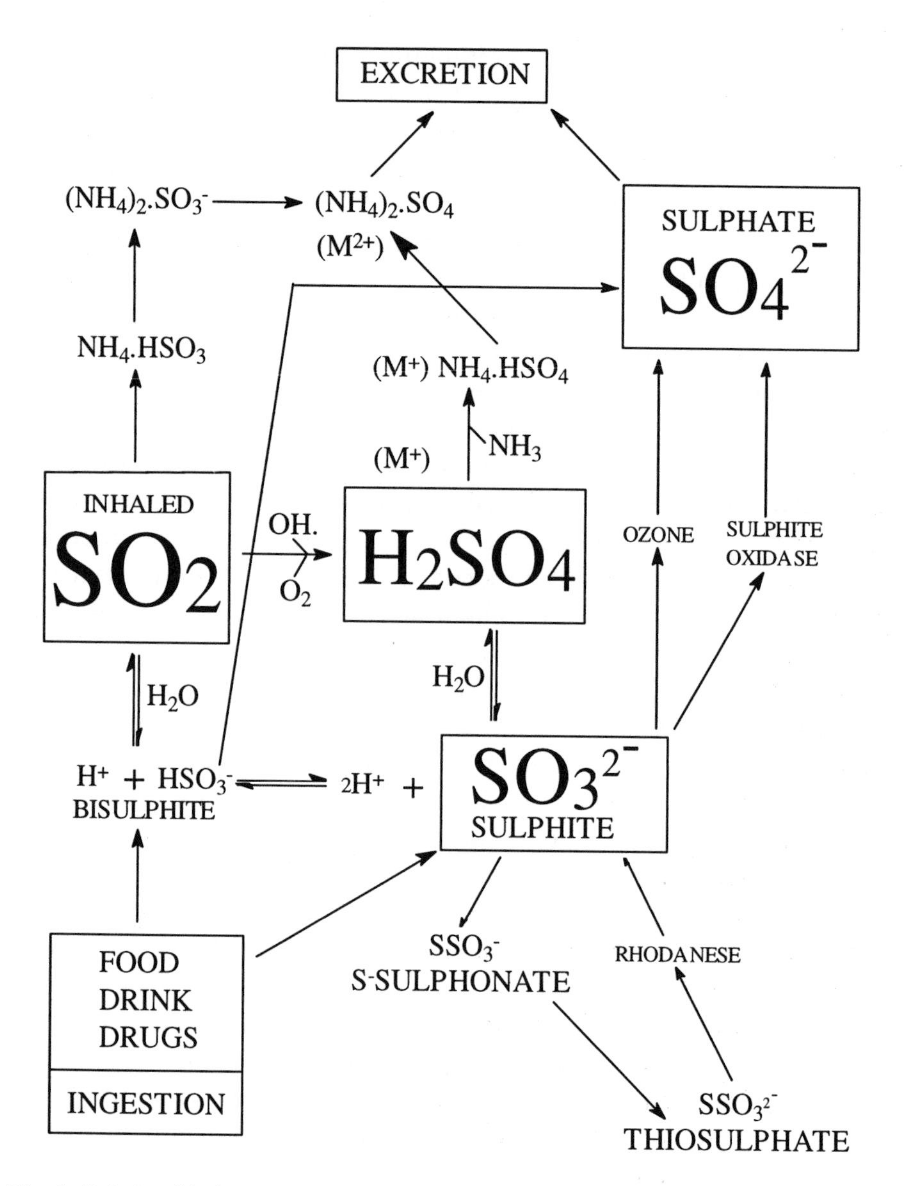

Fig. 4. Relationship between inhaled sulphur dioxide and other forms of sulphur that could occur systemically or within the ELF.

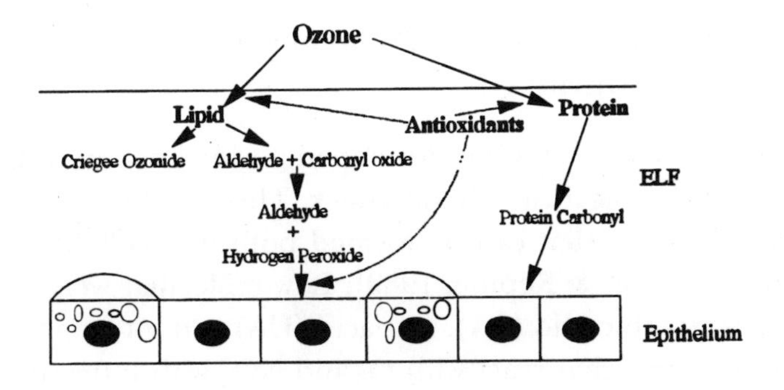

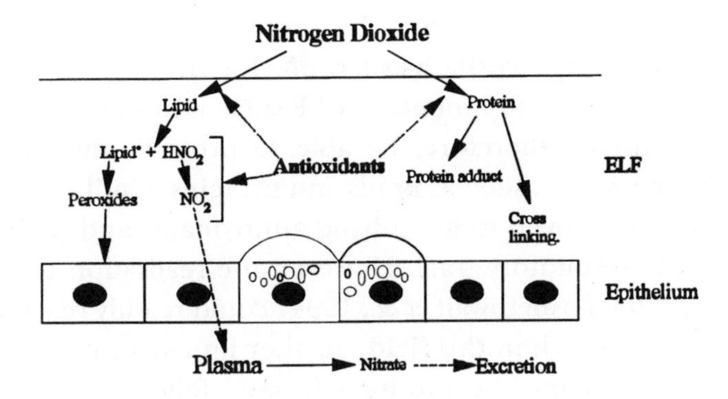

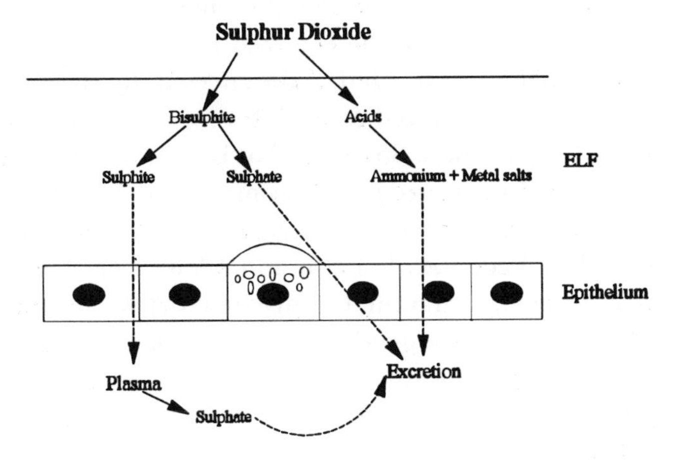

Fig. 5. Simplified schematic representation of the interaction of three pollutant gases with the respiratory tract.

5. Protection Against Inhaled Pollutants and Individual Sensitivity

The entire respiratory tract has developed a wide array of antioxidant defences to protect against oxidant attack. These include enzymic and non-enzymic systems that can be located both intracellularly and extracellularly (Heffner & Repine, 1989). Low molecular weight antioxidants such as ascorbic acid (AA), uric acid (UA) and α-tocopherol (AT) are electron donors that react with O_3 and NO_2 with intrinsically high reaction rates (Pryor, 1993), but to be effective in preventing O_3 and NO_2 induced damage, they must be present at high concentrations (Langford *et al.*, 1995; Postlethwait *et al.*, 1995).

The most reactive components of ELF such as low molecular weight antioxidants should, therefore, be able to protect other macromolecules that are present, such as lipids and proteins. Until recently, little attention has been given to identifying antioxidants and their specific location in the respiratory tract. However, the realisation that O_3 and NO_2 (Pryor, 1993; Postlethwait *et al.*, 1994) could readily react with ELF has promoted research in this field. Further impetus to such work has been provided because the products formed following pollutant gas interactions with ELF may mediate the subsequent toxicity to the underlying epithelia. Despite the fact that local dosimetry of the pollutant gases has been investigated, differences in the levels of antioxidants present in ELF obtained from different anatomical sites of the tract are poorly understood. Some of what is known of the distribution of antioxidants in human subjects is summarised in Table 1.

The major extracellular low molecular weight antioxidant in the nasal cavity is UA (Peden *et al.*, 1990) and AA is rarely detected unless subjects are receiving vitamin C enriched drinks or supplements (Housley, 1996). There are gender differences in UA content in nasal ELF, with males consistently showing higher levels than females (Housley, *et al.* 1996). Uric acid is present in bronchoalveolar lavage (BAL) fluids at lower levels than nasal lavage, but AA is substantially higher (Table 1). However, given the fact that the degree of dilution that occurs during a BAL (Rennard *et al.*, 1985) is larger than that

Table 1. Comparison of antioxidants present at different respiratory tract regions and in plasma of human subjects. All data expressed as μM/L unless otherwise stated. Data for nasal cavity (Housley, 1996) and conducting/respiratory regions expressed as median + interquartile range (Kelly *et al.*, 1997) except vitamin E (mean ± SEM) (Pacht *et al.*, 1986). Plasma values expressed as normal min-max data (Kelly *et al.*, 1995). All data represent lavage values except those in italics which represent possible *in vivo* concentrations (Kelly *et al.*, 1995; Housley 1996; with dilution factor for nasal lavage from Steerenberg *et al.*, 1995). N.D. = No data.

	Uric acid μM	Ascorbic acid μM	GSH μM	GSSG μM	Vitamin E μM
Nasal Cavity	Male 2.56; 0–4.225	0.59	0; 0–0.53	0.22; 0–0.507	N.D.
	Female 1.55; 0–2.253		0; 0–0.128	0.30; 0–0.5065	
	7.7–155	*2.2–9.7*	*3.2–24*	*2.2–26*	*N.D.*
Conducting Airways	0.22; 0.008–0.58	0.44; 0.4–0.6	0.45; 0.31–0.6	0.21; 0.06–0.3	N.D.
Respiratory Region	0.33; 0.11–0.47	0.5; 0.47–0.61	0.46; 0.31–0.68	0.17; 0.06–0.23	20.7 ± 2.4 ng/ml
	50–200	*50–300*	*30–300*	*N.D.*	*4*
Plasma	160–450	30–150	2–6	N.D.	10–40

occurring during nasal lavage (Steerenberg *et al.*, 1995), the *in vivo* concentrations of UA in these compartments are likely to be similar (Table 1). Again, differences in dilution occurring during sampling procedures are likely to mean that GSH concentration in lung-derived ELF is far higher than nasal ELF. Reduced glutathione (GSH) and its oxidised form (GSSG) have been detected in all respiratory tract ELF locations, although only 50% of subjects have detectable levels of GSH within nasal lavage (Housley, 1996), as opposed to 100% of BAL samples (Mudway *et al.*, 1996). Also, ascorbic acid could only be detected in 25% of nasal lavage samples (Housley *et al.* 1995) and 30% of BAL samples (Mudway *et al.*, 1996). As mentioned above, estimates of ELF concentrations of GSH in the lower respiratory tract show them to be much higher than those found in plasma indicating that the source of this antioxidant is likely to be (specific) cells lining the airways. Such a deduction does not apply to UA (Peden *et al.*, 1993) where plasma is the likely source. Few studies have been conducted for extracellular AT which may be an important cell membrane component or a constituent of pulmonary surfactant (Cross *et al.*, 1994).

Several investigators (Halliwell & Cross, 1994; Kelly *et al*,. 1995) have suggested that individual differences in the levels of these antioxidants may influence individual sensitivity to inhaled oxidants. It is thus argued that only after pulmonary antioxidant defences (extracellular and intracellular) have become depleted that the oxidative processes will give rise to damage of cellular components (Kelly *et al.*, 1995). Individuals who have low concentrations of antioxidants within the ELF and/or do not have the ability for rapid replenishment of these antioxidants would be more sensitive to inhaled oxidants. However, this generalisation may be too simplistic. This is because each oxidant gas reacts at different rates and with differing preferences with the antioxidants (Slade *et al.*, 1989). Thus, it has been found that GSH reacts favourably with NO_2 (Postlethwait *et al.*, 1995) but is less readily depleted by O_3 (ascorbic acid ≥ uric acid >> glutathione) (Mudway *et al.*, 1996). In attempting to identify sensitive individuals, it may be necessary to know their full antioxidant profile because an individual ELF that protects well against NO_2 may not be so efficient against O_3. Furthermore, little information is available about the variation in individual respiratory tracts with changing season or alterations in diet.

The exact concentration of O_3 or NO_2 and the duration of exposure required to compromise antioxidant defences is not known. However, in subjects exposed to very high levels of NO_2 (2 ppm for 4 hours), UA and AA were decreased 1½ hours after exposure. However, this effect was only transient and these antioxidants had increased above baseline 6 hours after exposure ceased (Kelly *et al.*, 1997). There is also limited information on the rate of replenishment of antioxidants in individuals, although the above study of NO_2 exposures has established that extracellular antioxidant concentrations can increase after exposure (Kelly *et al.*, 1997). Such a phenomenon equates well with the development of "tolerance" (Mustafa & Tierney, 1978) in that subjects become less sensitive to a second exposure to the same pollutant gas (Horvath *et al.*, 1981).

A number of *in vitro* studies utilising human derived ELF, have been performed to establish the selectivity and kinetics of antioxidant depletion by pollutant gases. Depletion of the major antioxidant UA in nasal lavage fluid was linearly related ($R^2 = 0.97$) to environmentally relevant O_3 concentrations. It was also established that a pure biochemical solution of UA (604 μM) was depleted at a rate of 30 nmol/l/hr/ppb O_3 whereas UA in diluted lavage was depleted at 18 nmol/l/hr/ppb O_3. This indicated that UA may account for some 60% of the antioxidant capacity of these nasal lavage fluids (Housley *et al.*, 1995). Studies using ^{18}O have confirmed extracellular deposition of O_3 in the nasal cavity (Hatch *et al.*, 1995) but it is not known if the degree of deposition correlates with the amount of UA present. For individuals with low nasal UA or for oral breathing subjects, the "scrubbing" effect of the nasal ELF may be inefficient. Moreover, females have approximately 50% of the nasal ELF UA concentration found in males and there is a report indicating that females are more sensitive to O_3 than males (Lauritzen & Adams, 1985). In contrast, other evidence indicates that a gender difference in sensitivity does not exist (Weinmann *et al.*, 1995). Subsequent *in vitro* studies of the depletion of UA, AA and GSH following exposure to O_3 (50–1000 ppb) have been performed by Mudway *et al.* (1996). Both AA and UA were consumed in a time- and O_3 concentration-dependent manner with rates of 1.7 ± 0.8 and 1.0 ± 0.5 pmol/l/s/ppb, respectively. In contrast, GSH was consumed at 50 ppb O_3 but the rate of consumption did not increase with O_3 concentration. The precise reason for the apparent low reactivity of GSH is still unclear, but previous overestimations of GSH reactivity may be partly responsible (Mudway *et al.*, 1996). Such *in vitro* depletion studies are not readily applicable to *in vivo* investigations where UA replenishment for nasal ELF may take immediate effect. Nevertheless, for most individuals, it seems probable that the antioxidant status of the nasal ELF is sufficient to sequester environmentally relevant concentrations of O_3. Despite this protection, responses to inhaled O_3 can be observed in volunteers, with approximately 10% of people currently believed to be sensitive to low levels of this oxidant gas.

6. Human Responses to Inhaled O_3

The three pollutant gases have been found to cause a wide range of effects in the respiratory tract, including alterations in the epithelia (Calderon-Garcidueñas *et al.*, 1994), influx of inflammatory cells (Koren *et al.*, 1990) and alterations in lung function (Lippman, 1989). A summary of some of the effects is shown in Table 2. Generally, O_3 causes a concentration-dependent decrease in exhaled volumes and flow. Specifically, these changes include decreases in forced expiratory volume (FEV), vital capacity (VC), total lung capacity (TLC) and tidal volume, but no changes in residual volume or functional residual capacity. Airway resistance and specific airway resistance increase as does respiratory frequency. Increases in bronchial responsiveness have also been observed. Other changes include airway inflammation and increased epithelial permeability (Leikauf *et al.*, 1995). Bhalla *et al.* (1986) proposed that the inflammatory changes are also important in causing epithelial damage following O_3 inhalation. Several studies have found greater responses when O_3 is inhaled in ambient air compared with purified air (Lippmann, 1989), suggesting that other components of ambient air are important. This indicates that further work on mixtures is required. Many of the published papers covering these controlled chamber studies have been well reviewed previously (Lippman, 1989;

Table 2. Comparison of some of the studies that have been performed to investigate alterations in lavage and lung function following exposure to pollutant gases.

Ambient concentrations of pollutant gases in the UK	O_3 30 ppb	NO_2 30 ppb	SO_2 15 ppb
Lavage studies	100 ppb for 6.6 hrs. 379% increase in PMN, 23% increase in protein, 58% increase in LDH, 74% increase in PGE_2, 236% increase in fibronectin and 393% increase in IL-6 (Devlin *et al.*, 1991).*	3 hrs 600 ppb. 47% increase in α_2-macroglobulin after 3.5 hrs, which resolved by 18 hrs. No change in cells (Frampton *et al.*, 1989).*	8000 ppb for 20 mins. 24 hrs after exposure, total BAL cells increased from 6.9 to 8.2×10^7/l. Lymphocytes increased from 6 to 20% and lysozyme positive macrophages from 5 to 18%. These had regressed toward normal by 72 hrs. No changes in lung function (Sandstrom *et al.*, 1989).*
Lung function studies	380 ppb for 2 hrs light exercise. 5% fall in FEV. This may decrease to 100 ppb in most sensitive subjects. With 6.6 hrs exposure and moderate exercise 70 ppb required to cause 5% fall in FEV (Dept of Health, 1991).*	600 ppb for 1 hr. No change in lung function (Adams *et al.*, 1987). 2000 ppb for 1 hr gave no change in lung function, but did show 20% increase in airway responsiveness (Mohsenin, 1988).	1000 ppb for 3 mins. 393% increaase in specific airway resistance (Balmes *et al.*, 1987). 2000 ppb for 30 mins caused changes in nasal airways but no changes in lung function (Bedi & Horvath, 1989).

*Exercise performed during exposure. PMN = polymorphonucleocyte; LDH = lactate dehydrogenase; PGE_2 = prostaglandin E_2; IL-6 = interleukin 6; FEV = forced expiratory volume and BAL = bronchoalveolar lavage.

Dept of Health, 1991). Unlike the other pollutant gases, measurable changes with O_3 can be observed at concentrations that are only 3–4 times above background average concentrations. However, changes occurring at environmentally relevant concentrations appear to be fully reversible. For example, volunteers exposed to 120 ppb O_3 for 6.6 hrs (2 × 3 hrs with 50 mins exercise per hr) showed an average FEV_1 change of 13%, but this waas fully resolved by 24 hrs post exposure (Folinsbee *et al.*, 1988). In a similar study, there was a high degree of variability in FEV_1, with changes ranging from 0% to 37% (Horstmann *et al.*, 1990). These functional changes have been noted to be proportional to concentration (C) and time (t) of exposure, although C is more important than t, i.e. $C^n t$, where $n > 1$ (Dept of Environment, 1993). Incorporation of minute volume to calculate total dose has also been used to relate functional changes to dose but again, the dose-response relationship is not linear (Dept of Health, 1991).

Very high levels of NO_2 (7500 ppb, 2 hrs) cause transient increases in airway resistance. Exposure to levels below 1000 ppb generally appear to cause no alterations in lung function, with the exception of a few studies indicating very small changes in forced vital capacity (Dept of Health, 1993).

Following inhalation of high concentrations of SO_2, normal subjects undergo bronchoconstriction, with asthmatics being more sensitive. These changes can be seen within minutes of exposure commencing. As with O_3, there appears to be wide variation in response among subjects. However, in normal subjects, SO_2 inhalation has not been associated with changes in lung function when inhaled at concentrations below 1000 ppb. Several studies have shown changes in asthmatic patients for concentration below 1000 ppb. Full details of the above studies have been discussed previously (Dept of Health, 1992).

7. Summary and Comment

The toxicity of the air pollutants O_3, NO_2 and SO_2 are probably mediated by reactions that occur within the epithelial lining fluids of the airway. For NO_2 and SO_2, however, products such as nitrite and sulphite may be able to diffuse to other sites such as plasma where they may undergo further conversion to other products. The composition of the ELF includes several antioxidants which are present at high enough concentration to afford protection to most individuals. This indicates that health effects following inhalation of these pollutants, even when present at concentrations several times above ambient, are likely to be transient. A number of investigations have been undertaken in an attempt to explain why about 10% of the population show an increased sensitivity to gases such as O_3. The mechanisms of sensitivity still remain unexplained. Moreover, whilst some asthmatics are susceptible to increasing levels of SO_2, there is no full explanation of how this reducing gas affects a compromised airway.

Few, if any, studies have attempted to explain why 90% of the population can tolerate episodic oxidant gas exposure with transient, minimal or no apparent health effects. Logic dictates that individuals with low, overall antioxidant status may represent those most at risk from respiratory effects. However, this may be a simplistic view until we have a better understanding of respiratory tract antioxidants with respect to their distribution, replacement, variation with age, gender, ethnic origin and season, asthmatic versus non-asthmatic subjects and whether individuals are known responders or non-responders on episode days. In addition, this review has focused on low molecular weight antioxidants whereas other non-protein and protein components such as albumin, transferrin, caeruloplasmin and bilirubin may also have important protective roles.

Finally, it should be noted that very few gaseous pollution episodes occur without substantial increases in ultrafine particles (PM_{10}) which in themselves, may react more than the gases or induce synergistic effects. It would seem particularly important to establish the effects of both summer smog ($O_3 + PM_{10}$) and a winter urban episode (NO_2 and/or $SO_2 + PM_{10}$) on individual antioxidant status.

References

Adams W.C. *et al.* (1987) *J. Appl. Physiol.* **62**(4), 1698–1704.

Aharonson E.F. *et al.* (1974) *J. Appl. Physiol.* **37**(5), 654–657.

Balmes J.R. *et al.* (1987) *Am. Rev. Respir. Dis.* **136**, 1117–1121.

Bascom R. *et al.* (1990) *Am. Rev. Respir. Dis.* **142**, 594–601.

Bedi J.F. & Horvath S.M. (1989) *JAPCA.* **39**(11), 1448–1452.

Ben-Jebria A. *et al.* (1995) In *Nasal Toxicity and Dosimetry of Inhaled Xenobiotics: Implications for Human Health,* ed. Miller F.J. pp. 386–389, Taylor Francis, Washington DC.

Bhalla P.K. *et al.* (1986) *J. Toxicol. Env. Health,* **17**(2–3), 269–283.

Calderon-Garcidueñas L. *et al.* (1994) *Environ. Health Perspec.* **102**, 1074–1080.

Cross C.E. *et al.* (1992) *Free Rad. Res. Comm.* **15**(6), 347–352.

Cross C.E. *et al.* (1994) *Environ. Health Perspec.* **102**(suppl. 10), 185–191.

Dahl A.R. (1990) *Toxicol. Appl. Pharmacol.* **103**, 185–197.

Dalçik H. *et al.* (1995) *Turkish J. Med. Sci.* **24**(Suppl), 88–89.

Department of the Environment (1993) *Third Report of the United Kingdom Photochemical Oxidants Review Group.* HMSO, London.

Department of Health (1991) *First Report: Ozone.* Advisory Group on the Medical Aspects of Air Pollution Episodes. HMSO, London.

Department of Health (1992) *Second Report: Sulphur Dioxide.* Advisory Group on the Medical Aspects of Air Pollution Episodes. HMSO, London.

Department of Health (1993) *Third Report: Nitrogen Dioxide.* Advisory Group on the Medical Aspects of Air Pollution Episodes. HMSO, London.

Devlin R.B. *et al.* (1991) *Am. J. Respir. Cell. Mol. Biol.* **4**, 72–81.

Folinsbee L.J. *et al.* (1988) *JAPCA.* **38**, 28–35.

Frampton M.W. *et al.* (1989) *Am. J. Respir. Cell. Mol. Biol.* **1**, 499–505.

Gerrity T.R. *et al.* (1988) *J. Appl. Physiol.* **65**(1), 393–400.

Gerrity T.R. & Weister M.J. (1987) *Experimental Measurements of the Uptake of Ozone in Rats and Human Subjects.* 80th meeting of APCA, June 21–26. New York.

Greenwood N.N. & Earnshaw A. (1995) *Chemistry of the Elements.* Cambridge University press.

Halliwell B. & Cross C.E. (1994) *Environ. Health Perspec.* **102**(suppl. 10), 5–12.

Hatch G.E. *et al.* (1994) *Am. J. Respir. Crit. Care Med.* **150**, 676–683.

Hatch G.E. *et al.* (1995) In *Nasal Toxicity and Dosimetry of Inhaled Xenobiotics: Implications for Human Health,* ed. Miller F.J. pp. 125–158, Taylor Francis, Washington DC.

Heffner J.E. & Repine J.E. (1989) *Am. Rev. Respir. Dis.* **140**, 531–554.

Horstmann D.H. *et al.* (1990). *Am. Rev. Respir. Dis.* **142**, 158–1163.

Horvath S.M. *et al.* (1981) *Am. Rev. Respir. Dis.* **123**, 496–499.
Hotchkiss J.A. *et al.* (1989) *Toxicol. Appl. Pharmacol.* **98**, 289–302.
Housley D.G. (1996) *Biochemical Characterisation of Human Nasal Lavage.* Ph.D thesis, University of Wales.
Housley D.G. *et al.* (1996) *Acta Otolaryngol. (Stockh).* **116**, 751–754.
Housley D.G. *et al.* (1995) *Int. J. Biochem. Cell Biol.* **27**(11), 1153–1159.
Hu S.C. *et al.* (1992) *J. Appl. Physiol.* **73**(4), 1655–1661.
Hu S.C. *et al.* (1994) *J. Appl. Physiol.* **77**(2), 574–583.
Karaöz E. *et al.* (1995) *Turkish J. Med. Sci.* **24**(Suppl), 91–92.
Kelly F.J. *et al.* (1997) *Am. J. Respir. Crit. Care Med.* **154**, 1700–1705.
Kelly F.J. *et al.* (1995) *Respir. Med.* **89**, 647–656.
Kikugawa K. *et al.* (1994) *Free Rad. Biol. Med.* **16**(2), 373–382.
Kimbell J.S. *et al.* (1993) *Toxicol. Appl. Pharmacol.* **121**, 253–263.
Kimbell J.S. (1995) In *Nasal Toxicity and Dosimetry of Inhaled Xenobiotics: Implications for Human Health,* ed. Miller F.J. pp. 73–84, Taylor Francis, Washington DC.
Koren H.S. *et al.* (1990) *Toxicology,* **60**, 15–25.
Langford S.D. *et al.* (1995) *Toxicol. Appl. Pharmacol.* **132**, 122–130.
Lauritzen S.K. & Adams W.C. (1985) *J. Appl. Physiol.* **59**(5), 1601–1606.
Leikauf G.D. *et al.* (1995) *Env. Health Perspec.* **103**(S2), 91–95.
Lippmann M. (1989) *JAPCA.* **39**(5), 672–695.
Manson M. (1980) Sulphur dioxide biochemistry. Internal report. Carshalton. Surrey. MRC Toxicology Unit.
Maynard R.L. & Waller R.E. (1993) In *General and Applied Toxicology,* eds. Ballantyne B., Marrs T. & Turner P., pp. 1227–1252, Stockton Press.
Menzel D.B. *et al.* (1986) *Adv. Exp. Med. Biol.* **197**, 477–492.
Miller F.J. *et al.* (1978) *Environ. Res.* **17**, 84–101.
Miller F.J. & Overton J.H. (1989) In *Atmospheric Ozone Research and Its Policy Implications,* eds. Schneider *et al.*, pp. 281–291, Elsevier Science, Amsterdam.
Miller F.J. *et al.* (1985) *Toxicol. Appl. Pharmacol.* **79**, 11–27.
Miller F.J. *et al.* (1993) In *Toxicology of the Lung,* eds. Gardner D.E., Crapo J.D. and McClellan R.O. pp. 485–526, Raven Press, New York.
Mohsenin V. (1988) *Arch. Environ. Health,* **43**(3), 242–246.
Mudway I. *et al.* (1996) *Free Rad. Res.* **25**(6), 499–513.
Mustafa M.G. (1990) *Free Rad. Biol. Med.* **9**, 245–265.
Mustafa M.G. & Tierney D.F. (1978) *Am. Rev. Respir. Dis.* **118**, 1061–1090.
Overton J.H. *et al.* (1987) *Toxicol. Appl. Pharmacol.* **88**, 418–432.
Pacht E.R. *et al.* (1986) *J. Clin. Invest.* **77**, 789–796.
Peden D.B. *et al.* (1990) *Proc. Natl. Acad. Sci.* **87**, 7638–7642.
Peden D.B. *et al.* (1993) *Am. Rev. Respir. Dis.* **148**, 455–461.

Postlethwait E.M. & Bidani A. (1989) *Toxicol. Appl. Pharmacol.* **98**, 303–312.
Postlethwait E.M. & Bidani A. (1990) *J. Appl. Physiol.* **68**(2), 594–603.
Postlethwait E.M. *et al.* (1990) *J. Appl. Physiol.* **69**(2), 523–531.
Postlethwait E.M. *et al.* (1991) *J. Appl. Physiol.* **71**(4), 1502–1510.
Postlethwait E.M. *et al.* (1991) *Toxicol. Appl. Pharmacol.* **109**, 464–471.
Postlethwait E.M. *et al.* (1992) *J. Appl. Physiol.* **73**(5), 1939–1945.
Postlethwait E.M. *et al.* (1994) *Toxicol. Appl. Pharmacol.* **125**, 77–89.
Postlethwait E.M. *et al.* (1995) *Free Rad. Biol. Med.* **19**(5), 5553–563.
Pryor W.A. (1981) *Science,* **214**, 435–437.
Pryor W.A. (1992) *Free Rad. Biol. Med.* **12**, 83–88.
Pryor W.A. (1993) *J. Lab. Clin. Med.* **122**, 483–486.
Pryor W.A. (1994) *Free Rad. Biol. Med.* **17**(5), 451–465.
Pryor W.A. & Church D.F. (1991) *Free Rad. Biol. Med.* **11**, 41–46.
Pryor W.A. *et al.* (1995) *Free Rad. Biol. Med.* **19**(6), 935–941.
Raphael G.D. *et al.* (1988) *Amer. J. Rhinol.* **2**(3), 109–116.
Rennard S. *et al.* (1985) *J. Appl. Physiol.* **60**(2), 532–538.
Sandström T. *et al.* (1989) *Am. Rev. Respir. Dis.* **140**, 1828–1831.
Slade R. *et al.* (1989) *Inhal. Tox.* **1**, 261–271.
Steerenberg P.A. *et al.* (1995) *Exp. Toxicol. Pathol.* **47**(4), 232–234.
Strandberg L.G. (1964) *Arch. Environ. Health,* **9**, 160–166.
Ultman J.S. (1995) In *Nasal Toxicity and Dosimetry of Inhaled Xenobiotics: Implications for Human Health,* ed. Miller F.J. pp. 59–72, Taylor Francis, Washington DC.
Uppu R.M. *et al.* (1995) *Arch. Biochem. Biophys.* **319**(1), 257–266.
Van der Vliet A. *et al.* (1995) *Arch. Biochem. Biophys.* **319**(2), 341–349.
Weinmann G.G. *et al.* (1995) *Am. J. Respir. Crit. Care Med.* **152**, 988–996.
Wright E.S. *et al.* (1990) *Toxicology Letters,* **51**, 125–145.
World Health Organisation. (1987) *Air Quality Guidelines for Europe. WHO Regional Publications, European Series No. 23.* Copenhagen: WHO.

CHAPTER 8

AIR POLLUTION POLICY IN THE EUROPEAN COMMISSION

R.L. Maynard & K.M. Cameron

The views expressed in this chapter are those of the authors and should not be taken as those of the UK Departments of Health and of Environment, Transport and the Regions or of the European Commission.

1. Introduction

This is a time of new developments in air pollution policy in the European Commission. This chapter describes what has been done in the past and the new approaches being taken to set air quality standards in Europe.

1.1. Why International Cooperation is Necessary in Europe

Air pollution is not a local issue. All air pollutants can be dispersed by wind and all potentially affect areas distant from their source. Thus, no local authority or individual government save, presumably, those of remote islands, can tackle air pollution effectively without cooperation from neighbouring authorities or states. For pollutants such as fine particles, dispersion is very effective and such particles may travel hundreds, perhaps thousands, of kilometres before being removed from the air by either dry deposition or by the action of rain and snow. Long-range transport of pollutants has been identified as a problem within the European area. Acidic materials produced in the UK are known to be transported by the prevailing south-westerly winds to northern Europe and damage to forests and acidification of lakes and rivers have been attributed to this process (Elsom, 1987). Damage to forests has also been reported from central and eastern Europe and here, again, deposition of transported acidic air pollutants has been blamed.

Ozone is produced by photochemical reactions in air masses loaded with primary pollutants, e.g. oxides of nitrogen and hydrocarbons, produced largely in urban areas (Photochemical Oxidants Review Group, 1987). Ozone is thus described as a secondary air pollutant. The generation of ozone takes time and the highest concentrations of ozone are often recorded at long distances from the sources of the primary pollutants. A classic example of high pollution levels occurring far from the source of the primary pollutants was noted in Southern England in the hot and sunny summer of 1976. Wind from the east carried polluted air from the densely industrialised areas of Northern Europe across the Low Countries towards the UK. As the polluted air mass moved westwards, the concentration of ozone rose and thus air with a high concentration of ozone "arrived" in London. Hourly concentrations of ozone in London exceeded 200 ppb each day for nearly two weeks (Hough & Derwent, 1990). These are exceptional concentrations — the more so because they were recorded in an area that is generally an ozone sink on account of the nitric oxide produced by high traffic density. In such areas, ozone reacts rapidly with nitric oxide, the main component of the mixture of oxides of nitrogen produced by motor vehicles, to produce nitrogen dioxide.

The easterly winds carried the polluted air mass further westward across the UK and a maximum hourly average concentration of 258 ppb was recorded at Harwell, Oxfordshire. The 1976 episode was exceptional and has not been repeated during the hot summers of the early-mid 1990s. The reasons for this are not completely understood and are discussed in the 4th Report of the UK Photochemical Oxidants Review Group published in 1997 (Photochemical Oxidants Review Group, 1997). It is clear that there is little chance that action taken unilaterally by the UK could have prevented the ozone episode of 1976.

A number of air pollutants of concern are released from point sources such as factories, power stations, etc. These may be transported over long distances but may also have effects in the area close to the source. In these instances, it may be necessary to limit the emissions from the source. Control of pollution costs money. Costs inevitably fall on industry and are passed on to the consumer as a rise in the price of the commodities offered for sale.

Modern industries have been sensitive to the changing climate of opinion regarding air pollution and technological developments have kept pace with requirements and standards. In addition, new industries aimed at developing equipment to control pollution have arisen and have been successful. To such industries, the imposition of tighter air pollution standards (either as emission standards or as general air quality standards) is often an advantage and an opportunity for further development and sales.

2. Air Pollution Legislation in the European Union

Historically, control of air pollution in Europe has been addressed using two approaches:

(i) legislation establishing limits on emissions from specified sources;
(ii) legislation establishing air quality standards.

This chapter focuses on the second of these approaches. It is interesting to look at the philosophy behind the two approaches. In the first type of legislation, emission limits are generally agreed on the basis of "best available technology" and "best practicable means". If used in isolation, these concepts should result in ever decreasing emissions as technology improves. More recently, a concept of balancing costs and benefits has been accepted, for example, in the recently adopted Integrated Pollution Prevention and Control (IPPC) Directive which establishes emission limits for a number of chemicals and an approach to integrated management of emissions to all media (Council Directive 96/61/EC, 1996). This approach allows the cost of pollution control to be taken into account and to be weighed against the benefits gained.

The second approach of using air quality standards as a means of managing air pollution has grown in importance. This approach requires the establishment and agreement of concentrations of air pollutants which do not cause harm to human health and/or the environment. As new information becomes available, the standards can be revised, although generally, the pressure is to revise downwards. In terms of air quality management, the next steps are to identify and then control the sources of pollution so that the standard is met. In principle, the two approaches should be complementary so that emission limits are set to meet air quality standards. This is, of course, much easier on a local scale where there is good understanding of sources contributing to local air pollution.

3. Developing European Directives on Air Pollution

In the past, directives setting air quality standards have been established to address sulphur dioxide and suspended particulate matter, lead, nitrogen dioxide and ozone (Council of EC, 1980; 1982; 1985; 1989; 1992). New directives addressing any issue are generally drawn up by staff from the European Commission with the assistance of experts from the member states and outside bodies. For technical issues, those drafting the legislation will draw on external scientific expertise. In the past, the European Commission has generally taken this approach to assist in the area of risk assessment for air pollutants. Conclusions of internationally recognised bodies such as the World Health Organisation (WHO) and specially commissioned reviews of the available literature have formed the basis of decisions made regarding the setting of air quality objectives.

Most of the existing directives on air quality were adopted many years ago and since then, more new information on the effects of air pollutants has become available. In the early 1990s, it was recognised that most of the early directives needed to be updated. In 1992, the Fifth Environmental Action Programme committed the Commission to developing new air quality standards based on WHO Guideline Values (*Towards Sustainability*, 1993).

4. Historical Approaches

Two examples are provided of how in the past the European Commission has developed limit values for air pollutants based on the work of WHO.

4.1. Sulphur Dioxide and Suspended Particulates

The Directive on sulphur dioxide and suspended particulate matter was adopted in 1980 (Council of EC, 1980). This Directive established the limit and guide values based on the conclusions reached by the World Health Organisation which were published in 1972 and updated in 1979 (WHO, 1972; 1979).

The first of these reports concluded that effects would be seen above the concentrations (24-hour exposure) shown in Table 1. The second report largely agreed with the first but was more certain regarding the level of sulphur dioxide at which the condition of patients with existing respiratory disease would be worsened and concluded that this would be seen at 250 $\mu g/m^3$ (24-hour average). These conclusions were largely based on data derived from studies on the effects of the smogs in London in the 1950s and 1960s (Lawther *et al.*, 1970). It was felt that the magnitude of the effects at these levels of exposure was small and that it was difficult to separate these from effects due to other factors such as weather or infections. For smoke, it was felt that further modification of the value chosen, determined from the studies

Table 1. Concentrations (24-hour mean) at which effects of exposure to sulphur dioxide and smoke will be seen.

Effect	24-hour mean ($\mu g/m^3$)	
	SO_2	Smoke
Excess mortality among chronically sick	500	500
Worsening condition of those with existing respiratory disease	250–500	250–500

mentioned above, would need new evidence in view of the uncertainties concerning differences in composition of this component of pollution from one area to another and the change with time even in one locality.

The limit values in the Directive were derived directly from the short-term values in Table 1 using a 98th and 95th percentile approach without the use of any uncertainty factors. This approach means that for only 2% or 5%, respectively, of the time (units specified e.g. hours, days etc. in the Directive), the limit value is allowed to be exceeded. More stringent guide values were set at approximately half of the values shown above for effects of short-term exposure on those with existing respiratory disease. It is worth noting that the problem of episodic pollution was also addressed, and the Directive considered to have been breached if concentrations set as the 98th percentile of all daily mean values taken throughout the year were exceeded on three or more consecutive days.

At the time at which these reviews of health effects were published and the EC Directive adopted, there was little information on the effects of exposure to sulphur dioxide or suspended particulate matter individually. Most of the information available related to exposure to both pollutants simultaneously as most of the exposures at that time also occurred in this way — the main source of both pollutants in urban areas than being the domestic burning of coal. The Directive set limit values for smoke independently but those for sulphur dioxide were set based on the levels of black smoke.

At the time of setting the Directive, there was also little information on effects or levels of particulate matter as measured by gravimetric methods. Limit values were set for countries which used gravimetric rather than black smoke methods for particulates and it was accepted at the time that these values were not equivalent and would be revised. Some revision came in the form of Directive 89/427/EEC (1989) but it was still the case that the approaches were not completely equivalent and a future review of these was promised. The limit values for sulphur dioxide and suspended particulate matter as measured by black smoke and gravimetric methods are shown in Tables 2–4.

4.2. Nitrogen Dioxide

The third Directive on air quality in the European Community set limit and guide values for nitrogen dioxide (Council of EC, 1985). The

Table 2. Limit values for sulphur dioxide (all values in $\mu g/m^3$).

Reference period	Limit value	Associated value for suspended particles	
		As black smoke (OECD method)	By gravimetric method
Year (median of daily mean values taken throughout the year)	80 120	> 40 ≤ 40	> 150 ≤ 150
Winter (1 October–31 March, median of daily mean values taken through the winter)	130 180	> 60 ≤ 60	> 200 ≤ 200
Year (98th percentile of all daily mean values taken throughout the year)	250 (1) 350 (1)	> 150 ≤ 150	> 350 ≤ 350

(1) Member states must take all appropriate steps to ensure that this value is not exceeded for more than three consecutive days.

Table 3. Limit values for suspended particulates (as measured by the black smoke method described in Annex III of Directive 80/779/EEC (Council of EC, 1980) expressed in $\mu g/m^3$.

Reference period	Limit value
Year (median of daily mean values taken throughout the year)	80
Winter (1 October–31 March, median of daily mean values taken through the winter)	130
Year (98th percentile of all daily mean values taken throughout the year)	250 (1)

(1) Member states must take all appropriate steps to ensure that this value is not exceeded for more than three consecutive days.

Table 4. Limit values for suspended particulates (as measured by the gravimetric method described in Annex IV of Directive 80/779/EEC (Council of EC, 1980) expressed in $\mu g/m^3$.

Reference period	Limit value
Year (median of daily mean values taken throughout the year)	150
Year (95th percentile of all daily mean values taken throughout the year)	300

Directive proposal was made in 1983 and the Directive adopted in 1985. The limit values proposed were based on two sources of information regarding the effects of nitrogen dioxide: the WHO review of oxides of nitrogen published in 1977 (WHO, 1977) and a subsequent review of the more recent literature by Wagner commissioned by the European Commission and published in 1985 (Wagner, 1985).

Few data relating to the effects of nitrogen dioxide on man were available at the time of these reviews. Both concluded that the lowest level at which effects (increased airway resistance, increased sensitivity to bronchoconstrictors and enhanced susceptibility to respiratory infections) would be expected in animals studied under long- and short-term exposures, was 940 $\mu g/m^3$. It was considered that the studies showing enhanced susceptibility to respiratory infections were particularly important and that these might have parallels in man.

The WHO report concluded that adverse reactions in human subjects occurred at approximately the same concentrations. Under controlled conditions, human subjects exposed to nitrogen dioxide at concentrations of 1300–3800 $\mu g/m^3$ for ten minutes experienced increased airway resistance. It had also been shown in one study that exposure of asthmatics to 190 $\mu g/m^3$ for one hour increased the bronchoconstrictor effect of carbachol although the WHO group felt that this study required confirmation (Orehek *et al.*, 1976).

Wagner did not find any evidence to disagree with the WHO conclusions. From the available human studies, he concluded that clear effects were only seen at exposures above 1880 $\mu g/m^3$. Slight but significant effects on lung elasticity, airway resistance and forced vital capacity observed at exposures between 940 and 1880 $\mu g/m^3$ were considered to have been within the normal range of variation or due to other pollutants. The question of increased sensitivity of the respiratory tract was considered not to have been answered and that as the findings indicating such an effect had not been duplicated, these would not form a sound basis for the development of thresholds or standards. Some consideration was given to studies of indoor air pollution and respiratory infections from the UK and US. These were considered inconclusive but not to be disregarded in the light of animal studies showing enhanced susceptibility to respiratory infections.

The WHO document recommended that an exposure limit based on a safety factor of 3–5 applied to the value of 940 $\mu g/m^3$ (giving approximately 320–190 $\mu g/m^3$) as a maximum hourly exposure, not to be exceeded more than once per month, would be acceptable.

Taking both sources into account the Commission proposed a limit value of 200 $\mu g/m^3$ as a 98th percentile of hourly means measured over one year. This is at the lower end of the WHO range but as a 98th percentile, the value can be exceeded for 175 hours per year (~ 14 hours per month) and thus in this respect, the requirement is less stringent than WHO. This value was subsequently adopted along with a guide value of 135 $\mu g/m^3$, also as a 98th percentile of hourly values recorded throughout the year.

5. WHO Air Quality Guidelines

At the same time as the Commission was considering the development of new legislation on air quality, WHO was updating and revising its air quality guidelines for Europe. The first edition was published in 1987 and provided an authoritative view of the major air pollutants affecting Europe (WHO, 1987). These guidelines are not standards in themselves but provide a starting point for countries to set standards, taking local (i.e. national) socio-economic factors into account. Thus, a national air quality standard might be set at a lower or higher level than specified in the appropriate guideline and not be incorrect. Also, these guidelines do not specify thresholds of concentrations which, if exceeded, will inevitably be followed by damaging effects on health. In fact, they specify concentrations at which effects are **not** expected even among sensitive groups in the population. This is often forgotten and small exceedances of WHO guideline concentrations are often assumed to be associated with damage to health.

This approach was followed in revising the guidelines in the 1990s. The priority pollutants for revision were identified in collaboration with the European Commission and WHO had published a series of interim guidelines between 1994 and 1996. The second edition of the guidelines was published in 1999.

In deriving the guidelines, WHO Expert Groups have tended to apply uncertainty factors to Lowest Observed (Adverse) Effect Levels and thus the guidelines incorporate a margin of safety. For some pollutants, this margin is large, while for others, it is comparatively small. It should also be recalled that in defining a Lowest Observed

(Adverse) Effect Level, WHO Expert Groups had tended to be cautious and concluded that effects likely on exposure to about this level of pollution are unlikely to be particularly marked or widespread in the general population. The WHO Air Quality Guidelines for Europe were set on the basis that, as guidelines, they would be applicable throughout Europe. Consideration of socio-economic factors was still intended in moving from these guidelines to standards.

WHO Air Quality Guidelines specify guidelines for most pollutants in terms of a concentration and an appropriate averaging time. However, for carcinogens and particulate matter, no simple numerical guidelines have been produced. Instead, risk estimates were calculated. In the case of carcinogens, including benzene, 1,3-butadiene and polycyclic aromatic hydrocarbons (PAHs) the additional risk of a specified disease, e.g. leukaemia or lung cancer, likely to be experienced as a result of lifetime exposure to unit concentration of the pollutant is specified. For example, in the case of benzene, the additional risk of leukaemia associated with lifetime exposure to 1 $\mu g/m^3$ is estimated to be 6×10^{-6}. For PAHs, the additional risk of lung cancer associated with lifetime exposure to a mixture of PAHs characterised by a concentration of benz-α-pyrene of 1 ng/m^3 is 8.7×10^{-5}. It will be appreciated that such guidelines cannot be simply read across to standards. Before a concentration and averaging time can be set, some agreement on the level of acceptable risk is needed. In practical terms, specifying the averaging time as an annual average is widely accepted as appropriate for carcinogenic air pollutants. Table 5 shows the interim WHO guidelines as are currently available.

In considering particulate matter, WHO Expert Groups meeting in 1996 concluded that current epidemiological evidence suggested that particulate matter should be regarded as a no-threshold pollutant for harm within the population as a whole and that it should be considered independently of sulphur dioxide. This represented a significant change from the first edition of the WHO Air Quality Guidelines for Europe where sulphur dioxide and particulate matter were treated together and guidelines produced were based on the concentrations of both. It has

Table 5. Revised WHO Air Quality Guidelines.

Compound	Guideline	Averaging time	Unit risk estimate (Unit concentration)
Arsenic			1.5×10^{-3} ($\mu g/m^3$)
Cadmium	5 ng/m^3	1 year	
Chromium (Cr^{iv})			4×10^{-2} ($\mu g/m^3$)
Fluoride	1 $\mu g/m^3$	1 year	
Lead	Not yet agreed		
Manganese	0.15 $\mu g/m^3$	1 year	
Mercury	No guideline proposed*		
Nickel			3.8×10^{-4} ($\mu g/m^3$)
Platinum	No guideline proposed*		
Benzene			6×10^{-6} ($\mu g/m^3$)
Butadiene	Not yet agreed		
Dichloromethane	3 mg/m^3 0.45 mg/m^3	24 hours 1 week	
Formaldehyde	0.1 mg/m^3	30 minutes	
PAHs (as BaP)			8.7×10^{-5} (ng/m^3)
Styrene	70 $\mu g/m^3$	30 minutes	
Tetrachloroethylene	0.25 mg/m^3	Not yet specified	
Toluene	0.26 mg/m^3	1 week	
Trichloroethylene			4.3×10^{-7} ($\mu g/m^3$)
Carbon monoxide	100 mg/m^3 60 mg/m^3 30 mg/m^3 10 mg/m^3	15 minutes 30 minutes 1 hour 8 hours	

Table 5 (*Continued*)

Compound	Guideline	Averaging time	Unit risk estimate (Unit concentration)
Nitrogen dioxide	200 μg/m³ 40 μg/m³	1 hour 1 year	
Ozone	120 μg/m³	8 hours	
Particulate matter	No single guideline recommended*		
Sulphur dioxide	500 μg/m³ 125 μg/m³ 50 μg/m³	10 minutes 24 hours 1 year	
PCBs	No guideline proposed*		
PCDDs and PCDFs	No guideline proposed*		

*Failure to recommend a guideline does not imply that the compound or compounds are of low toxicity. The second edition of the WHO Air Quality Guidelines should be consulted for detailed explanations.

already been noted that a similar approach was taken in the first EC Directive on air pollution (80/779/EEC) adopted in 1980 (Council of EC, 1980). In both cases, the recommended levels of sulphur dioxide and smoke were based upon contemporary epidemiological studies of the then common mixture of sulphur dioxide and smoke. In revising the WHO guidelines, Expert Groups concluded that the appearance of new epidemiological data and the general disappearance of classic coal smoke pollution warranted treating these pollutants separately. Detailed examination of the results of recent time-series epidemiological studies (see Chapter 5) showed that assumptions of a threshold of effect for particles were unsupportable. Instead, a series of coefficients defining the demonstrated association between daily average concentrations of particulate matter and daily number of health-related events, e.g. deaths and hospital admissions, were given. Tables of expected numbers of such events in specified populations at specified daily average concentrations of particles measured as PM_{10} and $PM_{2.5}$ (mass concentrations of particles of generally less than 10 and 2.5 μm diameter, respectively) thus form the latest WHO guideline. Here again, as with carcinogens, immediate adoption as a standard is impossible since an acceptable level of effect has to be defined.

6. Other Technical Input

Directives setting air quality standards also define how and where air pollutants should be monitored. In the past, the directives defined a particular method of measurement but allowed use of other methods if member states could demonstrate that their methods provided equivalent measurements to the reference method. Advice on measurement methods comes from technical experts within the member states and from the European Commission's Joint Research Centre at Ispra which also organises interlaboratory comparisons of different methods.

There has been a long-running debate about the siting of monitoring stations in order to comply with EC Directives. Some member states have taken the view that if they monitor where levels are likely to be highest, e.g. at the kerbside for pollutants emitted from motor vehicles, and at these points comply with the limit values in the Directive, then levels elsewhere will be below these values. Other member states have taken the view that measurements should be made where population exposure is likely to be highest. Even when directives have been complied with, there have been some difficulties when higher levels are recorded close to sources (Report: Commission on State of Implementation of Ambient Air Quality Directives, 1995).

7. The Air Quality Framework Directive

Following the European Commission's commitment to revise and update its existing air quality directives and to extend the list of pollutants for which Union-wide air quality objectives will be set, a new approach was developed. This consisted of establishing a framework directive defining principles followed by a series of daughter directives addressing individual pollutants. In 1996, the new Directive providing the framework for the assessment and management of ambient air quality was adopted (Council Directive, 1996). This Directive had four objectives:

(i) establishment of a framework for setting new standards for air pollutants;
(ii) assessment of air quality using common methods and criteria;
(iii) obtaining and disseminating information on air quality;
(iv) maintenance of good air quality and improvement where necessary.

This Directive set a timetable for the European Commission to come forward with new proposals on 12 air pollutants (see Table 6). Proposals on the first four pollutants in this list — sulphur dioxide, nitrogen dioxide, particulate matter and lead — were published in 1997 and are currently under discussion. Further proposals on ozone, benzene and carbon monoxide are currently being developed.

Table 6. Air pollutants which will be addressed in the European legislation.

Sulphur dioxide
Suspended particulate matter
Nitrogen dioxide
Lead
Ozone
Benzene
Carbon monoxide
Cadmium
Nickel
Arsenic
Mercury
PAH compounds

The Framework Directive required that a number of parameters be addressed in the daughter legislation and these include:

(i) the establishment of limit values and date for achievement of these;
(ii) the establishment of temporary margins of tolerance (see Fig. 1);
(iii) the establishment of alert thresholds as appropriate;
(iv) the establishment of criteria and reference techniques for measurement;
(v) the establishment of assessment requirements in relation to population density.

8. Development of Daughter Legislation

For each of the pollutants, a technical working group was set up comprising representatives from the member states, industry, environmental groups, WHO, European Environment Agency and the Commission. These groups considered the technical issues surrounding each of the pollutants and provided a report which was subsequently discussed in a meeting of all member states. The Commission then developed the daughter legislation based on this report and outcome of the discussions.

8.1. Limit Values

As indicated above, the Commission relied heavily on the work of WHO in providing the basis for the development of new air quality standards. The technical working groups were invited to make a critical evaluation of the conclusions reached by WHO. In general, these groups felt that the revised WHO guidelines represented the "ideal" and efforts should be made to achieve these. Therefore, rather than amending the guideline values to produce limit values, consideration was given to the date by which these could be complied with. This is considered in more detail below.

Limit values will consist of a concentration, averaging time and also, an allowed number of exceedances per year reflecting circumstances outside the control of member states, e.g. abnormal weather conditions and performance characteristics of measurement equipment. This approach was also used in the earlier air quality directives as described above. For example, Directive 85/203/EEC (Council of EC, 1985) on nitrogen dioxide specified an air quality limit value of 200 $\mu g/m^3$ as the 98th percentile of hourly mean concentrations of NO_2 recorded throughout the year. This means that up to 175 (2% of 365 × 24) hourly average concentrations can exceed 200 $\mu g/m^3$, to an undefined extent, without infringing the Directive. This form of specification of the Directive has proved difficult to grasp by the public and environmental pressure groups. It is generally not understood that until 175 hourly average concentrations have exceeded 200 $\mu g/m^3$ has the Directive been infringed: cries that a value of 400 $\mu g/m^3$ "breaches the Directive" are thus meaningless. In saying this, it should be understood that the specified acceptable degree of compliance is not based on an appreciation of the effects of NO_2 on health. Nobody would argue that exposure to concentrations of NO_2 in excess of 200 $\mu g/m^3$ on 174 occasions had no effect, but that for 175 occasions, an effect would be produced. The part of the limit value that is health-based is the 200 $\mu g/m^3$ hourly average concentration. Thus, it **would** be correct to say that on **each** occasion when a concentration in excess of 200 $\mu g/m^3$ is reported, this **is** undesirable in health terms but **not** that the Directive has been infringed.

Discussions to establish limit values for carcinogenic air pollutants are only just beginning. The additional difficulty of having to define an acceptable risk has already been described . Some European countries, for example The Netherlands, have already done this and established standards for carcinogens. It should be noted that the approach taken by WHO Expert Groups in dealing with carcinogens is not the only available approach. In the UK, for example, national non-mandatory air quality standards for benzene, 1,3-butadiene and PAH compounds have been set. These have been derived by identifying a level of exposure at which no or minimal effects have been recorded

Table 7. UK standards for carcinogens.

Pollutant	Standard (ann. av.)
Benzene	5 ppb
1,3-butadiene	1 ppb
PAHs	0.25 ng/m^3

from occupational studies and applying a series of uncertainty or safety factors. Thus, standards which do not guarantee complete absence of effects but rather suggest that such exposure of the UK population to the specified level is likely to "present an exceedingly small risk to health" have been recommended. The UK standards for carcinogens are shown in Table 7.

8.2. Alert Thresholds

Alert thresholds were introduced into European air quality legislation in the ozone Directive which set two values, namely, an information threshold and a warning threshold. The Framework Directive recognised that it may not be appropriate to set alert thresholds for all pollutants. These can have a number of functions, for example:

(i) establishment of a value which triggers a series of management actions, such as restrictions on traffic, industrial emissions, etc;
(ii) establishment of values where there are serious public health concerns and thus, sensitive groups within the population need to warned of possible consequence and actions they may take.

Often these are not clearly separated.

8.3. Managing Air Quality

The aim of the Framework Directive was to improve air quality rather than simply set limit values and prosecute member states when these

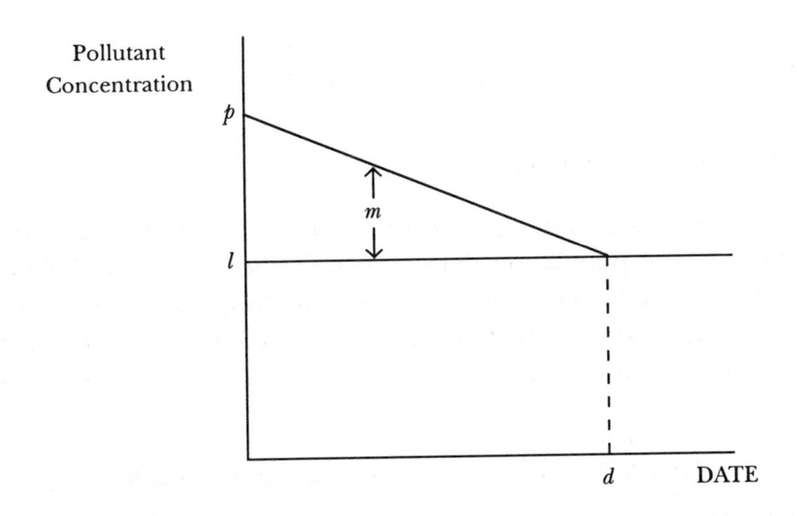

d: date by which limit value should be reached
l: limit value
p: current pollutant concentration
m: margin of tolerance, declining with time

Fig. 1. Framework for managing air quality.

are exceeded. In order to achieve this, the concept of the "margin of tolerance" was introduced. This is illustrated in Fig. 1. This concept allows time for the member states to develop and put into place a management plan which will enable them to improve air quality as necessary so that the limit value is complied with within the specified time. The margin of tolerance decreases with time so that the limit value is complied with by the attainment date. A further aspect of this concept is that it identifies those areas where air quality is worst and requires that detailed action be planned and taken so that efforts for improvement are concentrated in these areas.

Where air quality is above the margin of tolerance, member states must develop detailed action plans showing how the limit value will be achieved within the specified time. Where air quality is between the limit value and margin of tolerance, member states must report to the

Commission but must meet the limit value by the specified date. The margin of tolerance is defined for each air pollutant being addressed.

8.4. Assessment of Air Quality

Unlike earlier legislation, the effort required to be put into assessment is greater where population density is high and there is a likelihood of the limit value being exceeded. Upper and lower assessment thresholds are statistically defined so that the most intensive, high-quality measurement is used in the worst areas, whereas in areas where there

Table 8. Proposed requirements for assessing air quality.

Maximum pollution level in agglomeration or zone	Assessment requirements
1. Greater than upper assessment threshold	High quality measurement is mandatory. Data from measurement may be supplemented by information from other sources including air quality modelling
2. Less than upper assessment threshold but greater than lower assessment threshold	Measurement is mandatory but fewer measurements may be needed, or less intensive methods may be used, provided that measurement data are supplemented by reliable information from other sources
3. Less than lower assessment threshold	
a. In agglomerations only for pollutants for which an alert threshold has been set	At least one measuring site is required per agglomeration combined with modelling, objective estimation, indicative measurements
b. In non-agglomeration zones for all pollutants and in all types of zone for pollutants for which no alert threshold has been set	Modelling objective estimation and indicative measurements alone are sufficient

is little likelihood of exceedance, member states can use less accurate methods and modelling techniques. Table 8 sets out the proposed requirements in terms of upper and lower assessment thresholds. The Directive also proposes numbers of sampling points based on the population and the air pollutant concentrations in relation to the upper and lower thresholds. Guidance is also provided in relation to diffuse and point sources.

8.5. Economic Considerations

Article 130r of the Treaty on European Union (1996) states that the costs of action and inaction must be taken into account in developing new legislation on environmental issues. For new legislation on air quality, this was addressed through a Commission-sponsored study on the costs and benefits associated with meeting proposed limit values. This study determined whether additional action was needed beyond that already planned to meet proposed limit values, estimated costs using the most cost-effective solutions and assessed the additional benefits expected from meeting limit values.

Doing such a study on a European scale has considerable difficulties in addition to the usual problems of quantifying benefits in monetary terms. The approach chosen was to assess air quality on a regional scale and for cities where air quality information was available. Reference scenarios were developed for each pollutant, taking into account the effects of existing national, EC and international legislation together with Commission proposals up to the end of 1996. Costs could then be assigned to any additional measures necessary to achieve the proposed limit values. Carrying out such studies on a local basis may provide different results as local factors which might affect the outcome would be taken into account.

The Commission study employed the valuation of the statistical life approach which assesses the willingness-to-pay of individuals to reduce the risk of morbidity and mortality. The results are intended to indicate the importance attached to different types of risk. A number of studies employing this technique have been undertaken in Europe and have produced a range of values for a statistical life. The choice of which value to use is difficult but is obviously important if costs and benefits are to be compared.

A further difficulty when considering some air pollutants, for example, particulate matter, is that mortality appears to be greatest in those who are elderly or already chronically ill. The "willingness to pay" to avoid risks may not be the same in these groups as in the general population.

9. Current Situation

At the time of writing this chapter, the Commission's proposals for new limit values for sulphur dioxide, nitrogen dioxide, particulate matter and lead have been published. These were agreed at a meeting of the EU Environment Council in June 1998. The agreed figures are shown in Tables 9–12. Compliance dates for achieving the standards are set at 2005 for sulphur dioxide and lead, and 2010 for nitrogen dioxide. A two-stage approach is proposed for particulate matter which is described further below.

For sulphur dioxide, it can be seen that the Commission has proposed a one-hour limit value in order to minimise the monitoring burden on member states rather than a ten-minute value as recommended by WHO. This was developed from data provided by member states on short-term peaks in industrial areas. The Commission will require that both the one-hour and ten-minute data are provided so that the effectiveness of the one-hour limit value in protecting against exceedances of the WHO ten-minute guideline can be checked. The 24-hour value is directly taken from WHO. No annual value was set as the Commission felt, following examination of data from member states, that if the 24-hour value was complied with, there would be no need for an annual value. A further exception for sulphur dioxide is that it is the only pollutant for which an alert threshold was felt to be justified. This was based on the results of experiments in which asthmatic patients were exposed to sulphur dioxide when exercising

Table 9. Limit values for sulphur dioxide.

	Averaging period	Limit value	Margin of tolerance	Date by which limit value is to be met
1. Hourly limit value for the protection of human health	1 hour	350 $\mu g/m^3$ not to be exceeded more than 24 times per calendar year	150 $\mu g/m^3$ (43%) on entry into force of this Directive. Reducing on 1 January 2001 and every 12 months thereafter by an equal annual percentage to reach 0% by 1 January 2005	1 January 2005
2. Daily limit value for the protection of human health	24 hours	125 $\mu g/m^3$ not to be exceeded more than 3 times per calendar year	None	1 January 2005
3. Limit value for the protection of ecosystems	Calendar year and winter (1 October to 31 March)	20 $\mu g/m^3$	None	Two years from entry into force of the Directive

Limit values shall be expressed in $\mu g/m^3$. The volume must be standardised at the following conditions of temperature and pressure: 293 K and 101.3 kPa.

Table 10. Limit values for nitrogen dioxide and nitric oxide.

	Averaging period	Limit value	Margin of tolerance	Date by which limit value is to be met
1. Hourly limit value for the protection of human health	1 hour	200 $\mu g/m^3$ NO_2 not to be exceeded more than 18 times per calendar year	50% on entry into force of this Directive. Reducing on 1 January 2001 and every 12 months thereafter by an equal annual percentage to reach 0% by 1 January 2010	1 January 2010
2. Annual limit value for the protection of human health	Calendar year	40 $\mu g/m^3$ NO_2	50% on entry into force of this Directive, reducing on 1 January 2001 and every 12 months thereafter by an equal annual percentage to reach 0% by 1 January 2010	1 January 2010
3. Annual limit value for the protection of vegatation	Calendar year	30 $\mu g/m^3$ NO_2	None	Two years from entry into force of the Directive

Limit values shall be expressed in $\mu g/m^3$. The volume must be standardised at the following conditions of temperature and pressure: 293 K and 101.3 kPa.

Table 11. Limit values for particulate matter.

	Averaging period	Limit value	Margin of tolerance	Date by which limit value is to be met
Stage 1				
1. 24-hour limit value for the protection of human health	24 hours	50 $\mu g/m^3$ PM_{10} not to be exceeded more than 35 times per year	50% on entry into force of this Directive, reducing on 1 January 2001 and every 12 months thereafter by an equal annual percentage to reach 0% by 1 January 2005	1 January 2005
2. Annual limit value for the protection of human health	Calendar year	40 $\mu g/m^3$ PM_{10}	20% on entry into force of this Directive, reducing on 1 January 2001 and every 12 months thereafter by an equal annual percentage to reach 0% by 1 January 2005	1 January 2005
Stage 2*				
1. 24-hour limit value for the protection of human health	24 hours	50 $\mu g/m^3$ PM_{10} not to be exceeded more than 7 times per year	[to be derived from data and to be equivalent to the Stage 1 limit value]	1 January 2010
2. Annual limit value for the protection of human health	Calendar year	20 $\mu g/m^3$ PM_{10}	50% on 1 January 2005 reducing every 12 months thereafter by an equal annual percentage to reach 0% by 1 January to 2010	1 January 2010

*Indicative limit values to be reviewed in the light of further information on health and environmental effects, technical feasibility, and experience in the application of Stage 1 limit values in the member states.

Table 12. Limit values for lead.

	Averaging period	Limit value	Margin of tolerance	Date by which limit value is to be met
Annual limit value for the protection of human health	Calendar year	0.5 $\mu g/m^3$*	100% on entry into force of this Directive, reducing on 1 January 2001 and every 12 months thereafter by an equal annual percentage to reach 0% by 1 January 2005 or by 1 January 2010, in the immediate vicinity of specific point sources, which shall be notified to the Commission	1 January 2005 or 1 January 2010, in the immediate vicinity of specific industrial sources, which are situated on sites contaminated by decades of industrial activities which shall be notified to the Commission by the date referred to in Article 12.** In such cases, the limit value from 1 January 2005 shall be 1.0 $\mu g/m^3$

*The review process for this Directive foreseen in Article 10 will consider supplementing or replacing the limit value by a deposition limit value in the immediate vicinity of point sources.
**This notification shall be accompanied by an appropriate justification. The area in which higher limit values apply shall not extend beyond 1000 m from such specific sources.

and is aimed at this sensitive group within the population. Although short-term exposures to nitrogen dioxide and particulate matter have been associated with adverse effects, there are no clear thresholds on which to base alert thresholds for each.

As noted above, WHO did not recommend a guideline value for particulate matter. In proposing a standard, the Commission adopted a risk management approach which seeks to identify concentrations at which effects on the population as a whole would be small. The Commission is proposing that limit values are introduced for PM_{10} in two stages. Limit and action levels are shown in Table 11. Recognising that there are doubts over the size fraction of particulate matter associated with effects on morbidity and mortality, that much research is underway and that information on concentrations of this in Europe is limited, the Commission has committed to reporting to the European Council and Parliament by the year 2003 on developments in the scientific and technical understanding of particulate matter.

In conclusion, this is currently a time of new developments in air quality assessment and management in Europe. A new framework has been established and new limit values are being agreed which will result in improvements in air quality in Europe over the next ten years.

References

Council Directive 96/61/EC concerning integrated pollution prevention and control (1996) *Off. J. Eur. Communities,* L257/26.

Council Directive 96/62/EC on ambient air quality assessment and management (1996) *Off. J. Eur. Communities,* **L296**, 55.

Council of the European Communities (1980) Directive on Air Quality Limit Values and Guide Values for Sulphur Dioxide and Suspended Particulates (80/779/EEC). *Off. J. Eur. Communities,* **L229**, 30–48.

Council of the European Communities (1982) Directive on a Limit Value for Lead in the Air (82/884/EEC). *Off. J. Eur. Communities,* **L378**, 15–18.

Council of the European Communities (1985) Directive on Air Quality Standards for Nitrogen Dioxide (85/203/EEC). *Off. J. Eur. Communities,* **L87**, 1–7.

Council of the European Communities (1989) Council Directive Amending Directive 80/779/EEC on Air Quality Limit Values and Guide Values for

Sulphur Dioxide and Suspended Particulates (89/427/EEC). *Off. J. Eur. Communities*, **L201**, 53–55.

Council of the European Communities (1992) Directive on Air Pollution by Ozone (92/72/EEC). *Off. J. Eur. Communities*, **L297**, 1–7.

Elsom D. (1987) *Atmospheric Pollution.* Blackwell's Ltd., Oxford.

Hough A.M. & Derwent R.G. (1990) *Nature*, **344**, 645–648.

Lawther P.J. *et al.* (1970) *Thorax*, **25**, 525–539.

Orehek J. *et al.* (1976) *J. Clin. Invest.* **57**, 301–307.

Photochemical Oxidants Review Group (1987) *Ozone in the United Kingdom.* Department of the Environment, London.

Photochemical Oxidants Review Group (1997) *Ozone in the United Kingdom. Fourth Report of the Photochemical Oxidants Review Group.* Department of the Environment, Transport and the Regions, London.

Report from the Commission on the State of Implementation of Ambient Air Quality Directives (1995) Com(95) 372 Office for Official Publications of the European Communities, Luxembourg.

Towards Sustainability (1993) A European Community programme of policy and action in relation to the environment and sustainable development. Office for Official Publications of the European Communities, Luxembourg.

Treaty on European Union (1996) *Treaty Series No. 29.* HMSO, London.

Wagner H.M. (1985) *Update of a Study for Establishing Criteria (Dose-Effect Relationships) for Nitrogen Oxides. Environment and Quality of Life Series.* EUR 9412

World Health Organisation (1972) *Air Quality Criteria and Guides for Urban Air Pollutants. WHO Technical Report Series No. 506.* World Health Organisation, Geneva.

World Health Organisation (1977) *Oxides of Nitrogen. Environmental Health Criteria Series, No 4.* World Health Organisation, Geneva.

World Health Organisation (1979) *Sulfur Oxides and Suspended Particulate Matter Environmental Health Criteria Series, No. 8.* Geneva: World Health Organisation.

World Health Organisation (1987) *Air Quality Guidelines for Europe. WHO Regional Publications, European Series No. 23.* World Health Organisation, Copenhagen.

CHAPTER 9

RISKS, ESTIMATION, MANAGEMENT AND PERCEPTION

M. Jantunen

The ability to sense and avoid harmful environmental conditions is necessary for the survival of all living organisms. Survival is also aided by an ability to codify and learn from past experience. Humans have an additional capability that allows them to alter their environment as well as to respond to it. This capacity both creates and reduces risk. (Slovic, 1987). The goal of a risk assessment is to produce measures of risks, which contain the probabilities and severities of the possible adverse outcomes, the uncertainties involved, and to allow a maximum degree of comparability between different hazards and risks.

In modern West European and North American urban areas, the health risks of air pollutants — mainly from traffic, heat and power production and atmospheric chemistry — extend over large areas and consequently affect a great proportion of populations. Compared to pollutants from large industrial point sources, urban air pollution is relatively uniformly spread. High pollution episodes may be caused by unfavourable meteorological conditions, and such episodes may affect much larger areas than single cities.

It is important to realise that although significant associations have been observed between ambient urban air pollution levels and mortality and morbidity in the populations, most of the exposure occurs in indoor environments. Buildings both modify the outdoor air pollutants and add new pollutants from indoor sources into the exposure mixture. Unlike the rather smoothly dispersed urban outdoor air pollutants, air pollutants from indoor sources are very unevenly distributed among the spaces and individuals. At this stage, the most obvious needs for urban air pollution risk assessment and management rationale focus on non-cancer endpoints, complex and variable exposure mixtures and general populations in everyday urban environments.

Risk management should be an organised effort to collect information and to control risks. The public cannot and never will achieve risk-free air, water, or food. Risk management is fundamentally a

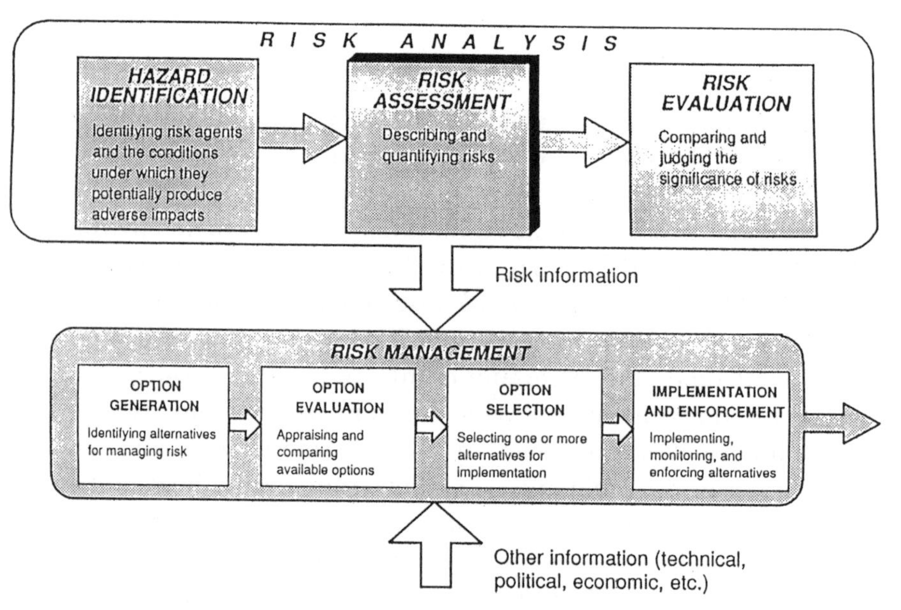

Fig. 1. The stages of risk analysis — hazard identification, risk assessment, and risk evaluation — provide key information for risk management (Covello & Merkhover, 1993).

question of values. In a democratic society, there is no acceptable way to make these choices without involving the citizens who will be affected by them (Moghissi, 1993).

The mutual relationships of the various steps in risk assessment and risk management according to Covello and Merkhover (1993) are presented in Fig. 1. Another view is provided later in Fig. 2.

1. Definitions of Risk

For a researcher, risk is *the magnitude of an unwanted consequence related to some activity multiplied by its probability*. This definition stems from insurance mathematics.

This is hardly the definition of risk likely to be given by members of the general public. Webster's Unabridged Dictionary defines *risk* as it is defined in mathematical risk analysis, but also as a synonym of *danger, peril, jeopardy, hazard.* Even in a framework limited to health risk assessment, risk has had many definitions. The following are found in relatively recent literature:

* A function determined by three variables: (a) the likelihood of a particular hazard causing harms to the exposed individuals, (b) the magnitude (severity) of the harms or their consequences, and (c) the number of people exposed to the hazard (e.g. Cox & Tait, 1991).
* The probability that an event will occur, e.g. that an individual will become ill or die within a stated period of time or age (Last, 1988).
* The magnitude of an unwanted consequence related to some activity multiplied by its probability (Morgan, 1993).
* A characteristic of a situation or action wherein two or more outcomes are possible, the particular outcome that will occur is unknown, and at least one of the possibilities is unwanted (Covello & Merkhofer, 1993).
* ... the probability of an adverse effect or condition ... and detriment (is) a product of probability and severity (Chadwick *et al.*, 1995).

* ... incidence and severity of the adverse effects likely to occur in a human population ... (in CEC Commission Directive 93/67/EEC, 1993, definition of "risk characterisation").

These definitions do not only differ in wording but also in substance. Sometimes, the type of application accounts for the differences in definitions, for example, if only fatal outcomes are considered, the severity element can be dropped. Often the differences between the definitions are even more fundamental. Lay people have different, broader definitions of risk, which in important respects, can be more rational than the narrow ones used by experts (Morgan, 1993). The risk perception of lay people depends on lack of previous knowledge of the risk, on dread of consequences (e.g. asbestos or radiation) outside their control, and on whether the risk has been imposed by an anonymous body outside of one's control, or whether it has been voluntarily taken for a practical or emotional purpose.

In a new American magazine, ICON (Zabriskie *et al.*, 1997), 40 individuals — public figures in arts, athletics, business, politics, media and public service, but none from science — were interviewed about "What is risk?" in their own lives. Not one of those interviewed was talking about risk as how a professional risk assessor defines it. They were discussing active risk-taking rather than passive risk-acceptance, mostly about calculated risks, risks that were searched for, and of the personal emotional rewards related to taking and overcoming those risks. No wonder there are considerable differences between the risk assessments made by many scientific assessors, let alone between the risk perceptions of the public and scientists. These may simply reflect the fact that the two sides mean two or more different things by the same word *risk*.

What is then the role of a researcher? Should the researcher reject the given mathematical definition and accept a broader definition of *risk* to better communicate with the public and politicians? Assuming that we still need *risk assessment* that produces *risk characterisations* which can be critically reviewed and compared, we still need a definition for *risk* that is unambiguous and enables *risk quantification* and *comparison*.

However, in communication with the general public, the researcher should always define exactly what he means by risk, and understand and accept the fact that the other party may not use this term in exactly the same way.

2. Risk Assessment

Risk assessment is the scientific and technical method for systematically evaluating the risk associated with an action or an existing condition (Moghissi, 1993). Risk assessment has also been described as a way of examining risks so that they can be better avoided, reduced, or otherwise, managed (Wilson & Crouch, 1987). Environmental policy-making has become more dependent on formal, quantitative risk assessment because of increasing attention paid to the prevention of human health damage from harmful exposures. Risk assessment helps set priorities for regulation of the very large number of chemicals that are of potential concern and helps direct limited social and governmental resources to the most significant risks (Russel & Gruber, 1987).

The starting point which determines the risk assessment approach in a specific case can be an *emission* (e.g. diesel exhaust), *environment* (e.g. a heavily industrialised town), *substance* [e.g. sulphur dioxide (SO_2)], *recipient* (e.g. individuals with asthma) or an *effect* (e.g. lung cancer). While the starting point defines the emphasis of a risk assessment, the whole chain from emissions to effects needs to be covered in a comprehensive risk assessment of any air pollution hazard.

The range of *uncertainty in risk assessment* is relatively small for well-known risks like cigarette smoking or high levels of SO_2 and fine particles in the air. The range is much broader for less known risks like those from low level exposures to dioxins from waste incinerators, or platinum and rhodium from catalytic converters of cars. Risk comparisons based on the high ends of probability distributions may overestimate a little known risk by several orders of magnitude in relation to a well-known risk, making the whole comparison outright misleading.

Risk assessment should be a scientifically based exercise and should only be limited by the normal ethical rules for experimental research and the protection of individuals' rights. In scientific risk assessment, the risk analyst should not attempt to overstate or understate threats, but rather to give the best estimate and the range of uncertainty. It is the role of decision makers to choose the proper amount of conservatism in setting the standard. The various governmental and international agencies ought to coordinate their risk assessment processes so that they arrive at similar estimates for a particular hazard (Lave, 1987).

While public trust in risk management has declined, ironically, the discipline of risk assessment has matured. It is now possible to examine potential hazards in a rigorous, quantitative fashion and thus to give people and their representatives facts on which to base essential personal and political decisions (Morgan, 1993).

2.1. Risk Comparisons

The purpose of *risk assessment* is to help in making decisions about the hazards causing risks, and so it is important to gain some perspective about the meaning and magnitude of the risk. *Risk comparisons* can be useful. We are not born with an instinct about what a risk of one in a million per lifetime means. It is particularly helpful to compare risks that are calculated in a similar way. Another common procedure is to compare exposures only. In some cases, it is also useful to contrast risks to indicate the different ways in which they are treated in society (Wilson & Crouch, 1987).

When more than two risks are compared with each other, the common method used is *risk ranking*. Risks may be ranked by using *risk statistics*, which are only available for certain clear-cut common risks like automobile driving or cigarette smoking, by using *animal toxicity* or *carcinogenicity* data, which may not provide appropriate ranking for humans (Ames *et al.*, 1987), or by asking a *panel of experts* for their combined professional judgement (Slovic *et al.*, 1980).

Risk comparison can be relatively straightforward when comparing well-known risks and similar outcomes, such as lung cancer risk from environmental tobacco smoke (ETS), radon and asbestos exposure. Such well-based risk comparison is, however, a rare luxury. Risk comparison is most useful and also, often most controversial, when comparing risks of competing or alternative services or techniques.

For risk comparison to be meaningful, it should use background data of comparable quality, or at least clarify the major differences between the data and their possible consequences for the risk assessment. The universal "yardsticks" that have been used in such comparisons include cases of death or disease, money and lost workdays. The most straightforward looking, yet most problematic, of these units are cases of death and money. Raw counts of deaths do not distinguish between the death of a paralysed 90-year-old, a 30-year-old working mother and a 10-year-old schoolboy. Direct and short-term monetary costs and benefits (of, e.g. catalytic converters and auto emissions reduction) are sometimes relatively easy to estimate, but they may be marginal in comparison to unestimated longer term (e.g. impact on climate change) or indirect costs (e.g. urban housing and transport development) and benefits (e.g. commuting to a home in a cleaner environment).

Comparison of dissimilar risks, such as traffic accidents versus sensory stress from irritating diesel exhaust, should make the best use of any relevant and straightforward comparison units, but should also provide risk managers and lay people with additional information, such as uncertainties, weights of evidence, conditions of analysis, and other factors restricting comparability, in a form which allows them to make as informed and responsible selections as possible. Simply incomparable qualities should be described but not compared.

Comparison of one risk estimate based on the exposure level that gives a certain outcome (cancer) probability per lifetime with risk estimates based on the lowest observed adverse effect level (LOAEL) is particularly tricky. In this case, the seemingly similar qualities (concentration units) are not comparable. Because different chemicals and pollution mixtures exhibit different dose-response curves (e.g. sigmoid versus linear, with and without threshold, etc.), it is essential that risk comparison is based on realistic and probable exposures. Just the apparently simple task of adding lifetime risks from compounds in a mixture can be extremely complicated, when the risks are characterised by different dose-response curves.

2.2. Epidemiological and Toxicological Risk Assessment Tools

The identification of the health risks of air pollutants in the 1990s has been dominated by the results of epidemiological studies, most importantly, the American Six Cities study (a 14–16-year, six-cohort, follow-up study) (Dockery *et al.*, 1993), and the European APHEA study (an 11-centre international daily mortality versus daily air pollution time-series study) (Katsouyanni *et al.*, 1997).

The epidemiological risk assessment process is based on numerical modelling of the relationship between the exposure (dose) of the harmful agent and the probability of a health outcome. The European Commission (EC) has published a comprehensive manual of study designs for Air Pollution Epidemiology (Katsouyanni, 1993). This process has two types of built-in errors. The first is practical, based on errors in measurement of exposure and health effect, or errors in the selected model. The second type of error is fundamental. *Verification and validation of numerical models in natural systems,* including the relationship between exposure and risk, *is impossible. This is because natural systems are never closed and because model results are always non-unique.* Measuring and correcting for confounders and incorporating effect modifiers into the model are scientific efforts to deal with this problem, but they are only effective in dealing with relatively small demographic, socio-economic, ethnic and cultural variations between the exposed groups and equally small variations in the total exposure mixtures. The validity of models can only be confirmed case by case by the demonstration of agreement between observation and prediction .... The most appropriate use of models is for *sensitivity analysis* (Oreskes *et al.*, 1994).

Although air pollution epidemiology studies of very different designs and magnitudes have been performed, certain features typical of urban air pollution epidemiology studies can be identified;

(i) complex exposures and rough exposure estimates, gradual ranges of exposures instead of exposed/non-exposed groups,
(ii) imprecise health effects measures such as overall or classified mortality, hospital admissions; only very simple physiological measurements, such as peak expiratory flow, and
(iii) weak associations between exposures and effects.

On the other hand, the studied populations and consequently, the number of observed events are often very large (especially in multi-centre time-series studies), giving the studies very high detection power. Careful treatment of confounders and effect modifiers is exceptionally important in air pollution epidemiology, because their impacts on the health outcomes may be larger than those of the air pollutants. These were discussed in Chapter 3. The roles of socio-economic and cultural factors in air pollution epidemiology were discussed in depth in a European Union report (Jantunen, 1997). The need to control the researcher's bias caused by a personal agenda in low-risk assessment is also important. This bias may link to error in both exposure evaluation, case selection or presentation of the results (Riylander, 1992).

On the counter-balance of all these sources of errors, epidemiological surveys study real populations in their real activities exposed to realistic levels of real pollution mixtures. An association observed in a well-designed environmental epidemiology study may contain large uncertainties, but it is definitely relevant for man in his environment.

In principle, the issues of uncertainty and bias apply also to experimental research on homogenous strains of laboratory animals. However, such research is conducted in significantly better controlled, closed systems than epidemiological research, making experimental results much more precise, repeatable and assignable to the assumed cause, in comparison with the statistical associations observed in epidemiological studies. Experimental studies are crucial for providing mechanistic information about cause and effect and thereby for establishing biological plausibility. These benefits are, however, compromised by the fact that experimental research is often conducted on artificial exposures (high and isolated) of homogeneous (age, strain, gender) groups of animals instead of men and women. Controlled experimental studies on humans, when ethically appropriate, are important final steps in risk assessment.

In practice, epidemiological and toxicological research both serve important and different purposes in risk assessment; toxicological experiments provide pharmacokinetic and mechanistic information and causal links between controlled exposures and effects, while epidemiological studies provide quantitative associations for heterogeneous human populations in the variable, complex and dynamic real-life exposure settings.

Reference populations

The reference population in managing industrial occupational environments comprises non-pregnant, healthy adults between 20 and 65 years of age, not exposed outside working hours. The reference population for the ambient air or indoor air quality guidelines is usually not specifically defined, because it has been assumed to be the general population with uninterrupted exposure. The increased responses of children, elderly, sensitive, health compromised, and other special social groups etc. need to be considered, when defining the reference population. However, in practical considerations excessively sensitive individuals have been excluded from the general population. This exclusion may need to be reconsidered or at least defined more clearly, because the fraction of the young population which is atopic and predisposed to allergies and asthma is already 5–7% and appears to be increasing by about 0.2% per year in many West European countries.

Most specific risk assessments deal with population risks, i.e. risks that would materialise in a sufficiently large population consisting of individuals of differing ages, predispositions, socio-economic and health status, and exposed to other unspecified hazards in a way which is comparable with the general population. The risks are expressed as expected excess cases of disease or death within this population, or as probabilities, e.g. 3×10^{-4} per year. Such a number is often used as if it would have a meaning at an individual level, e.g. claiming that living in a city with a PM_{10} concentration of X $\mu g/m^3$ will give an individual an excess mortality risk of Y%. In principle, such probabilities have a meaning, i.e. the validity of the statement can be tested, only at the population level. In practice, such an individual risk would depend strongly on other exposures (SO_2) and host factors such as genetic predisposition and health status (e.g. cardiopulmonary disease).

3. Systematic Approaches to Risk Assessment

In order to produce repeatable and comparable risk assessments, selection of a systematic approach or paradigm is needed. The first such approaches were used in assessing risks in and around nuclear facilities, and were based on the most exposed individual or worst-case scenarios, later on numerical analyses of the probabilities of the chains of independent events leading to the worst outcomes (WASH–1400 Report, 1975).

The International Agency for Research on Cancer (IARC) developed in the 1970s a systematic approach for cancer risk assessment of mostly individual compounds. This approach is based on a systematic compilation of existing data about sources of exposure, measured exposure levels, epidemiological and toxicological human data, experimental animal data, bioassay genotoxicity data and any other relevant data. These compilations are then critically reviewed and analysed by a panel of selected experts who produce a judgement as to the carcinogenicity class of the agent be it a compound or a mixture. This classification is not based on carcinogenic potency or quantitative estimation of risk to any population, but instead on the strength of evidence, i.e. IARC has evaluated toxicological properties of chemicals, not health risks of exposures.

IARC had, until 1998, classified 834 agents for carcinogenicity:

* Group 1. Proven human carcinogens 75
* Group 2A. Probable human carcinogens 59
* Group 2B. Possible human carcinogens 225

There are two other categories in the IARC classification of agents; Group 3 for agents which are not classifiable, and group 4 for agents which are probably not carcinogenic to humans (IARC, 1998). The ambient air pollutants that have been classified by IARC as carcinogens include asbestos, benzene and 1,3-butadiene.

The EU classification of carcinogens has similarities with that of IARC. Category 1 is for substances known to be carcinogenic to man, Category 2 for substances which should be regarded as if they were carcinogenic to man, and Category 3 for substances which cause concern with regard to possible carcinogenic effects in man, but for which the available information does not allow satisfactory assessment. It is interesting to note that while the IARC approach keeps a clear distance from practical decision making, the EU classification is more oriented towards the needs of administration.

In modern terminology, these cancer risk classifications are closer to hazard identification than risk assessment.

WHO has developed a similar approach for the classification of skin and airway sensitisers (WHO, 1997).

Needless to say, the IARC cancer risk assessment as well as the WHO classification of skin and airway sensitizers when compared with risk assessment of nuclear facilities, apply to entirely different types of risk and answer quite different questions. More universal risk assessment methodologies are needed to produce quantitative and comparable risk characterisations for the different risks associated with different air pollutants.

3.1. The NAS/NRC Paradigm, 1983

The U.S Environmental Protection Agency (EPA) announced its first cancer risk assessment guidelines in 1976, but "they lacked the intellectual construct that could serve to frame the thinking and discussion about risks" (Barnes, 1994). A new and comprehensive conceptual model or scheme for risk assessment was published by the U.S. National Academy of Sciences in 1983, known generally as the "NAS/NRC (1983) Paradigm". It separates risk assessment (RA), based on science, from risk management (RM), which in principle, follows from this independent assessment and has more practical goals and constraints. The risk assessment process was divided into four steps, namely:

* Hazard identification — Is this chemical harmful?
* Dose-response assessment — How bad is it?
* Exposure assessment — Who is exposed? What levels? How long?
* Risk characterisation — So...?

Originally, the NAS 1983 paradigm set a strict order (above) for these steps, and also separated RA from RM, so that RM would follow from a completed RA.

The NAS paradigm has finally succeeded in setting a widely agreed framework for risk assessment/risk management, and it has had broad applications from carcinogenic compounds to non-carcinogens, mixtures, radiations and other situations. Table 1 presents the logical framework of the NAS (1983) risk assessment model and a comparison with another model by Covello and Merkhofer (1993), and the CEC Directive 93/67/EEC (1993).

However, the wide application of the NAS paradigm has also exposed some basic weaknesses in it. Total separation of risk assessment and risk management does not exist in real life because it is not possible or even desirable in many cases. Instead of risk management following from risk assessment, risk assessment has often been explicitly guided by the gaps of information identified in the needs of RM. The NAS paradigm also favours the more manageable single chemical (e.g. asbestos or dioxin) approach over consideration of more realistic exposures to dynamic mixtures (e.g. traffic exhaust or the mixture of pollutants found in mouldy buildings). The current approach does not address the adversity of the effect and therefore helps little in comparison of multiple effects and different effects. While science (RA) refuses to compare apples and oranges, real life (RM) has to base some tough decisions on such comparisons.

3.2. Covello–Merkhofer, 1993

The Covello–Merkhofer (1993) risk assessment framework addresses a problem in the NAS paradigm, which has not been stated above, namely the role of hazard identification. Should hazard identification relate to effects observed in the experimental laboratory, or to exposures likely in the field? While the flow diagram of the NAS paradigm follows from an administrative viewpoint, the flow diagram of the Covello–Merkhofer approach for risk assessment is based on the causal flow of events from sources of pollutant to exposure, health consequences and scientific conclusions. Hazard identification can occur at any point of this flow or outside, for example, in the laboratory, and is considered as external in the Covello–Merkhofer approach.

The Covello–Merkhofer approach is particularly suitable for risk assessment where the development of risk is modelled from source to health outcome, such as diesel exhaust from tailpipe to ambient air, lungs and target organs. It does not apply as easily in cases where the risk assesssment focus is a sensitive subgroup, a particular environment, or a disease.

3.3. European Commission Directive 93/67/EEC, 1993

This directive lays down the principles for assessment of risks to man and the environment of chemicals (or substances notified in accordance with Council Directive 67/548/CEE). These principles as well as the terminology of this Directive are quite similar to those of NAS 1983. The directive requires that "the assessment of risks should be based on a comparison of the potential adverse effects of a substance with the reasonably foreseeable exposure of man", and "the assessment

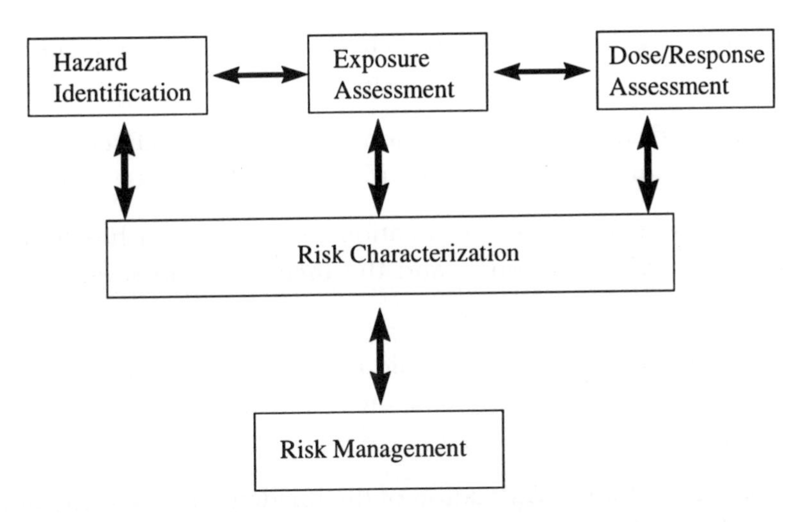

Fig. 2. The EU risk assessment framework (CEC Commission Directive 93/67/EEC).

of risks to man should take account of the physico-chemical and toxicological properties of a substance."

According to CEC Commission Directive 93/67/EEC, formal risk assessment is divided into four activities, which are defined as *hazard identification, dose (concentration)-response (effect) assessment, exposure assessment* and *risk characterisation,* which summarise the previous activities. Their roles and relations to each other are presented in Fig. 2.

Unlike the original NAS/NRC paradigm, in this EC framework, hazard identification, dose-response assessment and exposure assessment may all be connected with each other and with risk characterisation. Also, the connection between risk assessment and risk management can be bidirectional. In practice, these changes allow for risk assessment to follow from specific needs of RM. It also allows risk management to be based on, for example, hazard identification alone. Such risk management could lead to the prohibition of use of an identified carcinogen (e.g. asbestos in automobile brake linings) without going through comprehensive and time-consuming dose-response and exposure assessments.

The following definitions are used in the CEC Commission Directive 93/67/EEC:

Hazard identification — Identification of the adverse effects which a substance has an inherent capacity to cause.

Dose-response assessment — Estimation of the relationship between the dose or level of exposure, and the incidence and severity of an effect.

Exposure assessment — Determination of the emissions, pathways and transformations in order to estimate the concentrations/doses to which humans are or may be exposed

Risk characterisation — Estimation of the incidence and severity of the adverse effects. This may be accompanied with quantification of the likelihood of the detriment occurring and qualitative or quantitative estimation of uncertainties.

Risk management — Identification, selection and implementation of risk management alternative.

3.4. The Fundamental Steps of Risk Assessment

The most commonly accepted and used risk assessment framework is the one proposed by the National Academy of Sciences. The framework set by the CEC Commission Directive 93/67/EEC follows closely the NAS principles. The Covello–Merkhofer framework differs from the others in that it separates hazard identification from risk assessment. It is considered as an independent step that must precede risk assessment. Basic steps in each framework are summarised in Table 1.

The roles of dose/exposure-response assessment and exposure assessment in the chain of events from source to target organ effects are characterised in Fig. 3.

Table 1. The main steps of the three most commonly known risk assessment frameworks (Covello & Merkhofer, 1993; CEC, 1993; NAS/NRC, 1983).

Step	CEC Commission Directive 93/67/EEC, 1993	NAS/NRC, 1983	Covello–Merkhofer, 1993
Hazard identification	identification of whether a particular hazard/substance has the potential to cause adverse human health effects	determining whether a specified chemical causes a particular health effect	**external to the model:** identifying risk agents and conditions under which they may produce adverse consequences
Release assessment	—	—	quantifying the potential of a risk source to introduce risk agents into the environment
Dose-response assessment	determination of the relationship between the dose or level of exposure, and health consequences	determining the relationship between the magnitude of exposure and the probability of health effects	**Consequence assessment:** quantifying the relationship between exposures to risk agents and environmental consequences
Exposure sessment	quantification of the concentrations/ doses to which humans are or may be exposed	determining the extent of human exposure before and after application or regulatory controls	quantifying the as- exposures to risk agents resulting under specified release conditions
Risk characterisation	estimation of the incidence and severity and quantification of the magnitude of adverse effects, with qualitative or quantitative estimation of uncertainties	describing the nature and magnitude of human risk, including uncertainty	**Risk estimation:** estimating the likelihood, timing, nature, and magnitude of adverse consequences

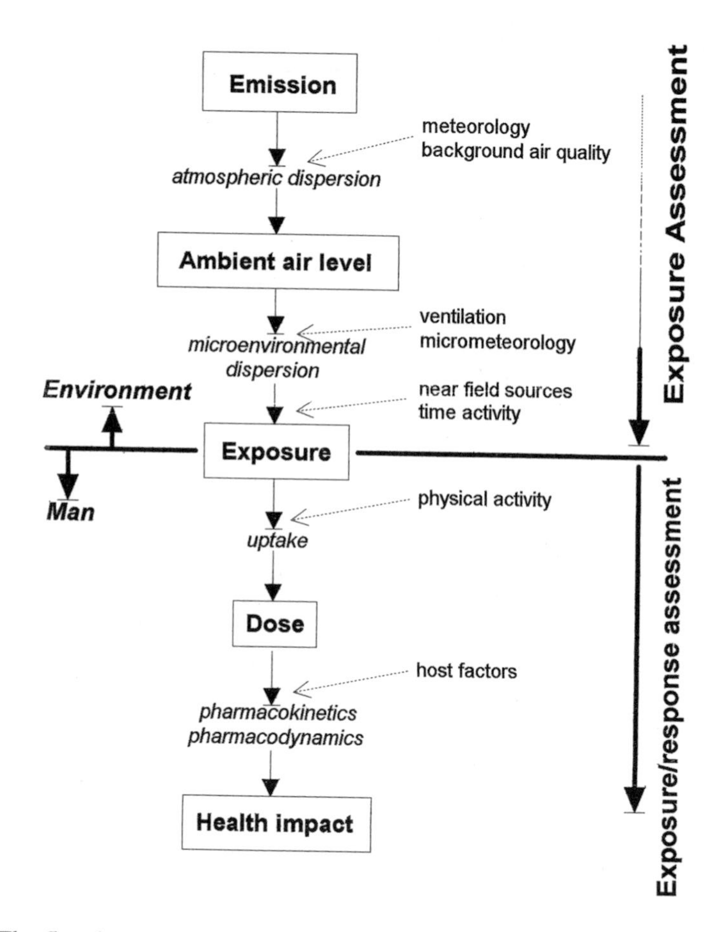

Fig. 3. The flowchart from emission to health impact showing **Exposure** as the interface between the human body and the environment, and presenting the roles of **Exposure (Dose) Assessment**, and **Exposure (Dose)-Response Assessment**, in **Risk Assessment**.

Hazard identification

Hazard identification may originate from air pollution epidemiology findings (e.g. an observed association of mortality with ambient air fine PM levels), occupational epidemiology (benzene and asbestos carcinogenicities), experimental toxicology (tetra-chloro-dibenzo-dioxin toxicity, 1,3-butadiene carcinogenicity) and even clinical observations (rapid increases of mortality during the air pollution episodes in Liege, 1930; Donora, 1948; London, 1952).

Exposure-response assessment

The exposure/dose-response assessment procedure depends very much on whether it is based on toxicological data from laboratory experiments or epidemiological data from field surveys. Toxicological data from well-designed and conducted experimental animal studies are quantitatively and qualitatively specific to the exposure of interest, but need multi-step extrapolation from homogeneous animal strains to heterogeneous human populations, and usually also from high to low exposures. Epidemiological data, especially when collected from the general population, often suffer from wide quantitative error margins and qualitative interpretation problems (other exposures, confounders and effect modifiers), but the results are relevant for humans under realistic exposure settings, and need little or no extrapolation. The European Union has published report on the methodologies for health effects assessment in air pollution epidemiology studies (Brunekreef, 1992).

For cancer risk assessment from experimental animal studies, the choice of the extrapolation model depends on the understanding of the mechanisms of carcinogenesis. For cancer initiating carcinogens, extrapolation based on a linear, non-threshold model has been used more frequently than models which assume an effect threshold. For cancer promoters, a threshold model is commonly assumed.

The risk associated with a lifetime exposure to a certain concentration of a carcinogen in the air has generally been estimated by non-threshold linear extrapolation and the carcinogenic potency expressed as the incremental unit risk estimate (WHO, 1987). The incremental unit risk estimate for an air pollutant is defined as the additional lifetime cancer risk occurring in a hypothetical population in which all individuals are exposed continuously from birth through life to a concentration of 1 $\mu g/m^3$ of the agent in the air they breathe.

Results expressed as unit risk estimates allow comparisons of the carcinogenic potencies of different agents and can help to set priorities in pollution control. By using unit risk estimates, any reference to the "acceptability" of risk is avoided in risk assessment. The decision on the acceptability of a risk should be made by national authorities in the framework of risk management.

For non-carcinogenic compounds, the common starting point is to define the lowest observed effect level (LOEL) or no observed effect level (NOEL). Whether the LOEL or the NOEL should be used is, in practice, mostly a matter of availability of data. The exposure of specific health concern relates better to the LOEL. In the case of sensory effects, the NOEL is more appropriate. Both approaches, however, make use of only one point in the dose-response curve. Methodologies have been developed which ensure the use of all dose-response information (e.g. benchmark approach) or incorporate mechanistic data to reduce uncertainties and account for variability (biologically based modelling) (Gaylor & Slicker, 1994). Comparing and summing the effects of different exposures with different exposure-response models and uncertainties can become very complicated, when the exposures of concern are variable mixtures (e.g. diesel exhaust or smoke from wood burning).

However, comparing and adding risks may not always be complicated. Crump *et al.* (1976) and Guess *et al.* (1977) have pointed out that if a pollutant produces cancer via the same mechanism by which background cancers occur, then there will be a linear incremental response to the incremental dose. Even if the biological dose response mechanism has a threshold or is non-linear, the existence of background cancers shows that the threshold is already exceeded by some background agent. Then, when a small amount of the same agent, or another agent operating in the same way as the first, is added, an incremental response linear with the incremental dose can result almost independently of the particular biological mechanism relating dose and response. It has also been argued (Crawford & Wilson, 1996) that this may also be true for non-cancer endpoints. The requirements are that (i) there exists a reasonably large background of the biological effect under consideration, and (ii) the pollutant acts in the same way as the background. It is evident that this might be satisfied by a large number of biological effects and pollutants. The generality of the argument suggests that linear dose-response relationship may be the rule, rather than the exception, at the low doses typical of exposures to air pollutants.

Exposure assessment

Exposure of an individual to a pollutant can be defined as the contact concentration of the pollutant experienced by the individual (Georgopoulos & Lioy, 1994) or as a coexistence of an individual and a pollutant in the same microenvironment (Ott, 1995).

The biological effects of exposures to toxicants result from the dose received by a sensitive target organ. The actual level of toxicant concentration in the target organ depends on the exposure of the individual, the toxicokinetics and metabolism of the substance in the organism, and accumulation and removal functions in the target organ. If there is a threshold concentration of the pollutant in the target organ in which an adverse biological effect will occur, two exposure level threshold limits can be defined. The lower is the highest exposure level that will never lead to a target organ dose threshold. The higher is the exposure level that will increase the target organ dose from zero to threshold within a specified integration time. Consequently, the lower exposure level threshold limit is independent of time, but the higher exposure level threshold limit is a function of the integration time, Δt. For many air pollutants, the individual differences between the threshold concentrations (susceptibilities depending on age, health, activity and genetics) as well as exposure scenarios (time-series depending on indoor, occupational and personal activity) vary significantly, or no threshold concentrations are known.

The full data of exposure to certain pollutant can be expressed only as the full time-series of instantaneous exposure concentration values experienced by each individual in the population. For almost any application, these are much more data than can be collected at any reasonable cost or utilised in any analysis. Consequently, an exposure assessment must be based on data which are very much reduced from the full time-series of the exposure of each individual in the population/sample. This data reduction should be done in a way that preserves all the data relevant for the risk assessment. The length of the exposure integrating/averaging time, Δt, should [ideally] be selected based on the physicokinetic mechanisms of corresponding health effects. The Δt should be defined "backwards", starting from considerations of the dose/response component of the exposure system (Georgopoulos & Lioy, 1994). The data reduction can proceed over three orders. The following reductions assume that the averaging time of interest is 24 hrs, but, of course, it can also be, e.g. 1 hr or 1 year.

An example of the full data set of 24-hr personal PM_{10} exposures covering 14 days and 14 non-smoking individuals in the THEES study in Phillipsburg, PA, USA (Lioy *et al.*, 1990) is presented in Fig. 4.

The first order data reduction combines all personal exposures for each day into daily frequency distributions. This is the time-series of the frequency distributions of personal exposures. It preserves the time-series data of the population, but all personal time-series data are lost, thus the data that relate to longer or shorter term health effects are lost.

Such a database can be used for identifying the days in which given personal exposure limits are exceeded within the population, and the daily percentages of the population exceeding such limits. For air pollutants which do not have significant indoor sources, the time-series of ambient air pollution levels measured at fixed air quality monitoring sites could be used as an approximation of the time-series of the mean or median personal exposures, but the exposure frequency distributions around these daily means must be obtained from other information sources.

In only three studies, THEES (Lioy *et al.*, 1990), LiiLa (Alm *et al.*, 1994; 1998) and Jansen *et al.* (1998), have the exposures of the same individuals been measured repeatedly to allow for estimation of the

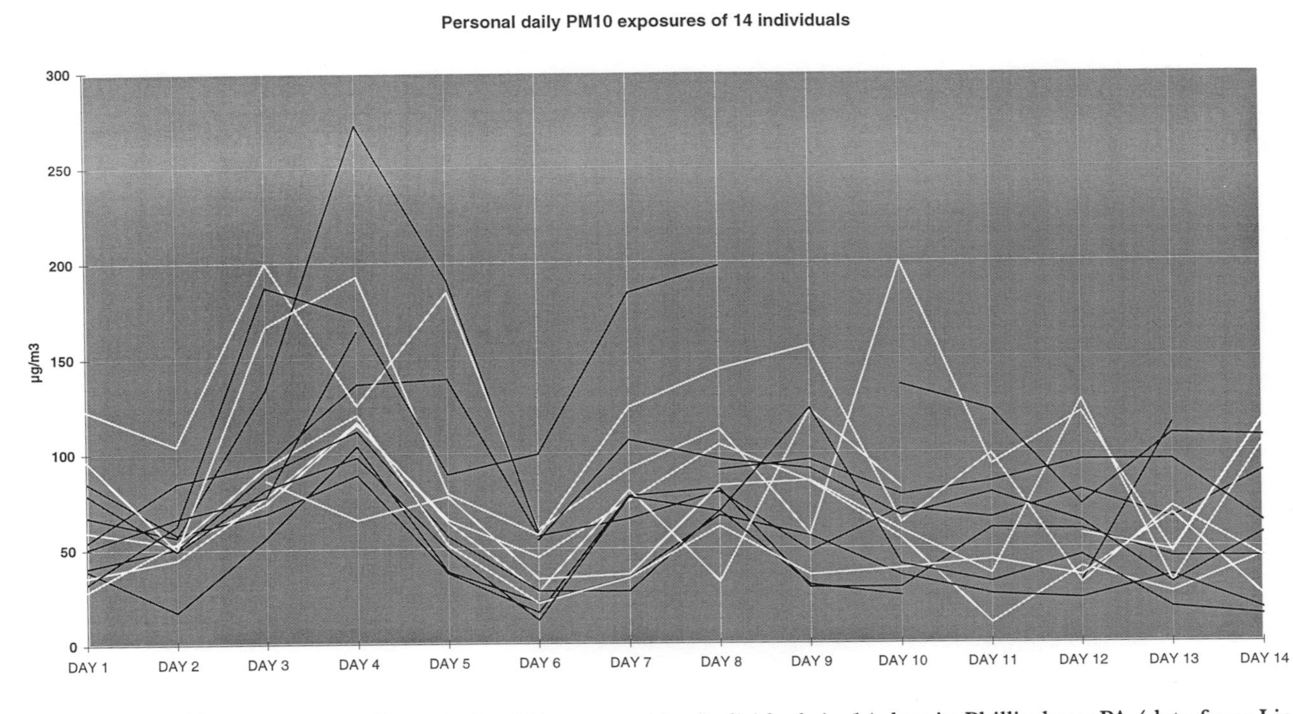

Fig. 4. Personal PM_{10} exposures (in $\mu g/m^3$) of 14 non-smoking individuals in 14 days in Phillipsburg PA (data from Lioy *et al.*, 1990).

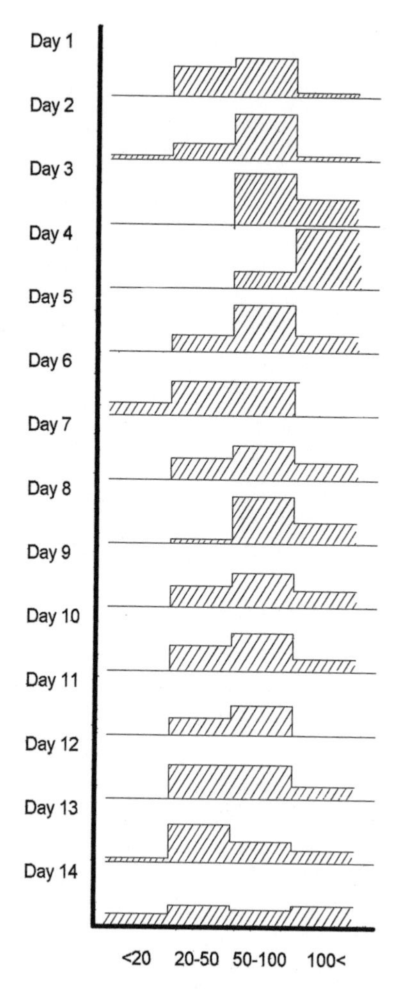

Fig. 5. A 14-day time-series of the frequency distributions of personal PM_{10} exposures (in $\mu g/m^3$) of 14 non-smoking individuals in Phillipsburg PA (data from Lioy *et al.*, 1990).

first-order time-series of the frequency distributions of personal exposures. A time-series of frequency distributions of personal exposures in the THEES data is presented in Fig. 5.

The time-series of the frequency distributions of personal exposures can be combined over time to form the second order, namely the frequency distribution of the (daily) personal exposures. Then, all time-series data are lost.

The second-order frequency distribution of personal exposures shows the percentages of the daily personal exposures in the whole population exceeding selected limits. It does not show how these exceedances are distributed between different individuals and times. Because the data are considerably reduced, the frequency distribution of the personal exposures allows direct graphical and statistical comparisons of different exposure data sets, e.g. comparison of the exposures of the suburban with downtown residents or residents of different cities. Data sets for frequency distributions of personal exposures have been produced in a number of air pollution exposure studies of representative population samples.

The second-order frequency distribution of the personal exposures in the THEES data is presented in Fig. 6.

If the frequency distribution of the personal exposures needs to be further reduced, it can be presented as the mean or median of the (daily) personal exposures. This can be accompanied by additional statistics, such as arithmetic or geometric standard deviation, minimum, maximum, etc. The frequency distribution data are now lost or reduced to a number but, on the other hand, an increased number of exposure data sets can be presented in a single table or histogram, e.g. as comparison of several population subgroups in several cities. If the mean personal exposure to ambient air concentration ratios have been determined for a representative population sample, the mean daily population exposures for a similar population in a similar environment can be estimated from similarly acquired ambient air quality data.

The mean of the daily PM_{10} exposures in the THEES study was 75 $\mu g/m^3$, with a standard deviation of 44 $\mu g/m^3$.

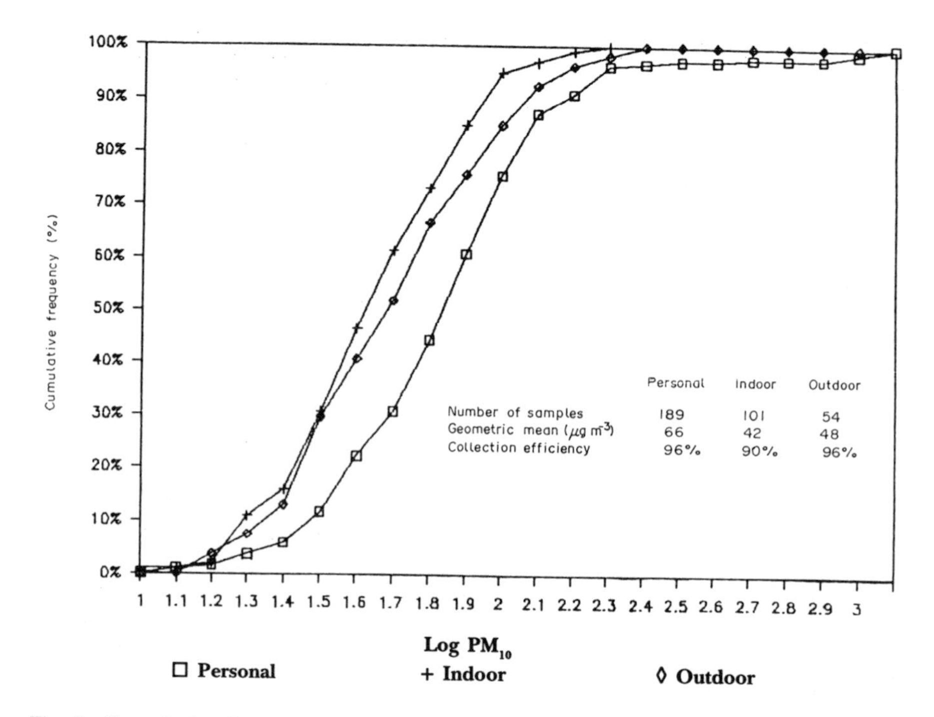

Fig. 6. Cumulative frequency distribution of 24-hr personal PM_{10} exposures, together with ambient air and indoor air PM_{10} levels (in $\mu g/m^3$) of 14 individuals in Phillipsburg PA (Lioy *et al.*, 1990).

The total human exposure to air pollutants is the sum of the exposures in different locations and times. People are exposed to outdoor air contaminants in the outdoor, indoor and transportation environments. The indoor environment is the major component of the total exposure because an overwhelming proportion of time is spent indoors. Table 2 presents the range of misclassifications which could result from using outdoor air concentration as an estimate for personal exposure. It is assumed that 85% of the time is spent indoors, 15% outdoors. The presence or absence of the some common indoor sources is also considered.

According to Table 2, individuals without gas appliances would have about half, while those with gas appliances would have almost

Table. 2. Human exposure at home for different pollutants (Ackermann *et al.*, 1997).

Pollutant ($\mu g/m^3$)	Indoor source	Indoor concentr.	Outdoor concentr.	Actual exposure	Assigned exposure	Ratio actual/ assigned
NO_2	—	12.5	25	14.37	25	0.57
NO_2	gas app.	50	25	46.25	25	1.85
PM	—	15	30	17.25	30	0.57
PM	tobacco	60	30	55.50	30	1.85
O_3	—	10	50	16	50	0.32

double the estimated NO_2 exposure, and very few people would actually be at the estimated exposure level. The case of particulate matter (PM) exposures of smokers and non-smokers is quite similar. For ozone or photochemical oxidants, with no indoor sources, the actual exposure of almost everybody is much lower than the estimated exposure. To take only risk estimates derived from epidemiological studies based on outdoor measurements would underestimate the risks of ozone exposure, because the observed effects have actually occurred at a lower than estimated ozone concentration.

Personal exposures can be measured directly or indirectly, or they can be modelled. Direct exposure measurements are performed by active or passive personal monitoring devices, which are carried by the subject over a day or a few days. Personal air pollution monitoring equipment needs to be independent of external power, silent, lightweight and robust. Indirect exposure measurements are made by measuring the pollutants in each microenvironment that the study subjects visit for any significant period of time, and summing the concentrations in each microenvironment weighted by the respective proportions of the time spent in the same microenvironments. The microenvironmental monitoring devices used in field studies need to be lightweight and robust. A fairly good variety of both personal and microenvironmental air pollution monitoring equipment is presently available. The European Union has published a report on the measuring and modelling methods available for air pollution exposure assessment (Williams, 1992).

Direct as well as indirect exposure measurements are costly (in the European EXPOLIS study the direct costs of each 48-hr personal and microenvironmental monitoring of PM2.5, VOCs, NO_2 and CO for one subject was approximately 1350 Ecu) and rather invasive for the subject. Presently, these facts reduce the maximum practical population sample size to a few hundred subjects, and the length of the monitoring period for each subject to a maximum of a few days, which gives extra weight to the need for careful population sampling and study design.

Exposure modelling is much cheaper and can be applied to much larger population samples. Different types of models have been developed:

Deterministic models are based on a mathematical construct of the physical phenomena that lead, on one side, from air pollution source emissions and dispersion, and on the other side, from human time-activity-patterns (Ackermann-Liebrich *et al.*, 1995) to the contact between the pollutants and the human receptors, i.e. exposures, e.g. SHAPE model (Duan, 1981; Ott, 1981).

Statistical models are based on a statistical construct (often a regression model) between the dependent variable, exposure, and a set of independent variables, which could include ambient air pollution levels, questionnaire-collected information about home, workplace, commuting and personal characteristics, and time-activity data.

Probabilistic models are usually based on a simple deterministic construct, but instead of using fixed input values, they apply full distributions of the input values and produce a more realistic probability distribution of exposures, and are therefore more capable of predicting the high and low ends of the population exposure distributions (Ott, 1984).

Deterministic and statistical models can be used for estimating exposures of specified individuals on specified days. Deterministic models are also applicable for extrapolation outside of the originally studied population, while statistical models are, in principle, only useful within the original studied population. Probabilistic model outputs are non-unique, and are therefore not suitable for estimating the unique exposure of a given individual on a given day. Although both the deterministic and probabilistic models are based on physical constructs, the constructs of probabilistic models are usually simpler and require less background data, because they do not try to estimate the actual exposures of given individuals and days, but more generally, the distributions of exposures within a larger population over a longer period of time.

Risk characterisation

Risk characterisation should be a decision-driven activity, directed towards informing choices and solving problems (Stern *et al.*, 1996). This risk assessment step produces an overview of the exposure-response assessment, the exposure assessment and other relevant information for the purposes of risk management. Risk characterisation should answer the questions: How harmful is the pollutant in question? What kind and how large is its public health impact? Which populations are most likely affected? and What sources and environmental factors are the most significant in increasing or controlling the extent and severity of the adverse health effects?

The analytic-deliberative process leading to risk characterisation should include early and explicit attention to *problem formulation*; representation of the spectrum of interested and affected parties at this early stage is imperative. The process should be mutual and recursive. Analysis and deliberation are complementary and must be integrated throughout the process, which benefits from feedback between the two. Those responsible for risk characterisation should begin by developing a provisional diagnosis of the decision situation so that they can better match the analytic-deliberative process leading to the characterisation of the needs of the decision (Stern *et al.*, 1996).

The previous paragraphs describe how far the NAS/NRC (1983) risk assessment paradigm has developed in 13 years from the original idea of keeping the risk assessment, including risk characterisation, a strictly scientific process, which, only after completion, should provide an input to risk management, where politics enters the process.

3.5 *Beyond the Traditional Physico-Chemical Risk Pathway*

All the above mentioned risk assessment paradigms and frameworks are strong in analysing physico-chemical and toxicological pathways for contaminant-specific health outcomes (e.g. mesothelioma from asbestos). They are much weaker in dealing with essentially non-specific

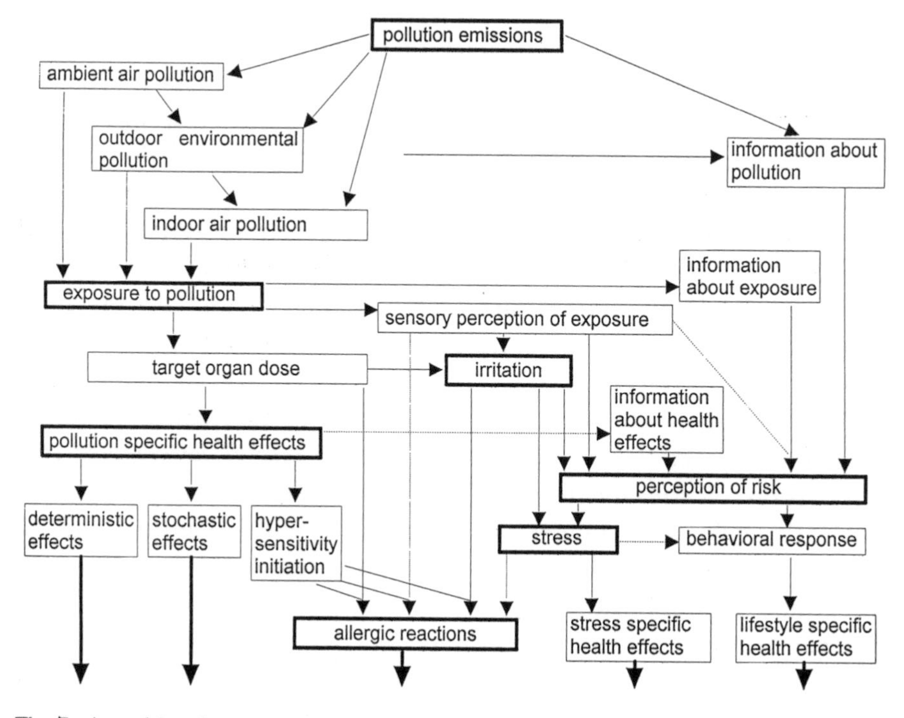

Fig. 7. A multi-pathway construct of the causal links from an air pollution source to various types of health outcomes. On the left are the traditionally-considered chemical-physiological pathways, in the centre, sensory perception and irritation pathways and on the right, psycho-social pathways. The specificity of the health effects by chemicals is high on the left of the figure, and is replaced by irritation, stress and lifestyle-specific health effects towards the right.

irritation and stress symptoms caused by irritating chemicals (e.g. a mixture of VOCs) in previously sensitised (e.g. pollen exposure) tissues and the perception of risk caused by sensory perception of potentially harmful exposures (e.g. odorous chemicals) or alarming information (about lung cancer risk from asbestos). Both irritation and risk perception can cause health consequences, which, although not via toxicological mechanisms specific to any chemical, are still causally linked to emissions of harmful pollutants.

The solution here could be a holistic approach, where all the different causal chains from pollution sources to different health outcomes are dealt with seriously, analysed, discussed and described separately, and any summing of the results is done with extreme caution and only to the extent necessary. The methods for assessing each causal risk chain should as much as possible, follow the same risk assessment paradigm, and preferably the paradigm described in the previous paragraphs.

Constructing a risk assessment model that will combine the toxicological, sensory-irritation and psycho-social chains leading from a source of contamination to the above discussed health effects is not as difficult as one might think. Figure 7 presents such a multi-pathway construct of the causal links from an air pollution source to different types of possible health outcomes, where all links can be subjected to scientific analysis.

4. Risk Management

Coping with a risk situation requires a broad understanding of the relevant losses, harms, or consequences to the interested and affected populations (Stern *et al.*, 1996). The public cannot and will never achieve completely risk-free air (Moghissi, 1993). Risk management is fundamentally a question of values. In a democratic society, there is no acceptable way of making these choices without involving the citizens who will be affected by them. Risk management is an organised effort to collect information about and to control risks (Morgan, 1993).

Risk assessment identifies significant hazardous effects, classifies chemicals and products according to health effect vs dose, and also identifies the significant routes and levels of exposures. From this information, it is possible to identify which specific stages in the manufacture, distribution, use or disposal of a substance give rise to particular risks. Risk management will draw an overall conclusion about the risks posed by a substance, set objectives and select means for reducing those risks. Risk management should provide a practical way of achieving the objective while minimising any adverse consequences of the measures being taken.

Drawing up a risk management strategy involves looking at the risks as identified by the risk assessment and considering how best to reduce those risks to an acceptable level. Two issues need to be considered:

* First, the options available for controlling the risks and these include a range of possibilities from labelling products to prohibiting use, and could include physical containment or the setting of emission, ambient air quality and exposure limits.
* Second, the administrative and legal framework within which any recommended action can be taken, including voluntary action by industry or citizens/consumers, or regulatory action (by EU or member states).

Risk managers can intervene at many points. They can prevent the hazardous process, reduce exposures, modify effects, alter perceptions or valuations through education and public relations or compensate for damage after the fact (Morgan, 1993).

There is no universally correct choice among the various criteria for making decisions about risks. Such decisions depend on the ethical and value preferences of individuals and society at large. It is, however, critically important that the decision framework is carefully and explicitly chosen and that this choice is kept consistent, especially in complex situations. Otherwise, inconsistent, and even incompatible approaches to the same risk may result (Morgan, 1993).

Selection of the set of rules

Formulation of any risk management practice depends strongly on which principal set of rules is selected. Most rules applied in risk management fall into one of three broad classes: utility-based, rights-based and technology-based. The first kind of rules attempts to maximise net benefits (cost/benefit analysis). Rights-based rules replace the notion of utility with one of justice. There are certain things that one party cannot do to another, regardless of costs or benefits. Technology-based criteria are concerned with the level of technology available to control risks. The key terms are: Best Available Technology (BAT), or As Low As Reasonably Achievable (ALARA) (Morgan, 1993).

An expression of the utility-based rules is the value set for life in road accident cost estimation. In 1990, this monetary value was $2 600 000 in the US, $900 000–$1 200 000 in Germany, New Zealand, Britain and Sweden, about $400 000 in France and Belgium, $130 000 in Holland, and $20 000 in Portugal.

An expression of the rights-based (or, in negative terms, *health risk*) rules is the attempt by many regulatory agencies to establish the highest acceptable lifetime risk level that a particular activity or chemical should impose on the exposed individual. According to many American policy decisions, this highest acceptable lifetime risk level is 10^{-4}–10^{-3} (Travis & Hattemer-Frey, 1988). The estimated health risk from air pollution in the eastern U.S. falls within this range (Wilson & Crouch, 1987). However, in California, the management of industrial carcinogenic emissions is based on the definition (Proposition 65) that an individual lifetime cancer risk in excess of 10^{-5} is significant, and the act requires businesses to give warning of exposures to listed toxic chemicals unless exposure can be shown to be insignificant, i.e. leading to a lifetime cancer risk of 10^{-6} or less (Ohshita & Seigneur, 1993).

An expression of the technology-based rules is the US National Emission Standards for Hazardous Air Pollutants Program, which aims at eventually controlling a list of 189 specified hazardous air pollutants by Maximum Available Technology Standards (Ohshita & Seigneur, 1993).

The role of trust and fairness

Economic theory suggests that hazardous facilities, even those with technological risks, will be supported by the public so long as the benefits sufficiently outweigh the risks. Revision to that theory suggests that non-market and unpriced values must be taken into account in cases where social, cultural and aesthetic values are involved. More recent research results imply that increased (decreased) trust in the managing organisation of the activity under concern will lead to lower (higher) levels of perceived risk and thus increase public support (opposition) for the activity. Understanding the dimensions of trust and their influence on risk perceptions may warrant much more attention in order to provide the insights needed to develop policy strategies that will gain public acceptance (Flynn *et al.*, 1992).

Economically and technically sound environmental policies may be doomed if people believe that they distribute benefits and burdens unfairly. Three concepts emerge from the philosophical and cultural basis of risk sharing: *parity*, in which each individual, group or country is treated equally; *priority*, in which the burden is given to those most deserving of it; and *proportionality*, in which burdens are shared according to need or contribution. Within this fairness triangle, no single concept of fairness ever wins the day, suggesting that there is something essential in plurality (IIASA, 1993).

To meet the challenge of the new scientific and social developments, risk analysts and managers will need to expand their agenda from evaluating dangers towards the general welfare; they will also have to adopt new communication styles and learn from the populace rather than simply trying to force information on them (Morgan, 1993).

4.1. Systematic Approaches to Risk Management

Probably, the most widely (geographically) applied and thoroughly tested risk management rationale is described in the ICRP principles for radiation protection (ICRP 60, 1990). It divides actions that may affect risks, into practices and interventions. The practices include the design, production and use of products which may cause harmful emissions or accidental damage. The interventions include all those actions in both on-going and accidental situations which aim at reducing emissions or risks/consequences. These principles lay a framework of rules, which the risk management needs to satisfy independently and simultaneously.

For practices (e.g. street traffic) the first rule is *justification*: each practice should produce enough benefits to the exposed individuals and/or the society to offset the detriment from emissions and exposures. The second is *optimisation*: the levels of exposures should be kept as low as reasonably achievable (ALARA). The third is *individual's rights*: each individual's exposure/risk should be limited below a set maximum.

For interventions (e.g. traffic reduction measures), the first rule states that any act of intervention should do more good than harm — all costs and effects considered. The second rule states that intervention (form, scale, duration) should be so optimised that its net benefit is maximised.

Yet another concept that has been discussed but not formally accepted, is called *de minimis*, or *level of no concern*. It states that if the measured or modelled level of a contaminant, exposure or health risk expected is below a set limit (e.g. a lifetime risk of less than 10^{-6}), the risk should be ignored. This rule is not based on the rationale that small risks are meaningless, but rather on the understanding that the intervention to reduce a small risk any further could — through possibly unaccounted pathways — add other risks and interfere with the rights of individuals or organisations.

The ICRP principles are designed for risks from exposures that can be reduced but not completely avoided and where the risk is assumed to grow linearly from zero exposure upwards. Therefore, they may be less applicable to, for example, such chemical risks which do have a threshold dose and do not accumulate in the exposed individual.

Cancer risk assessment, as in the IARC Cancer Risk Evaluation Monographs, applies, like the ICRP principles, to accumulating risks, which are not expected to have any threshold. Because IARC does not classify carcinogens according to potency but according to strength of evidence, its application in risk management also results in mostly qualitative actions, such as prohibiting the use of asbestos in car brake pads, or isolation of a hazardous waste treatment plant into a sparsely populated area.

The most widely applied risk management rationale is that adapted in the setting of occupational exposure limits (TLVs, MACs, etc.). They are mostly based on the rationale that each exposed individual is a non-pregnant, healthy adult between 20 and 65 years of age, does the same work and is exposed to the same harmful chemicals in a controlled environment 8 hrs per day, 40 hrs per week, that the number of chemicals of concern in any given case are few, and that the individual is not exposed outside of the working hours.

Occupational risk management rationale is not directly applicable to the population at large in the highly variable, complex and only poorly controlled daily environmental exposure situations. However, ambient air and drinking water pollution concentration guidelines and limits have been developed, critically reviewed, legislated and applied with varying success in non-occupational outdoor and indoor environments as well. Because the exposure times are not controllable outside the workplaces, different concentration limits are needed for different averaging times. Because also, the exposed individuals are not controllable outside the workplaces, the concentration limits are specified with regard to some reference population, which, however is not usually clearly defined, and may ignore the most susceptible or most exposed groups.

Ambient air pollution concentration limits are in existence for only a small number of pollutants. The questions of additivity of effects of different air pollutants or the effects of complex exposures (except PM_{10}) are not addressed at present in air quality standards or guidelines.

Air pollution risks of activities such as municipal waste incinerators often emerge in a context of emotional stress and outrage. Moreover, exposure to air pollutants is widespread and involuntary. While the health effects may be uncertain, risk acceptability of activities, which individuals perceive to be external to their personal needs, is very low. Children or elderly are often targeted as high risk populations. Complex scientific and technical data around such risks do not facilitate easy communication with the population. These facts set very specific requirements for both risk assessors and managers.

5. Risk Communication

Thomas Jefferson was right: The best strategy for assuring the general welfare in a democracy is a well-informed electorate. Implicit in the process of risk analysis and management is the crucial role of communication. If public bodies are to make good decisions about regulating potential hazards, citizens must be well informed. The alternative of entrusting policy to panels of experts working behind closed doors has proved a failure, both because the resulting policy may ignore important social considerations and because it may prove impossible to implement in the face of grass-roots resistance (Morgan, 1993).

Risk communication is a fundamental part of risk management. Successful communication will not raise undue concerns, but will suggest changes in the behaviour of the decision-makers and/or vulnerable individuals in a direction that effectively reduces avoidable risks, yet avoids non-essential interference in the lives of the individuals or in the choices of economic enterprises. Poor risk communication may not only fail to produce the wanted risk reduction, it may also unnecessarily limit individuals' options, add public and private costs, and generate new risks due to non-productive public concerns and potentially harmful behavioural changes. The objective of risk communication is to provide people with a basis for making an informed decision; any effective message must contain information that helps them in that task (Morgan, 1993). Scientific information about health and environmental risks is communicated to the public through a variety of channels, ranging from a public display of the present air pollution index to public meetings involving representatives from government, industry, the media, and the general public. Communication efforts can be frustrating for both risk communicators and for the intended recipients of the information. Government officials, industry representatives and scientists often complain that lay people do not accurately perceive and evaluate risk information. Representatives of citizen groups and individual citizens are often equally frustrated, perceiving risk communicators and risk assessment experts to be uninterested in their concerns and unwilling to take actions to solve seemingly straightforward health and environmental problems. The media often serves as transmitters and translators of risk information. But, the media have been criticised for exaggerating risks and for emphasising drama over scientific facts (Cohrssen & Covello, 1990). It is probably fair to say that while the media covers qualitatively all issues — including air pollution risks — of significant public interest, the quantitative content of the coverage is uncorrelated with the quantitative significance of the issues.

Risk implies uncertainty, therefore, risk assessment is largely concerned with uncertainty and hence, with a concept of probability that is hard to grasp. The results of even the simplest risk assessments need to be compared with similar assessments of commonplace situations to give them some meaning. We compare and contrast risk estimates to display their similarities and differences (Wilson & Crouch, 1987).

The essence of good risk communication is simple: learn what people already believe, tailor the communication to this knowledge and to the decisions people face and then subject the resulting message to careful empirical evaluation. Yet, almost no one communicates risks to the public in this fashion. People get their information in fragmentary bits through the media that often does not understand technical details and often chooses to emphasise the sensational. Those trying to convey information are generally either advocates promoting a particular agenda or regulators who sometimes fail to take a sufficiently broad perspective of the risks they manage. The surprise is not that opinion on hazards may sometimes force silly or inefficient outcomes; it is that the public does as well as it does (Morgan, 1993).

A common approach to risk communication has been public hearings before deciding upon general risk management measures, such as air quality guidelines, or specific issues like siting a hazardous waste facility. The setting in a town hall public hearing becomes almost unavoidably confrontational, where a smaller but more influential business interest party is confronted by a larger citizens interest party. Both parties bring in their supporting arguments and experts. This adversarial process is controlled by a logic which bears little resemblance to the logic of risk assessment and risk management. An approach, which has been more successful (but also consumes much more expert time), is an open house policy, where citizens can discuss with the experts in small groups over a much longer time period (Castle, 1993).

Based on the experiences in risk communication successes and failures, the US EPA has published a set of rules and guidelines for risk communication. In introducing the rules, the EPA noted that there are no easy prescriptions for effective risk communication. The EPA also noted that many of the rules may seem obvious, but are continually and consistently violated in practice:

Rule 1: *Accept and involve the public as a legitimate partner*
Rule 2: *Plan carefully and evaluate performance*
Rule 3: *Listen to your audience*
Rule 4: *Be honest, frank and open*
Rule 5: *Coordinate and collaborate with other credible sources*
Rule 6: *Meet the needs of the media*
Rule 7: *Speak clearly and with compassion*

(Cohrssen & Covello, 1990).

6. Risk Perception

In making judgements about uncertainty, including ones about risk, experimental psychologists have found that people unconsciously use a number of heuristics. Usually the rules of thumb work well, but under some circumstances, they can lead to systematic bias or other errors. As a result, people tend to underestimate the frequency of very common causes of death — stroke from hypertension, cancer from smoking, road accidents — by roughly a factor of 10. They also overestimate the frequency of very uncommon causes of death (e.g. leukaemia from nuclear radiation) by as much as several orders of magnitude. These mistakes apparently result from the so-called *heuristics of availability* — people often judge the likelihood of an event in terms of how easily they can recall (or imagine) examples (Morgan, 1993).

Public risk perception has been studied by psychometric methods developed for studying expressed preferences by analysing questionnaire data. Public perceptions of numerous risks were plotted on a coordinate system with two axes, the *controllability axis* and the *observability axis* (Fig. 8; Slovic *et al.*, 1985). The most feared risks are those that are perceived as unobservable and uncontrollable (DNA technology, radioactive waste, nuclear reactor accidents and weapons fallout), while the least feared are perceived observable and controllable (bicycles, boating and skiing). The risks of air pollution from fossil fuels and traffic are perceived to be near the middle of the risk space. They are feared more than smoking, but less than satellite crashes (*sic*!).

Perhaps the most important message from psychometric risk perception research is that there is wisdom as well as error in public attitudes and perceptions. Lay people sometimes lack certain information about hazards. However, their basic conceptualisation of risk is much richer than that of the experts and reflects legitimate concerns that are typically omitted from expert risk assessments. As a result, risk communication and risk management efforts are destined to fail unless they are structured as two-way processes. Each side, expert and public, must respect the insights and intelligence of the other (Slovic, 1987).

Whereas psychometric research implies that risk debates are not merely about statistics, some research implies that some of these debates may not even be about risk. Risk concerns may provide a rationale for actions taken on other grounds or they may be a surrogate for other

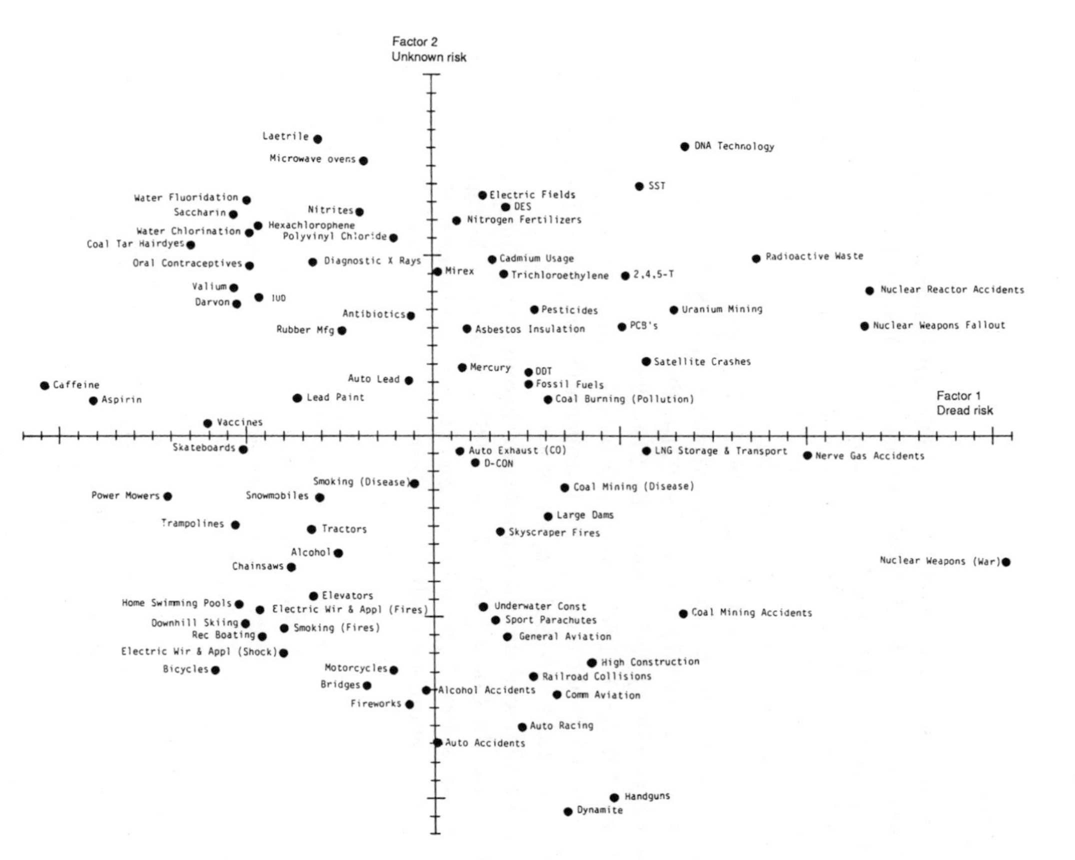

Fig. 8. Location of 81 hazards on a two-dimentional psychometric chart (Slovic *et al.*, 1985).

social or ideological concerns. When this is the case, communication about risks is simply irrelevant to the discussion. Hidden agendas need to be brought to the surface for discussion (Slovic, 1987).

References

Ackermann-Liebrich U. *et al.* (eds.) (1995) *Time Activity Patterns in Exposure Assessment.* p. 91, EUR 15892. Office for Official Publications of the European Communities, Luxembourg.

Ackermann U. *et al.* (1997) *Proceedings of International WHO and UN/ECE Workshop "Health Effects of Ozone and Nitrogen Oxides in an Integrated Assessment of Air Pollution".* Eastborne, UK.

Alm S. *et al.* (1994) *Atmos. Environ.* **28**, 3577–3580.

Alm S. *et al.* (1998) *J. Expos. Anal. Environ. Epidem.* **8**, 79–100.

Ames B.N. *et al.* (1987) *Science,* **236**, 271–280.

Barnes D.G. (1994) *Risk Analysis,* **14**, 219–223.

Brunekreef B. (ed.) (1992) *Health Effects Assessment.* p. 70, EUR 14346. Office for Official Publications of the European Communities, Luxembourg.

Castle G.J. (1993) *Air & Waste,* **43**, 963–969.

CCEE (Le Counceil de la Communauté Économique Européenne) (1967) Directive du Conseil du 27 juin 1967 concernant le rapproachement des dispositions législatives, réglementaires et administratives relatives à la classification, l'emballage et l'etiquetage des substances dangereuses (67/548/CEE). *Journal Officiel des Communautés Européennes* No. 196, 16.8.1967.

CEC (Commission of the European Communities) (1993) Commission Directive 93/67/EEC of 20 July 1993 laying down principles for assessment of risks to man and the environment of substances notified in accordance with Council Directive 67/548/EEC. *Official Journal of the European Communities* No. L 227, 8.9.93.

Chadwick M.J. *et al.* (1995) *Comparative Human Health and Environmental Risk Assessment: A Research Proposal Submitted to MISTRA.* p. 3, Stockholm Environment Institute, Stockholm.

Cohrssen J.J. & Covello V.T. (1990) *Risk Analysis: A Guide to Principles and Methods for Analysing Health and Environmental Risks.* pp. 99–104, U.S. Council for Environmental Quality, Washington, D.C.

Covello V.T. & Merkhofer M.W. (1993) *Risk Assessment Method; Approaches for Assessing Health and Environmental Risks.* p. 318, Plenum Press, New York.

Cox S.J. & Tait N.R.S. (1991) *Reliability, Safety and Risk Management.* Butterworth-Heineman, Oxford.

Crawford M. & Wilson R. (1996) *Human and Ecological Risk Assessment,* **2,** 305–330.

Crump K.S. *et al.* (1976) *Cancer Res.* **36**, 2973–2979.

Dockery D.W. *et al.* (1993) *NEJM,* **329**, 1753–1759.

Duan M. (1981) *Microenvironment Types: A Model for Human Exposure to Air Pollution.* Tech. Rep. No. 47 Department of Statistics, Stanford University, Stanford CA.

Gaylor D.W. & Slicker W. Jr. (1994) *Risk Analysis,* **14**, 333–338.

Georgopoulos P.G. & Lioy P.J. (1994) *J. Expos. Anal. Environ. Epidem.* **4**, 253–286.

Guess H. *et al.* (1977) *Cancer Res.* **37**, 3475–3483.

Flynn J. *et al.* (1992) *Risk Analysis,* **12**, 417–429.

IARC (1998, May 5) List of IARC Evaluations. http://193.51.164.11/monoeval/grlist.html

ICRP Publication 60 (1990) *Recommendations of the International Commission on Radiological Protection,* pp. 25–32, Pergamon Press, Oxford.

IIASA (1993) *Risk and Fairness Workshop Summary.* Laxenburg, Austria.

Jansen N.A.H. *et al.* (1998) *Occup. Environ. Med.* **54**, 888–894.

Jantunen M. (ed.) (1997) *Socioeconomic and Cultural Factors in Air Pollution Epidemiology.* p. 117, EUR 17510. Office for Official Publications of the European Communities, Luxembourg.

Katsouyanni K. (ed.) (1993) *Study Designs.* p. 166, EUR 15095. Office for Official Publications of the European Communities, Luxembourg.

Katsouyanni K. *et al.* (1997) *BMJ,* **314**, 1658–1663.

Last J.M. (ed.) (1988) *A Dictionary of Epidemiology.* Second edition. Oxford University Press, New York.

Lioy P.J. *et al.* (1990) *Atmos. Environ.* **24B**, 57–66.

Lave L.B. (1987) *Science,* **236,** 291–295.

Moghissi A.A. (1993) *Environ. Int.* **19**, 311–312.

Morgan M.G. (1993) *Scientific American,* **July**, 24–30.

National Research Council (1983) *Risk Assessment in the Federal Government: Managing the Process.* National Academy Press, Washington, D.C.

Ohshita S.B. & Seigneur C. (1993) *Air & Waste,* **43**, 723–728.

Oreskes N. *et al.* (1994) *Science,* **236**, 641–646.

Ott W.R. (1981) *Computer Simulation of Human Air Pollution Exposure to Carbon Monoxide.* Paper No. 81–57.6. Presented at the 74th Annual Meeting of the Air Pollution Control Association, Philadelphia.

Ott W.R. (1984) *J. Toxicol. Clin. Toxicol.* **21**, 883–887.

Ott W.R. (1995) *J. Expos. Anal. Environ. Epidem.* **5**, 449–472.

Riylander R. (1992) *Rev. Epidem. et Santé Publ.* **40**, 383–390.

Russel M. & Gruber M. (1987) *Science*, **236**, 286–290.

Slovic P. *et al.* (1980) In *How Safe is Safe Enough*, eds. Schwing R. & Alberts W. A. Jr. pp. 181–216, Plenum Press, New York.

Slovic P. *et al.* (1985) In *Perilous Progress: Managing the Hazards of Technology*, eds. Kates R.W., Hohenemser C. & Kasperson J.X. pp. 91–125, Westview, Boulder, CO.

Slovic P. (1987) *Science*, **236**, 280–285.

Stern P.C. & Fineberg H.V. (eds.) (1996) *Understanding Risk; Informing Decisions in a Democratic Society*. pp. 2–10, National Academy Press, Washington, D.C.

Travis C.C. & Hattemer-Frey H.A. (1998) *Environ. Sci. & Technol.* **22**, 872–876.

WASH-1400 Report (1975) *Reactor Safety Study: An Assessment of Accident Risks in U.S. Commercial Nuclear Power Plants.* Nuclear Regulatory Commission, Washington D.C.

WHO (1987) *Air Quality Guidelines for Europe.* p. 426, *WHO Regional Publications, European Series No. 23.* WHO, Copenhagen.

WHO (1997) *Criteria for Classification of Skin — and Airway-Sensitizing Substances in the Work and General Environments.* pp. 1–21, World Health Organisation, Copenhagen.

Williams M. (ed.) (1992) *Exposure Assessment.* p. 98, EUR 14345. Office for Official Publications of the European Communities, Luxembourg.

Wilson R. & Crouch E.A.C. (1987) *Science*, **236**, 267–270.

Zabriskie P. *et al.* (1997) *ICON*, **October**, 87–95.

CHAPTER 10

AIR POLLUTION AND INFORMATION RESOURCE

Geoff LeGouais, Joanna Carrington, Matthew Johnson, Charlotte Obhrai, Peter Webley & Peter Brimblecombe

1. Introduction

Literature represents the accumulation of human knowledge. This vast information entity, in its variety of printed and electronic guises, represents human thought and work in recorded form. In any study, it should be the initial resource for the scientist to build on rather than to duplicate.

An ideal world would see the researcher able to immediately call upon all knowledge that the scientific community has accumulated on the subject in question; it would be as if they could simply draw the relevant file from the well-ordered filing cabinet of human knowledge. The real world is one in which information is scattered between books, journals, governmental publications and a whole, array of electronic formats. These formats usually maintain their own systematisation when viewed separately, but as a whole appear incoherent and disorganised.

The late 20th century has become the information age and has seen the development of the World Wide Web. Although an increasing amount of material now appears on the Web, and it is certainly a positive step towards ease of access, its arrangements for retrieving that material are not always constructive. Furthermore, air pollution is a scientific field, which must by nature, bridge the gap between the respective literature of the pure sciences and government policies. Therefore, in this respect, the air pollution scientist finds the filing cabinet in chaos.

Like all areas of science, there has been a profound growth in air pollution literature. Activity was already regarded as high in the 1950's when it had reached some 250 journal articles a year (Halliday, 1961). It is hard to estimate current levels of output because of difficulties in defining the various areas of the literature. The *Scientific Citation Index* indicates that about a third of the material on pollution deals with air pollution. Abstracting (e.g. *Pollution Abstracts*) services are typically identifying about ten thousand pollution publications annually, thus, we can expect many thousands of significant documents pertaining to air pollution each year. Accessing this amount of literature is not an easy task and, as we will see, the researcher has to contend with a widening body of grey and Internet-based literature.

Nevertheless, a good relationship with the literature is required by all scientists and professionals. Even at the end of the 20th century the library remains the primary information collection centre. Journals, encyclopaedias, newspapers and CD-ROMs all reside there. The facilities have increasingly widened to encompass the newer media. The physical volume of materials is such a problem that a range of space-saving formats are used in many libraries. In particular, *Microfiche*, *Microfilm* and *MicroPrint*, whereby the material is photographed onto the film medium and stored in files organised in chronological order, have been popular. Thus, these formats are often found to be useful for lengthy serial records such as newspapers or government reports. The microfilmed records can be viewed using an appropriate projection screen in the library and can be copied. However, such records are not always easy to work with for long periods of time.

This chapter discusses some of the sources of information available and concepts of how it is organised. It focuses first on traditional collections of books and periodicals and then more on to newer information sources.

2. Cataloguing Knowledge

Cataloguing enables the user to discover what is present in a collection and where the material is located (Stibic, 1982). The art of cataloguing is to list with a clear plan a collection of materials in relation to the subject (Cutter, 1965; Hunter, 1985). Indexes are usually an alphabetical organisation of the knowledge within a subject at some detail.

2.1. Catalogues and Indexes

Libraries have their collections arranged into a classification system for accessibility of organisation and retrieval. There are three main systems in operation: the *Dewey Decimal Classification* (DDC), the *Universal Decimal Classification* (UDC) and the *Library of Congress Classification* (LCC). The first of these classification systems, DDC, is used worldwide and has been translated into many languages including Chinese. The DDC is favoured in public libraries and classifies knowledge as a whole into ten subdivisions. The subdivisions are further subdivided. However, it shows the signs of its 19th century origin as a cataloguing system for liberal arts libraries, which means that science is not well treated (e.g. Foskett, 1973). Interestingly, air pollution is usually located at 363.7, which makes it part of the social sciences, particularly, social pathology and services. The UDC is a development of the DDC, aimed at retrieval rather than shelf arrangement. If labelled and filed correctly, this system allows information to be found from whatever subject interest it is sought. It also has relating and connecting signs which are not present in the DDC, and has extensive expansions, especially in science (class 5) and technology (class 6). Despite the improvements, UDC is complex and schedule development in the sciences has been slow.

The LCC is the system favoured by academic libraries and is a system of coordinate series of special classifications. Each subject area is defined separately with internal structure for each general classification. Air pollution material is most typically found in the technology class at TD 883. A reader may find it useful to browse through this area of a library while trying to gain a sense of the holdings on air pollution. However, most information searches cannot be this serendipitous.

2.2. Encyclopaedias, Dictionaries and Handbooks

The encyclopaedia offers an efficient source from which to acquire general background information on a wide range of subjects. The *Encyclopaedia Britannica* is probably the most famous English encyclopaedia in publication. It is now published by the University of Chicago and groups subjects into a series of volumes with internal arrangement typically alphabetical, although not in the *Propaedia*. The alphabetical index volume of encyclopaedias is a much neglected resource even though it can often point to relevant information that would otherwise be missed. The information in encyclopaedias is often brief and concise but can be fairly detailed in some subject areas.

The traditional bookshelf-based encyclopaedia is only renewed once every decade or so, which can be a serious drawback as the science of air pollution is constantly evolving. Some encyclopaedias, such as the CD-ROM or on-line encyclopaedias, are easier to revise. On-line encyclopaedias can be updated constantly, and in theory, provide more up-to-date material but readers would be advised to check the date to ascertain. Computerised versions offer more effective subject searches by incorporating search programs which ask the user for a search word or expression and then scan the encyclopaedia for matches.

Scientific encyclopaedias, such as the *McGraw-Hill Encyclopaedia of Science and Technology* are more focused. Encyclopaedias of this type can provide the basis for a good scientific grounding in a subject. The main feature that distinguishes it from general encyclopaedias is the extent and quality of internal cross-referencing and bibliographical detail. This provides appropriate further reading and extra sources of information but may be dated.

The meanings of terms are explained in specialist scientific dictionaries. Many environmental dictionaries are now published. Good dictionaries will provide an accurate definition of a term for the reader. The difficulty for the specialist is that the definitions are sometimes written by compilers who do not have specific knowledge of the word use within the confines of atmospheric science. Some dictionaries provide almost encyclopaedia-type entries but these are most often meant for the general reader. Particularly valuable works in this category are volumes such as the *Dictionary of Organic Compounds* (also on CD-ROM), which lends itself very clearly to the dictionary form.

Handbooks are a more specialised source of information and prove useful for formulation, chemical names, and data such as physical and chemical constants, of which the Chemical Rubber Company handbooks are perhaps the best known. They may come from regular publishers or learned societies and other professional groupings. Although most are for reference rather than reading, some can have articles similar to encyclopaedias.

These reference forms can provide a rapid and easily accessible information source ideal for background understanding even though the source materials are often neglected. On the other hand, the above forms of literature are not appropriate if more detailed knowledge is required, so catalogues and indexes become a central feature in searching for detailed material on specific subjects.

2.3. Indexes, Bibliographies and Subject Guides

These books contain reference material under subject headings. Indexes are often updated annually to include references to the most recent material from the publications covered.

A general index covering all subjects, including the sciences and technology, is the *British Humanities Index (BHI).* The subjects are broad, much like the *Encyclopaedia Britannica,* but focus on bibliographic information. A more specialised index designed for science subject headings is the *General Science Index.* It is superb for science subjects and leads to references that are more specific than those offered by *BHI* and encyclopaedias. However, the suggested materials are still of a non-specialised nature relative to the expectations of most researchers. Thus, individual disciplines of science have subject guides which introduce the reader to the literature. These are well-developed in traditional disciplines such as chemistry and biology but their presence will no doubt be increasingly felt in the environmental sciences.

Policy-related research can use newspapers and magazines as a source of information. Major newspapers offer annual indexes, which include *The Financial Times, The Times, The New York Times, Guardian* and *Independent.* These indexes are now available in the CD-ROM format. These are often only a month or so out of date, compared with the bound hardcopy annual indexes. Each newspaper's CD-ROM index differs slightly in layout. The *Guardian* index, for instance, has a more aesthetically pleasing layout than that of the *Independent.* Information retrieval processes typically allows the user to call up a list of appropriate articles by specifying subject words, author or newspaper section. The CD-ROM format allows searches for several subject words to occur at once, and generally introduces a much more efficient search process than those used previously. This ease of search, coupled with rapid updating, makes the CD-ROM the best choice for recent materials whenever they are available. There is no doubt that parallel Internet searches will become more widely available.

2.4. Abstracts

Abstracts play a very special role in the sciences. They allow the researcher rapid access to the essential ideas available in a large volume of literature. Perhaps the most famous is the Chemical Abstracts Service (CAS) which produce the most comprehensive database of chemical information. It is particularly useful in air pollution research where chemistry plays such an important part. The *Chemical Abstracts* and *Registry* cover some 14 million abstracts and more than 18 million substances. In a more refined way, *Pollution Abstracts* from Cambridge Scientific Abstracts covers journals, conference proceedings and some reports. It treats the emission, detection, modelling, monitoring and analysis of air pollution among general pollution issues, and adds more than ten thousand abstracts a year.

3. Books

Indexes and catalogues will direct the researcher to printed literature. Books represent a major source of these forms of literature, but in many fields, they are overshadowed by periodicals and even more ephemeral materials. The difficulty with books is that key cataloguing elements such as author, title and subject may give little guidance to the exact contents. The subject will often be presented in just a few words yet larger books (especially multi-author ones where only the editor is listed by cataloguers) can have many chapters and numerous authors on a great range of specialised topics. In such cases, browsing the contents pages and indexes of relevant individual works may be the only way to find material.

In the case of recent books, information on those in print is available from *Whitaker's Bookbank.* This is a CD-ROM and on-line facility which provides information about book price, editions, publisher information and availability. It is useful for determining the cost and relevance of a book before purchase. Publishers usually have websites with search engines to find relevant material. Books of interest can be described in some detail, along with reviews and other information. Publishers currently see these sites as marketing tools, so they are not necessarily optimised for researchers.

In addition, journals such as the *New Scientist, Scientific American* and *Nature* have indexes to their various sections and articles, including the Book Review where scientific books are assessed. If the CD-ROM index format is employed, general search words or specific author and title searches can again be used to access reviews. These reviews cover books aimed at all levels from school texts to texts suitable for postgraduates, researchers and those in industry, but they are of variable quality. Although there are general indexes of reviews, these often cover the specialised scientific literature in a limited way. Specialist journals also review books on specific subjects relevant to air pollution but these are often small sections that may appear irregularly.

4. Scientific Journals

Serials are an important area of library collections and in science, it is the journal that occupies the prime place. The number of scientific journals has grown exponentially (Zsindeley *et al.*, 1997). Initially, the content of the journals had a more universal character, but with the development of modern scientific trends, more and more specialised journals appeared. This is especially true for the subject of air pollution. There are several general scientific journals which include air pollution issues within their scope, such as *Science, Nature* and *Journal of Geophysical Research.* At the same time, there are many highly specific journals, such as *Indoor Air* and the *Journal of Aerosol Science.* It should be noted that many of the journals, such as the *Journal of Air and Waste Management Association,* are produced by societies while others arise from commercial publishing houses.

4.1. Air Pollution Journals

The most comprehensive list of the titles and descriptions of journals related to air pollution can be found in the *Ulrich's International*

Periodical Directory under the heading of "Environmental Pollution". This list includes over two hundred titles of which only a few dozen are specific to air pollution. As well as listing the titles of all the journals, the Directory also includes the ISSN numbers, the language(s) of the texts, the dates first published, the editors' names and addresses, the publishers' addresses, website addresses and many more. The *Science Citation Index* (SCI) of the Institute of Scientific Information (ISI, Philadelphia, USA) processes information on over 3430 journals. The criteria for a journal to be included in the SCI listings are partially based on the journal's average citation rate, i.e. the impact factor (to be discussed later). The SCI is not as comprehensive as the *Ulrich's Directory*.

Again, most of the publishers of journals have web sites which list all the journals that they publish by subject headings. Journals related to air pollution are usually listed under the heading of "Environmental Science", but their search engines at most sites allow a rapid review of what they have.

Journals are usually very specific about their intended audience and such information is often contained in the journal's scope, typically printed within individual issues, but now also available at many web sites. As an example of audience specificity, Elsevier's *Air Pollution Consultant* is concerned with providing guidance to industry on changes in air pollution legislation while *Atmospheric Environment* draws on a more academic audience.

5. Influence of a Journal

It is useful to have an indication of the importance of journals to the development of air pollution. Parallel to this, there is much concern over the relative quality of scientific journals and the articles they publish. In academic publishing, citation has become an increasingly important indicator of the impact published articles have. Each year the ISI publishes a science citation report which lists *impact factors* for the journals it covers. *Impact factors* attempt to assess the relative importance of a journal in its field of research by determining the average frequency at which articles from the journal are cited. A journal's *impact factor* is the ratio between the number of that journal's articles cited elsewhere and the total number of articles published. The impact factor $I_j(X)$ for a journal j in year X is calculated using the following formula:

$$I_j(X) = \frac{C_{(X-1)} + C_{(X-2)}}{P_{(X-1)} + P_{(X-2)}},$$

where $C_{(\text{Year})}$ is the number of articles cited in other SCI listed journals and $P_{(\text{Year})}$ is the number of articles published. $X-1$ and $X-2$ represent the two previous years. Thus, the impact factor for a particular year is based on performance in the previous two years.

This is a useful determinant of quality but not necessarily the size of the impact. Some small journals may have articles that achieve a high citation rate but publish only a few articles. Some larger journals have a high influence by publishing a greater number of articles albeit a lower rate of citation. In particular, one should note the *Journal of Geophysical Research* which publishes a very large number of articles (not all on the atmosphere, of course) and has a very high impact factor.

Figures 1 and 2 show the impact factors and influence of some of the important journals on air pollution. Impact factors for many journals can be found in the *Journal Citation Report*. However, it is possible to calculate alternative impact factors using the World Wide Web (WWW).

Factors other than impact and influence may affect a journal. In some subject areas, citation is less frequent than others and poorly cited journals may have a surprising influence. In other cases, it is the rate of publication that is relevant (particularly for authors) and on-line or journals using camera-ready copy can ensure short publication times. Cost and availability can also affect a journal's stature.

The SCI impact factor is a measure of the relative importance of each journal across the entire range of sciences in which it is used. It

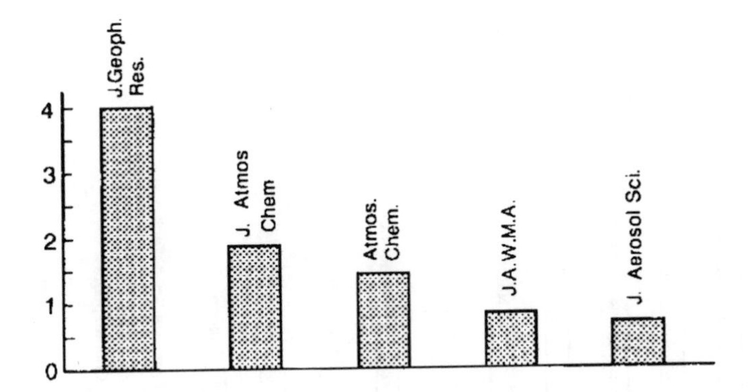

Fig. 1. Average impact factors 1985–94 (from data of Zsindeley *et al.*, 1997). NB: The *Journal of the Air Pollution Control Association* became the *Journal of the Air Waste Management Association* in 1990, and so they are combined under this heading.

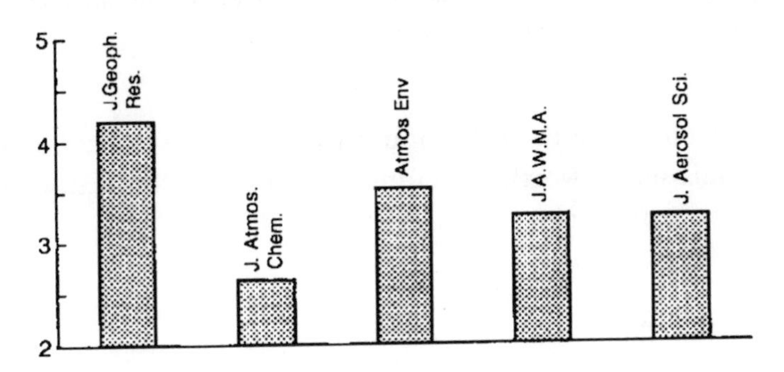

Fig. 2. Logarithm of influence (impact factors × articles published) 1985–94, using the impact factors from Fig. 1.

does not reveal a journal's relative importance within the field of air pollution. One indication of significant journals can be established by counting the number of times individual journals were cited in a prominent journal which deals with air pollution. Figure 3 shows the logarithm of the number of citations to some relevant journals in articles from the 1997 volume of *Atmospheric Environment.* The understandable bias towards citation for the journal being examined is very

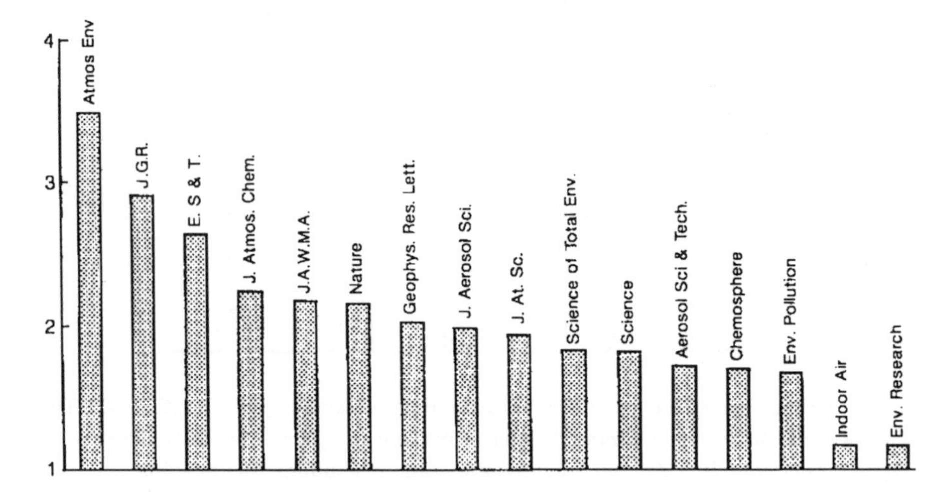

Fig. 3. Logarithm of the number of citations to some relevant journals in articles from the 1997 volume of *Atmospheric Environment.*

clear but the choice of the other journals gives some indication of what other journals see as worthy of citing. It is another indicator of which journals have influence.

6. Grey Literature

Grey literature (Auger, 1989) occupies an intermediate position between published books and periodicals (i.e. white literature) and ephemeral materials (e.g. bus schedules, theatre programmes, menus). It ranges from technical reports, theses, conference proceedings and committee reports to newsletters and product brochures. The Interagency Gray Literature Working Group's *Gray Information Functional Plan*, 1995 (Soule & Ryan, 1996) suggested that grey literature is "open source material that usually is available through specialised channels and may not enter normal channels or systems of publication, distribution, bibliographic control or acquisition by booksellers or subscription agents". This view places special emphasis on the difficulty of collecting grey material. It is further exacerbated by cataloguing problems which means that although publicly available, it is not easy to find. It is a popular form of publishing because it is rapid and flexible but this ultimately creates problems of wider access.

There are many different producers of grey literature, covering organisations such as local or national government, professional institutions, charities and pressure groups. The variety of document types and their sources add to the difficulty of acquiring the information. Non-standard formats frequently complicate cataloguing and storage and then exacerbate the problems of retrieval. Grey material rarely provides adequate information for cataloguing or retrieval. Information, such as author, title, place and date of publication and publisher, is often absent.

The lack of any broad indexes or sources of grey literature is a major problem in its use. A researcher who wishes to use the grey literature often encounters it in peripheral ways and must then obtain copies from the originating organisations. The potential of a *Gray Literature On-Line Catalog* has been addressed by Soule and Ryan (1996) and would provide a guide to some of the literature. As discussed in the section below, the Age of the Internet has changed the way in which much grey literature is published.

One route to grey literature is to access it through the originating organisation. Thus, one might go to the government department responsible for the field being researched. Individual organisations, even non-governmental and public interest groups, will often publish annual lists of their public and internal documents. These lists are also often part of the grey literature and have to be obtained by application to the organisation concerned but represent an excellent form of indexing (e.g. the UK's *Government Publications*). In some cases, however, especially with non-governmental organisations, such indexing may only cover recent publications, rendering access to older material almost impossible.

Grey literature varies greatly in quality as it is typically unrefereed. Much arises from specific organisations where documents may reflect institutional biases. This can limit their usefulness in terms of an objective scientific study unless some effort is made to understand the nature of the originating organisation. Industries can produce more detailed and technically advanced literature which may be suitable for research purposes, but there is often bias in terms of the industrial standpoint. Pressure groups will often take an entirely different view of the same industrial developments. However, there are some areas where more neutral sources of information can be obtained. The various professional institutions and some international, non-governmental organisations produce reports that merit examination in a scientific study of air pollution.

The grey literature is sometimes viewed as low in quality. This stems from the unrefereed nature of the materials rather than a true assessment of its worth. Some grey literature can be of the highest quality. In the last few years, the Intergovernmental Panel for Climate Change (IPCC) has published several quite comprehensive reports that are held by bookshops and libraries in most cities and represent possibly the most readily available example of the best of the grey literature. Much excellent material has remained in the grey literature simply because it has been too extensive and of such narrow interest range that it cannot be published in the commercial sense. The UK Department of Environment annual reports on *Air Pollution in the UK* list the measurements and assessments of air pollution at UK monitoring sites. These are extensive compilations that are increasingly found on the Internet. As a newsletter, *Air Quality Management* in the UK is a most informative publication.

The expansion of the WWW in recent years has seen many of the above institutions producing sites mainly as public relations exercises with only some providing useful indexes and contact details. Until the WWW sites of the various organisations responsible for producing grey literature do become comprehensive and well indexed, a lot of research in this area will rely on personal contacts and lists of names and addresses found in existing publications. This places a significant burden on the researcher who will have to investigate the various sources and then collate any information obtained themselves. A solution along the lines of the *Gray Literature On-Line Catalog* may be a hopelessly utopian notion, however attractive it seems. A fully catalogued grey literature may be a contradiction in terms because if it became this well-organised, it might no longer be defined as grey.

Dissertations and conference proceedings are a part of the grey literature that has increasingly effective catalogues. Major abstracting services, which tend to focus on US materials, create databases such as *Dissertation Abstracts On-Line*, and in the UK, Aslib publishes an *Index to Theses*. Conference proceedings are often more difficult to obtain, but increasingly, major national libraries, such as the British Library, have tried to collect and provide catalogues of such materials.

7. The Internet

The arrival of the Internet has brought with it enormous potential for global sharing of information. It is a rapid and easy way to publish materials, and this is in a sense expanding the grey literature. Virtually all the forms of scientific literature mentioned in this chapter are represented in some form on the Internet. The Internet has transformed the way that researchers work by revolutionising global information management. The Internet is growing at an astonishing rate. It doubles itself in just ten months. With a medium this fluid, changes occur very rapidly and sites and their addresses change frequently.

7.1. World Wide Web

The main on-line source is the World Wide Web. The WWW is an Internet-based navigational system and part of a global network that offers users access to information and documentation. It enables the user to navigate the Internet, moving with point-and-click ease from one location or one document to another. The WWW integrates different forms of information: still images, audio and video. A WWW user can jump effortlessly between locations, system applications and information formats.

Information can be accessed from the Internet using WWW Servers. Using the Hypertext Transfer Protocol (HTTP), these servers enable the user to access hypertext and hypermedia information. The addresses are in the form of uniform reference locators (URLs) which take the form:

http://www.uea.ac.uk/menu/acad_depts/env/

protocol domain name location

The domain name is a way of naming computers. These have a hierarchy of names. There are *top-level domain names* (TLDs) which are generic forms (e.g. EDU, COM, NET, ORG, GOV, MIL and INT) and often have two-letter country codes from ISO-3166. Generic forms can also be added as a second level in some countries. The example above shows the domain name www.uea.ac.uk, where ac represents the academic community of the United Kingdom, and further, the domain is divided into institutions (here, the University of East Anglia). Further subdivisions within the domain name are possible. The location gives the file and directory specifications of the resource on the server. The information available can be in the form of a text or file.

The File Transfer Protocol (FTP) is the means by which any type of file may be transferred between remote computers. These files could be documents, digital images, sound files or programs. The text file can be accessed using the relevant server. File transfer means an electronic copy of the document is obtained and this can be viewed, edited or printed at will. The Protocol makes information transfer easier for the user and the corresponding output more reliable. *Adobe Acrobat* (http://www.adobe.com), for example, allows you to view, navigate and print out files across all major computing platforms.

7.2. Search Engines

Search engines are the WWW equivalent of a library catalogue. They work by scanning the Internet for relevant information on the user's selected topic. The "index" is derived from programs called "spiders" that search the web on a periodic basis by following a trail of links from page to page. They then catalogue information on the sites visited. Depending on the searcher, the information collected may contain only the name and address of home pages, or it can also include information about the contents of a page, the words in a text-based document or information about multimedia files.

Search engines allow the user to enter keywords or phrases, and present a list of pages relating to the specific search terms used. There are many search engines to choose from and many different strategies for the search terms used. The search engines are still developing, and so here, we can only examine a few that typify those presently available.

Some search engines provide general information on hundreds of topics. *Hot Bots* directory is one of these. It divides on-line content into many thousands of categories. This form of WWW search is very good when you are looking for broad information on an air pollution topic such as indoor air pollutants. However, when specific information is needed on recent materials or text of conference papers, it is very tedious and not efficient as compared with other search engines.

In some cases, it may be better to use a search engine which is news based, e.g. *Infoseek*. Such searchers analyse the news on the WWW in newspapers and magazines as well as that from newsgroups. This form of searching is a useful form of collecting topical information. Air pollution is a branch of science which often deals with issues which arouse the public interest and so news based search engines may produce a surprisingly large number of hits. This form of search engine is best for scanning through the world's news but for other search tasks, it is poor.

There are search engines which only search sites that they have reviewed. These include *Excite* and *Lycos* . These are very useful search engines but as the Internet is changing so rapidly, they cannot fully review and analyse all the new incoming sites. These search engines depend upon the objective reviews and may not include in their search a specialist site which does not hold up to their review.

Each of the search engines mentioned so far has had some drawbacks but there are some searchers which avoid many of these pitfalls by using several search engines themselves. Examples are *Savvy Search* and *Altavista.*

Savvy Search arose as an experimental system based at Colorado State University. It is designed to query multiple Internet search engines simultaneously. Over the last few years, it has become a recognised search engine. *Savvy Search* takes out the problem of bias to specific sites as it encompasses 19 different search engines. These range from *Altavista,* through *Excite* and *Lycos,* to *Yahoo.* As time passes, the search systems include many more search engines and one can envisage further improvement with time. Enthusiasts regard *Savvy Search* as the best search engine around but it is experimental and not familiar to many users, particularly those who use the search facilities that come with many browsers. By contrast, *Altavista* is one of the most well-known search engines in the USA and is also one of the most informative.

Altavista is a fast, smart and powerful search engine. When search words are entered, the results snap up in less than a second with the best matches at the top. This also provides a search through *Yahoo* and *Look Smart,* which are two smaller but very useful web guides. The majority of search engines are biased towards American sites. *Savvy Search* and *Altavista* do not escape this pitfall. This bias creates a problem when looking for non-American information. While certain technical information from the USA is relevant and transferable abroad, policy issues will differ enormously globally and it is often best to use a search engine which is specific to the relevant country.

Services outside the US have been improved by establishing more local mirror sites. They are generally versions of the American search engines that search the relevant domain. *Altavista* UK searches in Northern Europe for English documents and it has mirror sites in Asia, Australia, Canada, Latin America, Northern Europe (including UK) and Southern Europe. *Excite* also has mirror sites in Europe and Asia.

Many search engines have the tendency to neglect newspapers, magazines and books. A good, all-encompassing web searcher is *Electronic Library.* It is designed for use by students and families and provides

Table 1. Types of WWW search engines.

Search engine	Address
Altavista (US)	http://altavista.digital.com
Saavy Search	http://www.savvysearch.com
Electronic Library	http://www.elibrary.com
Excite (US) (UK)	http://www.excite.com http://www.excite.co.uk
Infoseek	http://infoseek.go.com
Yahoo	http://www.yahoo.com
Looksmart	http://www.looksmart.com
Lycos	http://www.lycos.com
Hot Bots	http://www.hotbot.com

a database of over 150 newspapers, 800 magazines and thousands of books. *Electronic Library* provides a useful means of accessing air pollution literature in newspapers, magazines and books on the subject.

If none of these prove suitable, there is now a range of software which allows you to combine other search engines in a way that optimises searches to your particular needs.

7.3. Search Words and Strategies

Apart from the obvious strategy of selecting the most appropriate search engine, there are other strategies which will help to focus the search onto the most relevant information.

It is important to choose characteristic and unambiguous search words. It is also important to make use of the built-in logic that search engines employ. Within the search expression, many engines will understand Boolean logical terms like "and" and "or" so, words can be linked to narrow the search. The "not" expression may be useful if a certain search expression is repeatedly revealing documents related to a certain topic in the same area which is not of interest (e.g. rainbow not rainbow trout). If multiple words are entered as a search expression, then the search engine may be told to treat them as a single expression with all the words together in that order in the text or as individual words which may appear in any order in the document. Some search engines allow you to define the material you are looking for, e.g. images, software, etc.

In literature searches on the WWW, there are clear needs to approach the task with some prior knowledge. In addition, one has to consider the time taken for the search. The Internet becomes sluggish for UK users in the afternoon, possibly as a result of US or Transatlantic traffic at this time. Searches are often more efficient during the week-end or early morning. Where a site is known, it is possible to use a program such as *Web Buddy* 1.0 from DataViz which will download material at a time when the links are more efficient.

There is much concern about volatility of material on the WWW. Material can change over ill-defined time periods and older versions often vanish. Even more troublesome are sites that vanish completely. There are some well-maintained sites, usually from governments or large organisations where there is a sense of greater permanence. Nevertheless, the problem is a very real one. One solution is the idea of PURLs or persistent uniform resource locators (Weibel *et al.*, 1995) and they appear similar to URLs:

http://purl.oclc.org/OCLC/PURL/FAQ

protocol resolver address name

Functionally, they are URLs but instead of pointing directly to the location of an Internet resource, a PURL points to an intermediate resolution service. This service then associates the PURL with the actual URL and returns the URL to the user. This URL can then be used in the normal way.

7.4. Sites

Publishers and information providers are beginning to see the Internet as a key element in future strategies. It is likely that subject-specific sites will develop rapidly. Such a site could become an excellent starting point for information on any air pollution topic. These developments lie in the future, so the current specific sites remain very important.

The quality of sites has improved markedly over recent years. Many of the organisations relevant to air pollution research have produced WWW pages with useful information. A good example of a basic resource page is the UK Department of the Environment, Transport and the Regions (the DETR) at *http://www.detr.gov.uk/* which includes links to their air quality site but more importantly to a publications directory. There are an increasing number of similar sites containing monitoring data from around the world. Agency sites often provide the researcher with access to an on-line library of government acts, press releases and even transcripts of recent parliamentary debates. They often incorporate their own search engines that provide access to a lot of government documentations, including legislative acts, statutory instruments, etc. This is useful to both research and the application of legislation.

European integration means it is increasingly necessary to consider various forms of European legislation as well. The European Document Repository at *http://www.eudor.com* provides a similar level of functionality to the DETR but it also only provides the summary and heading lists on-line. Full document must be ordered. Although there are almost 50 UK European Documentation Centres (also *http://aleph.tau.ac.il:4500/ALEPH/eng/ATA/EEC/EEC/SCAN*), many are at UK universities which hold much of this documentation.

Non-governmental organisations, such as the National Society for Clean Air, produce a number of on-line fact sheets covering a variety of topics aimed primarily at the layperson. The NSCA site also has links to a number of other groups including Friends of the Earth and Greenpeace. The majority of the mainstream environmental pressure groups produce material that is designed primarily for lay-people and as such may not provide useful information for an air pollution researcher. There is also an amazing range of "crank" websites that will probably provide a challenge, if not information.

The Internet also provides the possibility of making contact with individuals who are identified as experts in the air pollution fields. There are academic directories at many university sites and a range of search engines (e.g. the *Internet Address Finder* http://www.iaf.net/ and Yahoo's *Four 11 White Pages* http://people.yahoo.com/) for finding e-mail addresses. These provide a universal service to find e-mail addresses as well as other important contact information.

7.5. Quality

There is also no guarantee as to the quality of the information on the web. The credibility which has traditionally been given to the printed word by the presence of publishers and referees simply does not apply on the web. Anyone can be a web publisher and the web can be as much an instrument of misinformation as one of information. The user should pay careful attention to the domain name and the information provided on the nature of the originating body within the site. These are hardly a perfect guarantee of reliability but they represent a good starting place.

7.6. Mailing Lists

In addition to the web, another extremely useful Internet information tool is the use of mailing lists and discussion groups. These are e-mail lists which people can join and are able to send and receive information to or from a group of people with similar research interests. There are also lists that have restricted authorship but the information on them can be read by anyone. Publishers' mailings can be very useful and some, such as Elsevier's *Contents Direct*, electronically mail free of charge journal contents some weeks before publication.

The World Wide Web is becoming the prime resource for accessing information including that concerning air pollution. Non-computer-based resources will continue to be used but wider availability of the Internet will place increasing reliance on it. The ability to access the WWW for relevant information will not be optional in the future.

8. Conclusion

The amount of information globally available on air pollution is staggering. In order to retrieve something useful, relevant and timely from the pool of sources available, it is not sufficient just to browse. The researcher must have a reasonable knowledge of the resources available and an understanding of how best to exploit them. The shape of information management has changed rapidly in the last few years and is likely to continue to do so. Fortunately, in an increasingly information conscious world, the increasing power of tools to search for information has made global access almost a reality. As always, though, the best research will be done by those who use the right tools in the right way.

References

Auger C.P. (1989) *Information Sources in Grey Literature*, Bowker-Saur, London.

Cutter C.A. (1965) *Rules for a Dictionary Catalog.* The Library Association, London.

Foskett A.C. (1973) *The Universal Decimal Classification: The History, Present Status and Future Prospects of a Large General Classification Scheme.* Bingley, London.

Halliday E.C. (1961) In *Air Pollution.* pp. 9–37, WHO, Geneva.

Hunter E.J. (1985) *Computerized Cataloguing.* Clive Bigley, London.

Internet Assigned Number Authority (http://www.iana.org/)

Soule M.H. & Ryan R.P. (1996) *Gray Literature*, http://www.dtic.mil/summit/tb07.html

Stibic V. (1982) *Tools of the Mind. Techniques and Methods for Intellectual Work.* North-Holland Pub. Co., Amsterdam.

Weibel S. *et al. PURLs: Persistent Uniform Resource Locators* (http://purl.oclc.org/OCLC/PURL/SUMMARY)

Zsindely S. *et al.* (1997) *Idojaros*, **101**, 93–103.

INDEX